D0054020

A Bibliography of Noise

1965 - 1970

A Bibliography of Noise

1965 - 1970

by

Mary K. Floyd

The Whitston Publishing Company
Incorporated
Troy, New York
1973

Library of Congress Catalog Card Number 72-87107

ISBN 0-87875-029-0

PREFACE

This is a six-year, near-complete world bibliography of noise and its physiological, psychological, sociological, and cultural effects covering the period, 1965-1970. Annual supplements will appear beginning with that of 1971 and in the fall of each succeeding year. The bibliography is divided into two sections: a title section arranged alphabetically by subject and alphabetically by title within subjects. Titles and subjects are provided so that researchers who do not trust subject assignments, which are arbitrary and unique to the individual compiler, can conduct their own to-some-extent original literature search. Especial care has been taken to develop subject headings meaningful to the user: entries are broken down into almost 280 subject categories, and they have been allowed to arise from the nature of the material, and not, that is, imposed on the matter by the Library of Congress, or other standard subject heading lists. In addition, adequate "See" and "See Also" entries are included.

The bibliography is designed to serve the needs of different kinds of researchers, from medical scientists to those with only passing interest in this increasingly severe social problem. It should be of service to the undergraduate, graduate, and professional student of medicine, psychology, sociology, and the physical and applied sciences; the architect, the educator, and the musician.

The following bibliographies, serials indexes, and abstracts have been searched in compiling this bibliography: APPLIED SCIENCE AND TECHNOLOGY INDEX; ART INDEX; BIBLIOGRAPHIC INDEX; BRITISH HUMANITIES INDEX; BUSINESS PERIODICALS INDEX; CANADIAN PERIODICAL INDEX; CUMULATIVE BOOK INDEX; CUMULATIVE INDEX TO NURSING LITERATURE; EDUCATION INDEX; INDEX MEDICUS; INDEX TO LEGAL PERIODICALS; INTERNATIONAL NURSING INDEX; LAW REVIEW DIGEST; LIBRARY OF CONGRESS CATALOG: BOOKS: SUBJECTS; THE MUSIC INDEX; PUBLIC AFFAIRS INFORMATION SERVICE; READERS GUIDE TO PERIODICAL LITERATURE; SOCIAL

SCIENCES AND HUMANITIES INDEX.

In using the subject section of this bibliography, one should note the following reminders: 1. Noise as vibration is generally not included unless it involves the physiological, the psychological, or the emotional. 2. Noise-induced pulsations, for example in oil burners, are not treated. 3. No attention has been paid to acoustic stresses to metals, generally. 4. No engine noise, as in the instance of an internal combustion engine, has been included when that noise pertains purely to engine or metal fatigue. 5. White noise, except in two instances, has been excluded. 6. Although industrial noise and machinery noise have, generically, their subject heads, they are also broken down into specific sub-components. 7. "Children," "Students," and "Youth" are separate subject heads.

Thus, this bibliography concerns noise and its effects on people, their culture, and cultural artifacts, such as architecture, cities, music, vehicles, whether automobiles or airplanes, and the like. It is concerned with acoustic stress on man, the animal, his physical and mental and emotional well-being. And thus it must be concerned with the noise of transformers and air-conditioners, with that of apartments, of dental drills, and of rock-bands; but not with, for example, radio noise, white noise, and with most forms of masking noise.

LIST OF ABBREVIATIONS

AAUW J	American Association of University Women Journal (Washington)
AIAA J	American Institute of Aeronautics and Astronautics Journal (New York)
ANJ or Amer J Nurs	American Journal of Nursing (New York)
AORN J	Association of Operating Room Nurses, Inc., Journal (New York)
ASHRAE J	American Society of Heating, Refrigerating and Air-Conditioning Engineers, Inc. (New York)
Acoustical Soc Am J	Acoustical Society of America Journal (New York)
Acta Biol Exp	Acta Biologiae Experimentalis (Warsaw)
Acta Biol Med German	Acta Biologica et Medica Germanica (Berlin)
Acta Cardiol	Acta Cardiologica (Brussels)
Acta Chir Acad Sci Hung	Acta Chirurgica Academiae Scientiarum Hungaricae (Budapest)
Acta Genet	Acta Geneticae Medicae et Gemcllologiae (Roma)
Acta Otolaryng	Acta Oto-Laryngologica (Stockholm)
Acta Otorhinolaryngol Belg	Acta Oto-Rhino-Laryncologica Belgica (Brussels)
Acta Otorinolaryngol Iber Am	Acta Oto-Rino-Laryngologica Ibero-Americana (Barcelona)
Acta Paediatr Scand	Acta Paediatrica Scandinavica (Stockholm)
Acta Physiol Pharmacol Neerl	Acta Physioligica et Pharmacologia Neerlandica (Amsterdam)
Acta Physiol Pol	Acta Physiologica Polonica (Warsaw)
Activ Nerv Sup	Activitas Nervosa Superior (Prague)
Adhesives Age	Adhesives Age (New York)
Adm Mgt	Administrative Management (New York)
Aeromed Rev	Aeromedical Reviews (Brooks Air Force Base, Texas)
Aeronautical J	Aeronautical Journal (London)
Aerosp Med	Aerospace Medicine (St. Paul)
Aerospace Tech	Aerospace Technology (Washington)
Agnes Karll Schwest	Agnes Karll-Schwester der Krankenpfleger (Frankfurt)

Air Cond Heat & Refrig N	Air Conditioning Heating and Refrigeration News (Birmingham, Michigan)
Air Cond Heat & Ven	Air Conditioning, Heating and Ventilating (New York)
Aircraft Eng	Aircraft Engineering (London)
Albany L Rev	Albany Law Review (Albany, New York)
Am Aviation	American Aviation (Washington)
Am Bar Assn J	American Bar Association Journal (Chicago)
Am City	American City Magazine (New York)
Am Dyestuff Rep	American Dyestuff Reporter (New York)
Am Federationist	American Federationist (Washington)
Am Gas Assn Mo	American Gas Association Monthly (Arlington, Virginia)
Am Home	American Home (New York)
Am J Optom	Amreican Journal of Optomerty and Archives of American Academy of Optomerty (Minneapolis)
Am Legion	American Legion Magazine (New York)
Am Lib	American Libraries (Chicago)
Am Mach	American Machinist (New York)
Am Sch Bd J	American School Board Journal (Evanston, Illinois)
Am Soc C E Proc	American Society of Civil Engineers, Proceedings (New York)
Am Soc Safety Eng J	American Society of Safety Engineers. Journal (Park Ridge, Illinois)
Amer Ass Industr Nurses J	American Association of Industrial Nurses Journal (New York)
Amer Industr Hyg Ass J	American Industrial Hygiene Association Journal (Southfield, Michigan)
Amer J Dis Child	American Journal of Diseases of Children (Chicago)
Amer J Ment Defic	American Journal of Mental Deficiency (Albany, New York)
Amer J Physiol	American Journal of Physiology (Bethesda, Maryland)
Amer J Psychol	American Journal of Psychology (Urbana, Illinois)
Amer J Public Health	American Journal of Public Health and the Nations Health (New York)
Amer Psychol	American Psychologist (Washington)
An Exp Odontoestomat	Anales Espanoles de Odontoestomatologia (Madrid)

Ann Laring	Annali di Laringologia, Otologia, Rinologia, Faringologia (Torino)
Ann NY Acad Sci	Annals of the New York Academy of Sciences (New York)
Ann Occup Hyg	Annals of Occupational Hygiene (Elmsford, New York)
Ann Oculist	Annales d'Oculistique (Paris)
Ann Otolaryng	Annales d'Oto-Laryngologie et de Chirurgie Cervico-Fociale (Paris)
Ann Ottal	Annali di Ottalmologia e Clinica Oculistica (Parma, Italy)
Ann Univ Curie Sklodowska	Annales Universitatis Mariae Curie-Sklodowska; Sectio D: Medicina (Lublin)
Appraisal J	Appraisal Journal (Chicago)
Arch Belg Med Soc	Archives Belges de Medecine Sociale, Hygiene, Medecine du Travail et Medecine Legale (Brussels)
Arch Design	Architectural Design (Glendora, California)
Arch Environ Health	Archives of Environmental Health (Chicago)
Arch Gen Psychiatry	Archives of General Psychiatry (Chicago)
Arch Ital Laring	Archivio Italiano di Otologia, Rinologia e Laringologia (Milan)
Arch Ital Otol	Archivio Italiano di Otologia, Rinologia e Laringologia (Milan)
Arch Klin Exp Ohr Nas Kehlkopfheilk	Archiv fuer Klinische und Experimentelle Ohrennasen- und Kehlkopfheilkunde (Berlin)
Arch Mal Prof	Archives des Maladies Professionvelles de Medecine du Travail et de Securite Sociale (Paris)
Arch Otolarng	Archives of Otolaryngology (Chicago)
Arch Phys Ther	Archives of Physical Therapy (Chicago)
Arch Rec	Architectural Record (New York)
Arch Sci Physiol	Archives des Sciences Physiologiques (Paris)
Architects J	Architects Journal (London)
Archives Environmental Health	Archives of Environmental Health (Chicago)
Arh Hig Rada	Arhiv za Higijenu Rada i Toksikologiju (Zagreb)

Arzneim Forsch	Arzneimittal-Forschung (Aulendorf)
Astronautics & Aeronautics	Astronautics and Aeronautics (New York)
Atlan	Atlanta (Atlanta, Georgia)
Atti Accad Fisiocr Siena	Atti dell'Accademia dei Fisiocritici di Siena (Siena)
Audio	Audio (Philadelphia)
Audio Eng Soc J	Audio Engineering Society. Journal (New York)
Audubon	Audubon Magazine (New York)
Aust Dent J	Australian Dental Journal (Sydney)
Aust Nurses J	Australian Nurses Journal (Glebe, Australia)
Automation	Automation (Cleveland, Ohio)
Automation & Remote Control	Automation and Remote Control (Pittsburg, Pennsylvania)
Automobile Eng	Automobile Engineer (London)
Automotive Eng	Transportation Engineer (Detroit, Michigan)
Av Instr	Audiovisual Instruction (Washington)
Aviation W	Aviation Week and Space Technology (New York)
Barron's	Barron's (New York)
Behav Sci	Behavioral Science (Ann Arbor, Michigan)
Beitr Gerichtl Med	Beitraege zur Gerichtlichen Medizin (Wien)
Bell Lab Rec	Bell Laboratories Record (Murray Hill, New Jersey)
Bell System Tech J	Bell System Technical Journal (Murray Hill, New Jersey)
Bet Hom & Gard	Better Homes and Gardens (Des Moines)
Biorheology	Biorheology (Elmsford, New York)
Biull Eksp Biol Med	Biulleten Eksperimentalndi Biologii i Meditsiny (Moscow)
Boll Mal Orecch Gola Naso	Bollettino della Malattie dell'Orecchio, della Gola, del Naso (Florence)
Boll Oculist	Bollettino D'Oculistica (Bologna)
Boll Soc Ital Biol Sper	Bollettino della Societa Italiana di Biologia Sperimentale (Napoli)
Boll Soc Medicochir Cremona	Bollettino della Societa Medico-Chirurgica di Cremona (Cremona)
Br J Psychol	British Journal of Psychology (New York)
Br J Ind Med	British Journal of Industrial Medicine (London)

Br Med J	British Medical Journal (London)
Brit Dent J	British Dental Journal (London)
Brit J Physiol Opt	British Journal of Physiological Optics (London)
Bsns W	Business Week (New York)
Build Res Sta Digest	Building Research Station Digest (London)
Builder	Builder (Springfield, Illinois)
Bul Atom Sci	Bulletin of the Atomic Scientists (Chicago)
Bull Inst Mar Med Gdansk	Bulletin of the Institute of Marine Medicine in Gdansk
Bull Inst Nat Sante	Bulletin de l'Institut National de la Sante et de la Recherche Medicale (Paris)
Bull NTRDA	Bulletin of the National Tuberculosis and Respiratory Disease Association (New York)
Bull NY Acad Med	The New York Academy of Medicine Bulletin (New York)
Bull Soc Sci Med Grand Duche Luxemb	Societe des Sciences Medicales du Grand-Duche de Luxembourg. Bulletin (Luxembourg)
C R Acad Sci	Comptes Rendus herdomadaires des Seances de l'Academie des Sciences; D: Sciences Naturelles (Paris)
C R Soc Biol	Comptes Rendus des Seances de la Societe de Biologie et de Ses Filiales (Paris)
Calif J Ed Res	California Journal of Educational Research (Burlingame, California)
Calif Med	California Medicine (San Francisco)
Can Bank	Canadian Banker (Toronto)
Can Bus	Canadian Business (Montreal)
Can J Chem Eng	Canadian Journal of Chemical Engineering (Ottawa)
Can J Public Health	Canadian Journal of Public Health (Toronto)
Can Min & Met Bul	Canadian Mining and Metallurgical Bulletin (Montreal)
Can Pers	The Canadian Personnel and Industrial Relations Journal (Toronto)
Canad Hosp	Canadian Hospital (Toronto)
Canad J Physiol	Canadian Journal of Physiology and

Pharmacol	Pharmacology (Ottawa)
Canad J Psychiat Nurs	Canadian Journal of Psychiatric Nursing (New Westminster, Canada)
Canad J Psychol	Canadian Journal of Psychology/Revue Canadienne de Psychologie (Toronto)
Canad Med Ass J	Canadian Medical Association Journal (Toronto)
Canad Nurse	Canadian Nurse (Ottawa)
Cath Nurse	Catholic Nurse (Bristol, England)
Cesk Hyg	Ceskoslovenska Hygiena (Prague)
Cesk Otolaryng	Ceskoslovenska Otolaryngologie (Prague)
Cesk Pediatr	Ceskoslovenska Pediatrie (Prague)
Cesk Stomat	Ceskoslovenska Stomatologie (Prague)
Chem & Eng N	Chemical and Engineering News (Washington)
Chem Eng	Chemical Engineering (New York)
Chem Eng Prog	Chemical Engineering Progress (New York)
Chem Industr R	Chemical Industry Report (Washington)
Chem Process & Eng	Chemical Processing and Engineering (Bombay)
Chem W	Chemical Week (New York)
Child Develop	Child Development (Chicago)
Chr Today	Christianity Today (Washington)
Clear House	Clearing House (Teaneck, New Jersey)
Clin Otorinolaring	Clinica Otorinolaringoiatrica (Rome)
Clin Pediat	Clinical Pediatrics (Philadelphia)
Coal Age	Coal Age (New York)
Colum L Rev	Columbia Law Review (New York)
Columbia J Law & Social Problems	Columbia Journal of Law and Social Problems (Charlottesville, Virginia)
Combustion	Combustion (New York)
Comp Air Mag	Compressed Air Magazine (Phillipsburg, New Jersey)
Comp Biochem Physiol	Comparative Biochemistry and Physiology (Elmsford, New York)
Composer	Composer (London)
Compr Psychiatry	Comprehensive Psychiatry (New York)
Concours Med	Concours Medical (Paris)
Confin Neurol	Confinia Neurologica/Borderland of Neurology (Basel)
Cong Q W Rept	Congressional Quarterly Service: Weekly Report (Washington)

Conn Med	Connecticut Medicine (New Haven)
Consumer Rep	Consumer Reports (Mount Vernon, New York)
Contemporary	Kortars/Contemporary (Budapest)
Country Life	Country Life (Appleton, Wisconsin)
Crawdaddy	Crawdaddy (New York)
Czas Stomat	Czasopismo Stomatologiczne (Warsaw)
DDZ	DDZ; Monatsschrift fuer Wissenschaftliche und Praktische Fortbildung des Zahnarztes (Munich)
Dent Dig	Dental Digest (Pittsburgh)
Deutsch Gesundk	Deutsche Gesundheitswesen (Berlin)
Deutsch Med Wschr	Deutsch Medizinische Wochenschrift (Stuttgart)
Deutsch Zahnaerztl Z	Deutsche Zahnaerztliche Zeitschrift (Muchen)
District Nurs	District Nurse (London)
Dom Eng	Domestic Engineering (Chicago)
Dtsch Tieraerzti Wochenschr	Deutsche Tieraerztliche Wochenschrift (Hannover-Waldhausen, W. Germany)
Dtsch Zahn Mund Kieferheilkd	Deutsche Zahn-Mund- und Kieferheilkunde (Leipzig)
Economist	Economist (London)
Elec Com	Electrical Communication (New York)
Elec Constr & Maint	Electrical Construction and Maintenance (New York)
Elec Mus R	Electronic Music Review (Trumansburg, New York)
Electr World	Electrical World (New York)
Electroencephalogr Clin Neurophysiol	Electroencephalography and Clinical Neurophysiology (Amsterdam)
Electronic N	Electronic News (New York)
Electronics	Electronics (New York)
Electro-Tech	Electro-Technology (Beverly Shores, Indiana)
Eng N	Engineering News (London)
Engineer	Engineer (New York)
Engineering	Engineering (London)
Environmental Control Mgt	Environmental Control and Safety Management (Morristown, New Jersey)
Ergonomics	Ergonomics (London)
Exceptional Child	Exceptional Children (Washington)
Exp Neurol	Experimental Neurology (New York)

Experientia	Experientia (Basel)
Eye Ear Nose Throat Mon	Eye, Ear, Nose and Throat Monthly (Chicago)
Factory	Factory (Bombay)
Farm J	Farm Journal (Philadelphia)
Fed Proc	Federation Proceedings: Federation of American Societies for Experimental Biology (Bethesda)
Feldsh Akush	Fel'dsher i Akusherka (Moscow)
Fin Post	Financial Post (Toronto)
Fiziol Norm Pat	Fiziologia Normala si Pathologica/Normal and Pathologic Physiology (Bucharest)
Fiziol Zh	Fiziolohichnyi Zhurnal (Moscow)
Fleet Owner	Fleet Owner (New York)
Folia Histochem Cytochem	Folia Histochemica et Cytochemica (Warsaw)
Folia Med	Folia Medica (Naples)
Folia Phoniat	Folia Phoniatrica (Basel)
Folia Psychiat Neural Jap	Folia Psychiatrica et Neurologica Japonica (Tokyo)
Forbes	Forbes (New York)
Forest Ind	Forest Industries (Portland, Oregon)
Forsvarsmedicin	Forsvarsmedicin (Stockholm)
Fortune	Fortune (Chicago)
Foundry	Foundry (Clevelnad, Ohio)
G Ig Med Prev	Giornale di Igiene e Medicina Preventiva (Genoa)
Gas	Gas (Houston, Texas)
Gaz Med France	Gazette Medicale de France (Paris)
German Med Monthly	German Medical Monthly (Stuttgart)
Gig Sanit	Gigiyena i Sanitariya (Moscow)
Gig Tr Prof Zabol	Gigiyena Truda i Professional'nyye Zabolevaniya (Moscow)
Glass Ind	Glass Industry (New York)
God Zborn Med Fak Skopje	Godisen Zbornik na Medicinskiot Fakultet vo Skopje
Good H	Good Housekeeping (New York)
Grad Res Ed	Graduate Research in Education and Related Disciplines (New York)
Graefe Arch Ophthal	Graefes Archiv fuer Klinische und Experimentelle Ophthalmoloige (Berlin)
Greater London Research Quarterly Bulletin	Greater London Research Quarterly Bulletin (London)

Abbreviation	Journal
Growth	Growth (Lakeland, Florida)
HNO	HNO Wegweiser fur die Fachaerztliche Praxis (Berlin)
Harv Civil Rights L Rev	Harvard Civil Rights-Civil Liberties Law Review (Cambridge)
Harv J Legis	Harvard Journal on Legislation (Cambridge)
Heating-Piping	Heating, Piping and Air Conditioning (Chicago)
High Fidelity	High Fidelity (New York)
Hippokrates	Hippokrates (Stuttgart)
Hlth Serv World	Health Services World
Hosp Top	Hospital Topics (Chicago)
Hospital	Hospital (Rio de Janerio)
Hospitals	Hospitals (Chicago)
House & Garden	House and Garden (New York)
House & Home	House and Home (New York)
House B	House Beautiful (New York)
Hum Factors	Human Factors (Baltimore, Maryland)
Hydraulics & Pneumatics	Hydraulics and Pneumatics (Cleveland, Ohio)
Hydrocarbon Process	Hydrocarbon Processing-Petroleum Refiner (Houston, Texas)
IEEE Proc	IEEE. Proceedings (New York)
IEEE Trans Com Tech	IEEE Transactions. Communication Technology (New York)
IEEE Trans Ind & Gen Applications	IEEE Transactions. Industry and General Applications (New York)
IEEE Trans Power Apparatus & Systems	IEEE Transactions. Power Apparatus and Systems (New York)
Ia L Rev	Iowa Law Review (Iowa City, Iowa)
Ig Mod	Igiene Moderna (Parma, Italy)
Image	Image (Chicago)
Ind Arts & Voc Ed	Industrial Arts and Vocational Education (Greenwich, Connecticut)
Ind Can	Industrial Canada (Toronto)
Ind Des	Industrial Design (New York)
Ind Finishing	Industrial Finishing (Wheaton, Illinois)
Ind Med Surg	Industrial Medicine and Surgery (Miami)
Ind Mktg	Industrial Marketing (Chicago)
Ind Res	Industrial Research (Beverly Shores, Indiana)
Ind W	Industrial Worker (Chicago)

Infirmiere	Infirmiere (Brussels)
Inst Mech Eng Proc	Institution of Mechanical Engineers. Proceedings (London)
Instrumentation Tech	Instrumentation Technology (Pittsburg)
Int Arch Arbeitsmed	Internationales Archiv fuer Arbeitsmedizin (Berlin)
Int Arch Gewerbepath	Internationales Archiv fur Gewerbepathologie und Gewerbehygiene (Berlin)
Int J Biometeor	International Journal of Biometeorology (Leiden, Netherlands)
Int J Nurs Stud	International Journal of Nursing Studies (Elmsford, New York)
Int Nurs Rev	International Nursing Review (Geneva)
Int Z Angew Physiol	Internationale Zeitschrift fur Angewandte Physiologie einschliesslich Arbeitsphysiologie (Berlin)
Interiors	Interiors (New York
Irish Nurs News	Irish Nursing News (Dublin)
Iron Age	Iron Age (Philadelphia)
Iron & Steel Eng	Iron and Steel Engineer (Pittsburg)
Iron & Steel Inst J	Iron and Steel Institute. Journal (London)
Izv Akad Nauk SSSR	Izvestia Akademii Nauk SSSR (Academy) of Science of the USSR. Bulletin. English Translation). (New York)
JAG J	Judge Advocate General Journal (Washington)
JAMA	Journal of the American Medical Association (Chicago)
J Abnorm Psychology	Journal of Abnormal Psychology (Washington)
J Acoust Soc Amer	Journal of the Acoustical Society of America (New York)
J Aircraft	Journal of Aircraft (New York)
J Amer Dent Ass	Journal of the American Dental Association (Chicago)
J Amer Osteopath Ass	Journal of the American Osteopathic Association (Chicago)
J Ap Mech	Journal of Applied Mechanics (New York)
J Appl Psychol	Journal of Applied Psychology (Washington)
J Basic Eng	Journal of Basic Engineering (New York)
J Cell Comp Physiol	Journal of Cellular Physiology (Philadelphia)

J Chronic Dis	Journal of Chronic Diseases (St. Louis)
J Clin Pathol	Journal of Clinical Pathology (London)
J Comm	Journal of Communication (Flint, Michigan)
J Comp Physiol Psychol	Journal of Comparative and Physiological Psychology (Washington)
J Egypt Med	Journal of the Egyptian Medical Association (Cairo)
J Egypt Public Health Ass	Journal of the Egyptian Public Health Association (Cairo)
J Eng Power	Journal of Engineering for Power (New York)
J Engin Psychol	Journal of Engineering Psychology (Margate, New Jersey)
J Environmental Sci	Journal of Environmental Sciences (Mt. Prospect, Illinois)
J Exp Anal Behav	Journal of the Experimental Analysis of Behavior (Bloomington, Indiana)
J Exp Syschol	Journal of Experimental Psychology (Washington)
J Florida Med Ass	Journal of the Florida Medical Association (Jacksonville)
J Formosa Med	Journal of the Formosan Medical Association (Taipei)
J Franc Otorhinolaryng	Journal Francais d'Oto-Rhino-Laryngologie et Chirurgie Maxillo-Faciale (Lyon)
J Gen Psychol	Journal of General Psychology (Provincetown, Massachusetts)
J Home Econ	Journal of Home Economics (Washington)
J Laryngol Otol	Journal of Laryngology and Otology (London)
J Med Lyon	Journal de Medecine de Lyon (Lyon)
J Mus Therapy	Journal of Music Therapy (Lawrence, Kansas)
J Nerv Ment Dis	Journal of Nervous and Mental Disease (Baltimore, Maryland)
J Neurol Sci	Journal of the Neurological Sciences (Amsterdam)
J Neurophysiol	Journal of Neurophysiology (Bethesda, Maryland)
J Nutr	Journal of Nutrition (Bethesda, Maryland)
J Occup Med	Journal of Occupational Medicine (New York)

J Opt Soc Am	Journal of the Optical Society of America (Lancaster)
J Otolaryng Jap	Journal of Otolaryngology of Japan (Tokyo)
J Otolaryng Soc Aust	Journal of the Oto-Laryngological Society of Australia (Melbourne)
J Personality Soc Psychol	Journal of Personality and Social Psychology (Washington)
J Pharm Sci	Journal of Pharmaceutical Sciences (Washington)
J Physiol	Journal of Physiology (London)
J Pract Nurs	Journal of Practical Nursing (New York)
J Proj Tech Pers Assess	Journal of Projective Techniques and Personality Assessment (Portland, Oregon)
J Prosth Dent	Journal of Prosthetic Dentistry (St. Louis)
J Psychosom Res	Journal of Psychosomatic Research (Oxford)
J Roy Coll Gen Pract	Journal of the Royal College of General Practitioners (Dartmouth, England)
J Roy Nav Med Serv	Journal of the Royal Naval Medical Service (Alverstoke)
J Sch Health	Journal of School Health (Columbus)
J Sci Instr	Journal of Physics E: Scientific Instruments (Formerly: Journal of Scientific Instruments). (London)
J Sci Med Lille	Journal des Sciences Medicales de Lille (Lille, France)
J Speech & Hearing Dis	Journal of Speech and Hearing Disorders (Washington)
J Speech Hearing Res	Journal of Speech and Hearing Research (Washington)
Jamacian Nurse	Jamacian Nurse (Jamaica)
Jap J Clin Med	Japanese Journal of Clinical Medicine (Osaka)
Jap J Hyg	Japanese Journal of Hygiene (Kyoto)
Kardiologiia	Kardiologiia (Moscow)
Klin Mbl Augenheilk	Klinische Monatsblaetter fur Augenheilkunde (Stuttgart)
Klin Med	Klinicheskaia Meditsina (Moscow)
Klin Oczna	Klinika Oczna: Acta Ophthalmologica Polonica (Warsaw)
L Rev	Jogtudomanyi Kozlony/Law Review (Budapest)

Lab Anim Care	Laboratory Animal Care (Joliet, Illinois)
Lakartidningen	Larkatidningen (Stockholm)
Lamp	Lamp (New York)
Lancet	Lancet (London)
Lang Speech	Language and Speech (Middlesex)
Laryngoscope	Laryngoscope (Collinsville, Illinois)
Life	Life (New York)
Life & Hlth	Life and Health (Washington)
Life Sci	Life Sciences (Oxford)
Lijecn Vjesn	Lijecnicki Vjesnik (Zagreb)
Lloyds Bank Review	Lloyds Bank Review (London)
McCalls	McCalls (Dayton, Ohio)
Mach	Machinery (New York)
Machine Design	Machine Design (Cleveland, Ohio)
Macl Mag	Maclean's Magazine (Toronto)
Manager	Management Today (Supersedes: Manager). (London)
Marine Eng Log	Marine Engineering/Log (New York)
Md Law R	Maryland Law Review (Baltimore, Maryland)
Mech Eng	Mechanical Engineering (New York)
Mech Illus	Mechanix Illustrated (New York)
Med J Aust	Medical Journal of Australia (Sydney)
Med Klin	Medizinische Klinik (Munchen)
Med Lavoro	Medicina del Lavoro (Milan, Italy)
Med Prac	Medecine Practicienne (Paris)
Med Pregl	Medicinski Pregled (Novi Sad)
Med Res Engin	Medical Research Engineering (Great Notch, New Jersey)
Med Serv J Canada	Medical Services Journal of Canada (Ottawa)
Med Trop	Medecine Tropicale: Revue du Service de Sante des Troupes de Marine (Marseille)
Med Welt	Medizinische Welt (Stuttgart)
Metallurgia	Metalurgia (Manchester, England)
Milit Med	Military Medicine (Washington)
Min Cong J	Mining Congress Journal (Washington)
Minerva Med	Minerva Medica (Torino)
Minerva Otorinolaring	Minerva Otorinotoringologica (Turin)
Minn L Rev	Minnesota Law Review (Minneapolis)
Miss & Roc	Missiles and Rockets (Washington)
Mod Hosp	Modern Hospital (Chicago)

Mod Manuf	Modern Manufacturing (New York)
Mod Materials Handling	Modern Materials Handling (New York)
Mod Plastics	Modern Plastics (New York)
Mod Textiles Mag	Modern Textiles Magazine (New York)
Monatsschr Ohrenheilkd Laryngorhinol	Monatsschrift fuer Ohrenheilkunde und Laryngo-Rhinologie (Vienna)
Motor B	Motor Business (London)
Mschr Kinderheilk	Monatsschrift fur Kinderheilkunde (Berlin)
Mschr Ohrendeilk	Monatsschrift fur Ohrenheilkunde und Laryngorhinologie (Wien)
Mschr Unfallheilk	Monatsschrift fur Unfallheilkunde, Versicherungs-, Versorgungs- und Verkehrsmedizin (Berlin)
Munchen Med Wschr	Muenchener Medizinische Wochenschrift (Munich)
Municipal	Municipality (Madison, Wisconsin)
Mus & Mus	Music and Musicians (London)
NM	Nursing Mirror (London)
NT or Nurs Times	Nursing Times (London)
NY Times Mag	New York Times Magazine (New York)
NYU L Rev	New York University Law Review (New York)
Naika	Naika (Tokyo)
Nat Parks	Nationla Parks Magazine (Washington)
Nat Safety Congr Trans	National Safety Congress. Transactions (Chicago)
Nation	Nation (New York)
Nations Bsns	Nation's Business (Washington)
Nation's Cities	Nation's Cities (Washington)
Nations Sch	Nation's Schools (Chicago)
Natural Resources J	Natural Resources Journal (Albequerque, New Mexico)
Nature	Nature (London)
Naval Eng J	Naval Engineers Journal (Washington)
Nederl Milit Geneesk T	Nederlands Militair Geneeskundig Tijdschrift (Hague)
Nederl T Geneesk	Nederlands Tijdschrift voor Geneeskunde (Amsterdam)
New Eng J Med	New England Journal of Medicine (Boston)
New Society	New Society (London)
New Statesman	New Statesman (London)
New York Dent J	New York Dental Journal (New York)

New York L Forum	New York Law Forum (New York)
New York Law Journal	New York Law Journal (Middletown, Connecticut)
New Zealand Nurs J	New Zealand Nursing Journal (Wellington)
Newsweek	Newsweek (New York)
Nord Hyg T	Nordisk Hygienisk Tidskrift (Stockholm)
Nord Med	Nordisk Medicin (Stockholm)
Nuovi Ann Ig Microbiol	Nuovi Annail d'Igiene e Microbiologia (Roma)
Nurs Clin North Am	Nursing Clinics of North America (Philadelphia)
Nurs J India	Nursing Journal of India (New Delhi)
Nurs Mirror	Nursing Mirror and Midwive's Journal (London)
Nurs Res	Nursing Research (New York)
Observer	Observer (London)
Occup Health	Occupational Health (Auckland)
Occup Health Nurs	Occupational Health Nursing (Thorofare, New Jersey)
Occup Health Rev	Occupational Health Review (Ottawa)
Occupational Psychology	Occupational Psychology (London)
Oeff Gesundheitswesen	Oeffentliche Gesundheitswesen (Stuttgart)
Office	Office (Stamford, Connecticut)
Oil & Gas J	Oil and Gas Journal (Tulsa, Oklahoma)
Oil Paint & Drug Rep	Oil Paint and Drug Reporter (New York)
Opt Soc Am J	Optical Society of America. Journal (New York)
Orv Hetil	Orvosi Hetilap (Bduapest)
Otolaryng Pol	Otolaryngologia Polska (Warsaw)
Otolaryngology	Jibinkoka (Tokyo)
Otorinolaringol Ital	Oto-Rino-Laringologia Italiana (Bologna)
Otorinolaringologie	Oto-Rino-Laringologie (Bucharest)
Pacific Law Journal	Pacific Law Journal (Seattle)
Paper Tr J	Paper Trade Journal (New York)
Parents Mag	Parents Magazine and Better Family Living (New York)
Parks & Rec	Parks and Recreation (Washington)
Pediat Akush Ginek	Pediatriia Akusherstvo i Gienkologiia (Kiev)
Percept Motor Skills	Perceptual and Motor Skills (Missoula, Montana)
Pflueger Arch	Pfluegers Archiv; European Journal of Physiology (Berlin)

Physiol Bohemoslov	Physiologia Bohemoslovoca (Prague)
Pipeline & Gas J	Pipeline and Gas Journal (Dallas)
Pit & Quarry	Pit and Quarry (Chicago)
Plant Eng	Plant Engineering (Barrington, Illinois)
Plastics Tech	Plastics Technology (New York)
Plastics World	Plastics World (Denver, Colorado)
Pol Med J	Polish Medical Journal (Warsaw)
Pol Tyg Lek	Polski Tygodnik Leharski (Warsaw)
Pop Sci	Popular Science Monthly (New York)
Postgrad Med	Postgraduate Medicine (Minneapolis)
Power	Power (New York)
Power Eng	Power Engineering (Barrington, Illinois)
Prac Lek	Pracovni Lekarstvi (Prague)
Practitioner	Practitioner (London)
Praxis	Praxis (Bern)
Presse Therm Climat	Presse Thermale et Climatique (Paris)
Probl Actuels Otorhin-laryngol	Problemes Actuels d'Oto-Rhino-Laryngologie (Paris)
Proc Roy Soc Med	Proceedings of the Royal Society of Medicine (London)
Proc Soc Exp Biol Med	Proceedings of the Society for Experimental Biology and Medicine (New York)
Product Eng	Product Engineering (New York)
Progress Arch	Progressive Architecture (Stamford, Connecticut)
Przegl Lek	Przeglad Lekarski (Krakow)
Psychiat Neurol Med Psychol	Psychiatrie, Neurologie und Medizinische Psychologie (Leipzig)
Psychol Forsch	Psychologische Forschung (Berlin)
Psychol Rep	Psychological Reports (Missaula, Montana)
Psychopharmacologia	Psychopharmacologia (Berlin)
Psychophysiology	Psychophysioloyg (Detroit, Michigan)
Public Health Rep	Public Health Reports (Washington)
Pulp & Pa	Pulp and Paper (New York)
Purchasing	Purchasing (New York)
Queensland Nurses J	Queensland Nurses Journal (Brisbane, Australia)
RIBA J	RIBA (Royal Institute of British Architects) London
RN	RN; National Magazine for Nurses (Oradell, New Jersey)
R Sci Instr	Review of Scientific Instruments (New York)

Rass Int Stomat Prat	Rassegna Internazionale di Stomatologia Pratica (Milan)
Read Digest	Reader's Digest (Pleasantville, New York)
Real Estate Appraiser	Real Estate Appraiser (Chicago)
Resen Clin Cient	Resenha Clinico-Cientifica (Sao Paulo)
Rev Asoc Odont Argent	Revista de la Asociacion Odontologica Argentina (Buenos Aires)
Rev Corps Sante Armees	Revue des Corps de Sante des Armees Terre Mer Air (Paris)
Rev Infirm Assist Soc	Revue de l'Infirmiere et de l'Assistante Sociale (Paris)
Rev Laryng	Revue de Laryngologie (Bordeaux)
Rev Med Psychosom	Revue de Medecine Psychosomatique (Toulouse)
Rev Med Suisse Rom	Revue Medicale de la Suisse Romande (Lausanne)
Rev Neurol	Revue Neurologique (Paris)
Rev Otoneuroophtalmol	Revue d'Oto-Neuro-Ophtalmologie (Paris)
Rev Paul Med	Revista Paulista de Medicina (Sao Paulo)
Rhode Island Med J	Rhode Island Medical Journal (Providence)
Right of Way	Right of Way (Los Angeles, California)
Riv Ital Ig	Rivista Italiana d'Igiene (Pisa)
Riv Med Aero	Rivista di Medicina Aeronautica e Spaziale (Rome)
Riv Otoneurooftal	Rivista Oto-neuro-oftalmologica (Bologna)
Rock Prod	Rock Products (Chicago)
Rocky Mtn Med J	Rocky Mountain Medical Journal (Denver)
Roy Aeronautical Soc J	Royal Aeronautical Society. Journal (London)
Roy Inst Brit Arch J	Royal Institute of British Architects. Journal (London)
Roy Soc Health J	Royal Society of Health Journal (London)
Rubber World	Rubber World (New York)
Ry Age	Railway Age (Bristol, Connecticut)
Sa Nurs J	South African Nursing Journal (Pretoria)
SAE J	SAE Journal (New York)
S Arf Med J	South African Medical Journal (Capetown)
SMPTE J	SMPTE Journal (New York)
Safety Maint	Safety Maintenance (Morristown, New Jersey)
Sales Mgt	Sales Management (New York)
Sat R	Saturday Review (New York)

Sborn Ved Prac Lek Fak Karlov Univ	Sbornik Vedeckych Praci Lekarske Fakulty Karlovy University (Hradec Kralove)
Scand J Psychol	Scandinavian Journal of Psychology (Stockholm)
Sch Mgt	School Management (Greenwich, Connecticut)
Schwest Rev	Schwestern Revue
Sci Amer	Scientific American (New York)
Sci & Cit	Scientist and Citizen (St. Louis)
Sci & Tech	Sciences et Techniques (Paris)
Sci Digest	Science Digest (New York)
Sci N	Science News (Washington)
Sci Tecah	Science Teacher (Washington)
Science	Science (Washington)
Sem Hop Paris	Semaine des Hopitaux de Paris (Paris)
Skyscraper Mgt	Skyscraper Management (Chicago)
So Calif L Rev	Southern California Law Review (Los Angeles, California)
Socialmed T	Socialmedicinsk Tidskrift (Stockholm)
Sotilaslaak Aikak	Sotilaslaaketieteellinen Aikakauslehti (Helsinki)
Space/Aeronautics	Space/Aeronautics (New York)
Spectator	Spectator (London)
Sr Schol	Senior Scholastic (New York)
Srpski Ark Celok Lek	Srpski Arhiv za Celokupno Lekarstvo (Beograd)
Stanford Research Institute. J	Stanford Research Institute. Journal (Menlo Park, California)
Statist	Statist (London)
Steel	Steel (Cleveland, Ohio)
Stero R	Stero Review (New York)
Stomatologiia	Stomatologiia (Moscow)
Stud Cercet Endocr	Studii si Cercetari de Endocrinologie (Bucharest)
Stud Gen	Studium Generale (Berlin)
Supervisory Mgt	Supervisory Management (New York)
Svensk Farm T	Svensk Farmaceutisk Tidskrift (Stockholm)
T Norsh Laegeforen or Tidsskr nor Laegeforen	Tidsskrift for den Norske Laegeforening (Oslo)
T Ziekenverpl	Tijdschrift voor Ziekenverpleging (Amsterdam)
Tech W	Technology Week (Washington)

Textile World	Textile World (New York)
Ther Gegenw	Therapie der Gegenwart (Berlin)
Ther Notes	Therapeutic Notes (Detroit, Michigan)
Therapeutique	Therapeutique (Paris)
Therapiewoche	Therapiewoche (Karlsruhe, W. Germany)
Three Banks	Three Banks Review (London)
Time	Time (Chicago)
Times Educ Sup	Times Educational Supplement (London)
Times R Ind & Tech	Times Review of Industry and Technology (London)
Today's Hlth	Today's Health Magazine (Chicago)
Town Plan Inst J	Town Planning Institute. Journal (London)
Town Plan R	Town Planning Review (Liverpool)
Traffic Q	Traffic Quarterly (Saugatuck, Connecticut)
Trans Amer Acad Ophthal Otolaryng	Transactions of the American Academy of Ophthalmology and Otolaryngology (Rodrester, Minneapolis)
Trans Ass Industr Med Officers and Trans Soc Occup Med	Society of Occupation Medicine. Transactions. (Formerly: Association of Industrial Medical Officers. Transactions). (London)
Trans NY Acad Sci	New York Academy of Science. Transactions (New York)
Tunisie Med	Tunisie Medicale (Tunis)
Turk Hemsire Derg	Turk Hemsireler Dergisi (Istanbul)
UN Med Canada	Union Medicale du Canada (Montreal)
UNESCO Courier	UNESCO Courier (Paris)
U Richmond L Rev	University of Richmond Law Review
US Air Force. Aerospace Med Res Lab	US Air Force Systems Command Aerospace Medical Research Laboratories Wright-Patterson Air Force Base (Ohio)
US Air Force. Sch Aerospace Med	US Air Force. School of Aerospace Medicine Brooks Air Force Base (Texas)
US Army Med Res Lab	US Army Medical Research Laboratory (Ft. Knox, Kentucky)
US Fed Aviat Agency. Office Aviat Med	US Federal Aviation Agency. Office of Aviation Medicine (Washington)
US NASA	US NASA
US Naval Submar Med Cent	US Naval Submarine Medical Center (Groton, Connecticut)
US News	US News and World Report (Washington)

Ukr Biokhim Zh	Ukrayins'kyi Biokhimichnyi Zhurnal (Kiev)
Ultrasonics	Ultrasonics (Surrey, England)
Urban Studies	Urban Studies (Edinburgh, Scotland)
Valsalva	Valsalva (Rome)
Vestn Akad Med Nauk SSSR	Vestnik Akademii Nauk SSSR (Moscow
Vestn Otorinolaring	Vestnik Otorinolaringologii (Moscow)
Virginia Med Monthly	Virginia Medical Monthly (Richmond, Virginia)
Vision Res	Vision Research (New York)
Vital Speeches	Vital Speeches (Southold, Long Island)
Voen Med Zh or Voennomed Zh	Voenno-Meditsinskii Zhurnal (Moscow)
Vogue	Vogue (New York)
Voj Zdrav Listy	Vojenski Zoravotnicke Listy (Czechoslovakia)
Vojnosanit Pregl	Vojnosanitetski Pregled (Beograd)
Vop Med Khim	Voprosy Meditsinskoi Khimii (Moscow)
Vop Pitan	Voprosy Pitaniya (Moscow)
Vrach Delo	Vrachebnoe Delo (Kiev)
WHO Chron	WHO Chronical (Geneva)
WHO Public Health Pap	World Health Organization Public Health Papers (Geneva)
Wall St J	Wall Street Journal (New York)
Water & Sewage Works	Water and Sewage Works (Chicago)
Week	Week (Nottingham)
Werk	Werk (Switzerland)
Wien Klin Wochenschr	Wiener Klinische Wochenschrift (Vienna)
Wien Med Wschr	Wiener Medizinische Wochenschrift (Vienna)
Wire & Wire Prod	Wire and Wire Products (New York)
Wireless World	Wireless World (London)
World Health	World Health (Geneva)
Yachting	Yachting (New York)
Yonsei Med J	Yonsei Medical Journal (Seoul)
Z Aerztl Fortbild	Zeitschrift fur Aerztliche Fortbildung (Jena)
Z Allgemeinmed	Zeitschrift fur Allgemeinmedizin: der Landarzt (Stuttgart)
Z Exp Angew Psychol	Zeitschrift fuer Experimentelle und Angewandte Psychologie (Goettingen, W. Germany)

Z Ges Hyg	Zeitschrift fur die Gesamte Hygiene und Ihre Grenzgebiete (Berlin)
Z Laryng Rhinol Otol	Zeitschrift fuer Laryngologie, Rhinologie, Otologie und Ihre Grenzgebiete (Stuttgart)
Z Mikr Anat Forsch	Zeitschrift fuer Mikroskopisch-Anatomische Forschung (Leipzig)
Zahnaerztl Mitt	Zahnaerztliche Mitteilungen (Cologne-Braunsfeld, W. Germany)
Zahnaerztl Welt or Zahnaerztl Rundsch	Zahnaerztliche Welt/Zahnaerzliche Rundschau (Heidelberg)
Zambia Nurs	Zambia Nurse (Kitwe)
Zbl Arbeitsmed	Zentralblatt fuer Arbeitsmedizien und Arbeitsschutz (Heidelberg)
Zbl Bakt	Zentralblatt fuer Bakteriologie, Parastenkunde, Infektionskrankheiten und Hygiene (Stuttgart)
Zh Nevropat Psikhiat Korsakov	Zhurnal Nevropatologii Psikhiatrii Im. S. S. Korsakova (Moscow)
Zh Ushn Nos Gorl Bolez	Zhurnal Ushnykh, Nosovyky i Gorlovykh Boleznei (Kiev)
Zh Vyssh Nerv Deiat Pavlov	Zhurnal Vysshei Nervnoi Deyatel'nosti Im. I. P. Pavlova (Moscow)

SUBJECT HEADINGS USED IN THIS BIBLIOGRAPHY

ATP: Treatment
Acoustic Accident
Acoustics: Brazil
Adenosinotriphosphoric Acid
Adolescents
Age
Aircraft Design
Aircraft Noise
Airlines
Airplanes
Airports
Air Conditioning
Air Movement
Alanime Aminotransferase
Alcohol
Aldolase
Aluminum Spars
Amplifiers
Amplitude Discrimination
Amylasuria
Apartment Noise
Architecture
Aspartic Aminotransferase
Attention
Automobile Noise
Barotrauma
Basal Metabolism
Bearing Noise
Behavior
Bekesy Typing
Bibliography
Biochemical Processes
Bioelectricity
Bioflavonoid
Biology
Brakes
Burners
CR-121
Camping Areas
Cams
Cardiovascular System
Children
Circuit-breakers
Cleft Palate
Cleregil
Clinical Aspects
Combustion Noise
Communications
Community Noise
Compressed Air Noise
Compressor Noise
Computer Noise
Concrete Breakers
Construction Industry
Cytochrome C
Dental Noise
Dentists
Diabetes
Diet
Digestive System
Directional Discrimination
Divers
Domestic Animals
EEG
EU-4200
Ear Injury
Earphones
Education
Electric Shock
Electrical Systems
Elevator Noise
Employees
Engines
Engines: Automobile
Engines: Diesel
Engines: Ducts
Environmental Health
Ethanol
Exhaust Systems
FAA
Fan Noise
Fertility

Fetus
Frogmen
Furnance Noise
Gamma Glutamyl
  Transpeptidas
Generator Noise
Glands: Adrenal
Glossaries
Glycogen Metabolism
Grinders
Head Injuries
Hearing Aids
Hearing Conservation
Hearing Field
Hearing Measurement
Heathrow Jets
Helicopter Noise
History
Hoarseness
Hospital Noise
Household Noise
Hunters
Hydraulic Engineering
Hygienic Studies
Hypacusis
Hypoxia
Impact Noise
Impulse Noise
Industrial Hygiene
Industrial Noise
Industrial Noise: Chemical
  Plants
Industrial Noise: Combustion
Industrial Noise: Concrete
Industrial Noise: Confectionery
Industrial Noise: Construction
Industrial Noise: Flax Mills
Industrial Noise: Forge Shops
Industrial Noise: Foundry
Industrial Noise: Glass-blowers
Industrial Noise: Iron Works
Industrial Noise: Krasnodar
  Factory
Industrial Noise: Metal Cutting
Industrial Noise: Metallurgy
Industrial Noise: Paper
Industrial Noise: Plastics
Industrial Noise: Plate Field
  Mills
Industrial Noise: Punch-presses
Industrial Noise: Quarries
Industrial Noise: Refinery
Industrial Noise: Shoe Factory
Industrial Noise: Synthetic Fiber
  Plant
Industrial Noise: Textile
Industrial Noise: Weaving
Industrial Noise: Welding
Industrial Noise: Yarn
Infrasonic Noise
Inner Ear
Internal Ear
Jet Noise
L-idotyl Dehydrogenase
Laboratory Noise
Lactic Dehydrogenase
Laryngectomy
Larynx
Laws and Legislation
Learning
Liability
Libraries
Machine Design
Machinery Noise
Machinery Noise: Cement Plants
Machinery Noise: Data Processing
Machinery Noise: Electrical
Machinery Noise: Farm Equipment
Machinery Noise: Hydraulic
Machinery Noise: Paper
Machinery Noise: Rotating
Machinery Noise: Turbo
Machinery Noise: Turbocompres-
  sors
Man
Masking Noise

TABLE OF CONTENTS

BOOKS

AIRCRAFT ENGINE NOISE AND SONIC BOOM; papers presented at the Fluid Dynamics Panel and Propulsion and Energetics Panel (33rd) Joint Meeting held at the Institut francoallemand de recherches, Saint-Louis, France, 27-30 May 1969. Neuilly-sur-Seine: NATO, 1969.

American Industrial Hygiene Association. INDUSTRIAL NOISE MANUAL, 2nd ed. Detroit: The Association, 1966.

Annus, Louise K., comp. AIRCRAFT NOISE AND SONIC BOOM. Washington: U.S. Federal aviation agency. Library services division, 1966.

Australia. House of representatives. Select committee on aircraft noise. INTERIM REPORT, JUNE 1970. Canberra: The Committee, 1970.

Aylesworth, Thomas G. THIS VITAL AIR, THIS VITAL WATER: man's environment crisis. Chicago: Rand McNally, 1968.

Baron, Robert Alex. THE TYRANNY OF NOISE. New York: St. Martin's Press, 1970.

Bazley, E. N. THE AIRBORNE SOUND INSULATION OF PARTITIONS. London: H.M.S.O., 1966.

Beales, Philip H. NOISE, HEARING AND DEAFNESS. London: M. Joseph, 1965.

Bell, Alan. NOISE: an occupational hazard and public nuisance. Genva: W.H.O.; London: H.M.S.O., 1966.

Berland, Theodore. THE FIGHT FOR QUIET. Englewood Cliffs: Prentice-Hall, 1970.

--NOISE--THE THIRD POLLUTION, 1st ed. New York: Public Affairs Committee, 1970.

Blachman, N. M. NOISE AND ITS EFFECT ON COMMUNICATION. New York: McGraw, 1967.

Boleszny, Ivan. CONTROL OF NOISE IN INDUSTRY. Res. serv. bibliogs, ser4, no85. Adelaide: State lib., 1967.

Bolt, Beranek and Newman, inc. NOISE ENVIRONMENT OF URBAN AND SUBURBAN AREAS. Developed under the technical studies program of the Federal Housing Administration, Department of Housing and Urban Development. Washington: G.P.O., 1967.

--A STUDY: INSULATING HOUSES FROM AIRCRAFT NOISE. Developed under the technical studies program of the Department of Housing and Urban Development, Federal Housing Administration. Washington: G.P.O., 1967.

Branch, Melville Campbell, et al. OUTDOOR NOISE AND THE METROPOLITAN ENVIRONMENT; case study of Los Angeles, with special reference to aircraft. Introd. by Vern O. Knudsen. Los Angeles: Dept. of City Planning, 1970.

British Standards Institution. METHOD OF RATING INDUSTRIAL NOISE AFFECTING MIXED RESIDENTIAL AND INDUSTRIAL AREAS. London: The Institution, 1967.

Burns, William. NOISE AND MAN. Philadelphia: Lippincott, 1969.

--and D. W. Robinson. HEARING AND NOISE IN INDUSTRY. London: H.M.S.O., 1970.

Chalupnik, James D., ed. TRANSPORTATION NOISES: a symposium on acceptability criteria. Seattle: University of Washington Press, 1970.

Chedd, Graham. SOUND: from communication to noise pollution. Garden City: Doubleday, 1970.

Chelsea & Kensington Action Committee on Aircraft Noise. A STUDY OF THE EFFECTS OF AIRCRAFT NOISE UPON THE ROYAL BOROUGH OF KENSINGTON & CHELSEA. London: The Committee, 1968.

Clark,D. M. SUBJECTIVE STUDY OF THE SOUND-TRANSMISSION CLASS SYSTEM FOR RATING BUILDING PARTITIONS. Ottawa: National Research Council of Canada, 1970.

Close, Paul Dunham. SOUND CONTROL AND THERMAL INSULATION OF BUILDINGS. New York: Reinhold, 1966.

Conference on Acoustic Noise and Its Control, London, 1967. PROCEEDINGS. London: Institution of Electrical Engineers, 1967.

Congres de la qualite de la construction, Brussels, 1967. SYMPOSIUM INTERNATIONAL "COMMENT COMBATTRE LE BRUIT." Bruxelles, 1967. Bruxelles: Communaute de l'isolation thermique et acoustique, 1967.

Day, Brian Frederick, R. D. Ford and P. Lord, eds. BUILDING ACOUSTICS. Amsterdam, New York: Elsevier, 1969.

Duerden, C. NOISE ABATEMENT. London: Butterworths, 1970.

Engineering Equipment Users Association. MEASUREMENT AND CONTROL OF NOISE. London: Constable, 1968.

Franken, Peter A. and David Standley. AIRCRAFT NOISE AND AIRPORT NEIGHBORS: a study of Logan International Airport; technical report. n.p., 1970.

French, Burrell O., et al. EFFECTS OF LOW FREQUENCY PRESSURE FLUCTUATIONS ON HUMAN SUBJECTS. Washington: National Aeronautics and Space Administration, 1968.

Galloway, William J., Welden E. Clark and Jean S. Kerrick. HIGHWAY NOISE: measurement, simulation, and mixed reactions. Washington: Highway Research Board, National Research Council, 1969.

Gilbert, P. AN INVESTIGATION OF THE PROTECTION OF DWELLINGS FROM EXTERNAL NOISE THROUGH FACADE WALLS. Garston, Eng.: Building Research Station, 1970.

Great Britain. THE LAW ON NOISE. London: Noise Abatement Society, 1969.

Great Britain Board of Trade. NOISE MEASUREMENT FOR AIRCRAFT

DESIGN PURPOSES INCLUDING NOISE CERTIFICATION PURPOSES. London: H.M.S.O., 1970.

Great Britain. Department of Employment and Productivity. NOISE AND THE WORKER, 2nd ed. London: H.M.S.O., 1968.

Great Britain. Parliament. House of Commons. Parlimentary Commissioner. SECOND REPORT OF THE PARLIAMENTARY COMMISSIONER FOR ADMINISTRATION: session 1967-68. London: H.M.S.O., 1967.

Greater London Council. TRAFFIC NOISE. London: The Council, 1966.

Harmelink, M. D. NOISE AND VIBRATION CONTROL FOR TRANSPORTATION SYSTEMS. Toronto: Dept. of Highways, Ontario, 1970.

Hayes, Charles D. and Michael D. Lamers. OCTAVE AND ONE-THIRD OCTAVE ACOUSTIC NOISE SPECTRUM ANALYSIS. Pasadena: Jet Propulsion Laboratory, California Institute of Technology, 1967.

Haywood, Maria R., compiler. AIRCRAFT NOISE AND SONIC BOOM; selected references. Washington: U.S. Dept. of Transportation. Library Services Division, 1969.

Heijbel, Carl Axel. PRACTICAL EXPERIENCE OF HEARING CONSERVATION IN INDUSTRY. Lecture held at the Industrial Noise Conference in London, December, 1961. Billesholm Sweden: Gullhogen, 1966.

Hildebrand, James L, ed. NOISE POLLUTION AND THE LAW. Buffalo: Hein, 1970.

Hines, William Arthur. NOISE CONTROL IN INDUSTRY, with applications to industrial, commercial, domestic and public buildings. London: Business Publications, 1966.

Ingemansson, Stig. CALCULATION OF SOUND INSULATION IN A BUILDING. Stockholm: Svensk byggtjanst, 1970.

International civil aviation organization. REPORT OF THE SPECIAL MEETING ON AIRCRAFT NOISE IN THE VICINITY OF AERODROMES. Montreal, 25 November-17 December, 1969. Montreal: The Organization, 1970.

International Conference on the Reduction of Noise and Disturbance Caused by Civil Aircraft, London, 1966. AIRCRAFT NOISE. London: H.M.S.O., 1967.

International Wrought Copper Council. INTRODUCTION TO THE STUDY OF NOISE IN INDUSTRY. London: I.W.C.C., 1968.

King, Arthur John. THE MEASUREMENT AND SUPPRESSION OF NOISE, with special reference to electrical machines. London: Chapman & Hall, 1965.

Kryter, Karl D. THE EFFECTS OF NOISE ON MAN. New York: Academic Press, 1970.

--REVIEW OF RESEARCH AND METHODS FOR MEASURING THE LOUDNESS AND NOISINESS OF COMPLEX SOUNDS. Springfield, Va.: Clearinghouse for Federal Scientific and Technical Information, 1966.

Los Angeles, California. Department of city planning. Graphics section. OUTDOOR NOISE AND THE METROPOLITAN ENVIRONMENT: case study of Los Angeles with special reference to aircraft. Los Angeles: The Department, 1970.

McClure, Paul T. INDICATORS OF THE EFFECT OF JET NOISE ON THE VALUE OF REAL ESTATE. Santa Monica: Rand Corp., 1969.

--SOME PROJECTED EFFECTS OF JET NOISE ON RESIDENTIAL PROPERTY NEAR LOS ANGELES INTERNATIONAL AIRPORT BY 1970. Santa Monica: Rand Corp., 1969.

McKennell, A. C. and E. A. Hunt. NOISE ANNOYANCE IN CENTRAL LONDON: a survey made in 1961 for the Building Research Station, 1st ed. reissued. London: H.M.S.O., 1968.

Middleton, David and P. J. F. Clark. ASSESSMENT AND DEVELOPMENT OF METHODS OF ACOUSTIC PERFORMANCE PREDICTION FOR JET NOISE SUPPRESSORS. Toronto: Institute for Aerospace Studies, University of Toronto, 1969.

Moore, John Edwin. DESIGN FOR NOISE REDUCTION. London: Architectural P., 1966.

Morgan, Candance D. AIRCRAFT AND INDUSTRIAL NOISE: a selected bibliography. Chicago: M.R. Library, 1968.

Navarra, John Gabriel. OUR NOISY WORLD, 1st ed. Garden City: Doubleday, 1969.

New Jersey. General assembly. Committee on air and water pollution and public health. PUBLIC HEARING ON ASSEMBLY BILLS NUMBERS 114 and 479 (noise control), held, Trenton, New Jersey, September 24, 1969. Trenton: The Committee, 1969.

New York, N.Y. Mayor's task force on noise control. TOWARD A QUIETER CITY: a report. New York: New York board of trade, 1970.

Nygaard, Kurt. REDUCTION OF NOISE IN CLOSED LOOP SERVO SYSTEMS. Nykoping: Aktiebolaget Atomenergi, 1970.

Olynyk, D. and T. D. Northwood. SUBJECTIVE JUDGMENTS OF FOOTSTEP-NOISE TRANSMISSION THROUGH FLOORS. Ottawa, National Research Council, 1969.

Parkin, Peter Hubert and H. P. Humphreys. ACOUSTICS, NOISE AND BUILDINGS. Foreword by Hope Bagenal, 3rd ed. London: Faber, 1969.

--LONDON NOISE SURVEY: [for the Building Research Station and the Greater London Council]. London: H.M.S.O., 1968.

Pearsons, Karl S. COMBINATION EFFECTS OF TONE AND DURATION PARAMETERS ON PERCEIVED NOISINESS. Washington: National Aeronautics and Space Administration, 1969.

Power, Joseph K. AN INVESTIGATION OF SONIC BOOM SIMULATOR TECHNIQUES AND MEASUREMENT DEVICES. n.p., 1968.

Purkis, Hubert John. BUILDING PHYSICS: acoustics, 1st ed. Oxford, New York: Pergamon, 1966.

Quigley, Hervey C., et al. FLIGHT AND SIMULATION INVESTIGATION OF METHODS FOR IMPLEMENTING NOISE-ABATEMENT LANDING APPROACHES. Washington: National Aeronautics and Space Administration, 1970.

Raes, Auguste C. ISOLATION SONORE ET ACOUSTIQUE ARCHI-

TECTURALE; problemes techniques et solutions pratiques. Paris: Editions Chiron, 1965.

Rettinger, Michael. ACOUSTICS; room design and noise control. New York: Chemical Pub. Co., 1968.

Rimskii-Korsakov, A. V., ed. PHYSICS OF AERODYNAMIC NOISE. Washington: National Aeronautics and Space Administration, 1969; Boston Spa: National Lending Library for Science & Technology, 1969.

Robinson, Douglas William. AN OUTLINE GUIDE TO CRITERIA FOR THE LIMITATION OF URBAN NOISE. London: H.M.S.O., 1970.

Rodda, Michael. NOISE AND SOCIETY. Edinburgh, London: Oliver & Boyd, 1967.

Salimivalli, Altti. ACOUSTIC TRAUMA IN REGULAR ARMY PERSONNEL. Clinical audiologic study. Turku, 1967.

Salmon, V. and S. K. Oleson. Noise control in the bay area rapid transit system. INTERIM REPORT. February, 1965. Palo Alto, California: Stanford Research Inst., 1965.

Sawyer, Richard H. and William T. Schaefer, Jr. OPERATIONAL LIMITATIONS IN FLYING NOISE-ABATEMENT APPROACHES. Washington: National Aeronautics and Space Administration, 1969.

Schubert, L. K. REFRACTION OF SOUND BY A JET: a numerical study. Toronto: Institute for Aerospace Studies, University of Toronto, 1969.

Skobtsov, Evgenil Aleksandrovich, A. B. Izotov and L. V. Tuzov. METHODS OF REDUCING VIBRATION AND NOISE IN DIESEL ENGINES. Translated by J. S. Shapiro, edited by B. J. Fielding. Boston Spa (Yorks): National Lending Library for Science and Technology, 1966.

Smetana, Ctirad. CORRECTION OF VALUES IN THE MEASUREMENT OF THE SPECTRA OF IMPULSE NOISE. Trondheim: Norger tekniske hogskole, Akustisk laboratorium, 1969.

Smith, Charles L. NOISE--UNWANTED SOUND: bibliography. (Bibliography no. 25), [61 San Mateo Road]Berkeley, California [94707]: 1969;

addendum, 1970.

Southampton, English University. Institute of Sound and Vibration Research. DEVELOPMENT OF AN AIRCRAFT FLYOVER NOISE RATING SCALE; technical report, SAE research project R-6.1. n.p., 1970.

Southwest Research Institute. Department of Applied Physics. A STUDY OF NOISE INDUCED HEARING DAMAGE RISK FOR OPERATORS OF FARM AND CONSTRUCTION EQUIPMENT. San Antonio, Tex.: The Institute, 1969.

Stephen, John E. and Lyman M. Tondell, Jr., LEGAL AND RELATED ASPECTS OF AIRCRAFT NOISE REGULATION; papers presented at the International Conference on the Reduction of Noise and Disturbance caused by Civil Aircraft, London, England, November 22-30, 1966. Washington: G.P.O., 1967.

Still, Henry. IN QUEST OF QUIET; meeting the menace of noise pollution: call to citizen action. Harrisburg: Stackpole Books, 1970.

Sutherland, L. C., ed. SONIC AND VIBRATION ENVIRONMENTS FOR GROUND FACILITIES; a design manual. Huntsville: Wyle Laboratories, 1968.

Symposium on Aerodynamic Noise. PROCEEDINGS OF AFOSR-UTIAS SYMPOSIUM, held at Toronto, 20-21 May 1968. Toronto: University of Toronto Press, 1969.

SYMPOSIUM ON MACHINERY NOISE; Los Angeles, 1969. New York: American Society of Mechanical Engineers, 1969.

SYMPOSIUM ON NOISE IN INDUSTRY; sponsored by the Departments of Public Health and Labour and Industry of South Australia, in collaboration with the University of Adelaide and the active support of a number of private industries; held at the University of Adelaide 28th and 29th February, 1968. Adelaide? 1968.

Symposium on the Sound of Outdoor Equipment, Detroit, 1967. PAPERS. New York: American Society of Heating, Refrigerating and Air-Conditioning Engineers, 1967.

Taylor, Rupert. NOISE. Harmondsworth: Penguin, 1970.

Tinney, E. Roy, ed. ANALYTICAL AND EXPERIMENTAL STUDIES OF SOUND SUPPRESSOR COMPONENTS FOR NASA ROCKET ENGINE TEST STANDS. Pullman: Technical Extension Service, Washington State University, 1967.

United States. Department of the Army. NOISE CONTROL FOR MECHANICAL EQUIPMENT. Washington: G.P.O., 1970.

--Department of housing and urban development. Office of metropolitan planning and development. Environmental planning division. AIRPORT ENVIRONS: land use controls. May 1970 35p. Washington: G.P.O., 1970.

--Department of transportation. First federal aircraft noise abatement plan: FY 1969-1970. Washington: G.P.O., 1970.

--Federal Council for Science and Technology. Committee on Environmental Quality. NOISE: sound without value. Washington: G.P.O., 1968.

--Office of science and technology. ALLEVIATION OF JET AIRCRAFT NOISE NEAR AIRPORTS: a report of the jet aircraft noise panel. Washington: G.P.O., 1966.

--Panel on Noise Abatement. THE NOISE AROUND US: findings and recommendations; report to the Commerce Technical Advisory Board. Washington: U.S. Dept. of Commerce, 1970.

--Small Business Administration. Technology Utilization Division. SELECTED ADVANCES IN INSULATION TECHNOLOGY. Washington: G.P.O., 1970.

Warring, R. H., ed. HANDBOOK OF NOISE AND VIBRATION CONTROL. Morden: Trade and Technical P., 1970.

Welch, Bruce L. and Annemarie S. Welch, eds. PHYSIOLOGICAL EFFECTS OF NOISE. New York: Plenum Press, 1970.

World health organization. NOISE: an occupational hazard and public nuisance. New York: United Nations, 1966.

Yerges, Lyle F. SOUND NOISE, and vibration control. New York: Van Nostrand Reinhold Co., 1969.

PERIODICAL LITERATURE

TITLE INDEX

"AICB-Congress in Baden-Baden," by L. Trbuhovic. WERK 53:sup 163, July, 1966.

"The AMA and noise-induced hearing loss," by G. D. Taylor. ARCH OTOLARYNG 90:543, November, 1969.

"Abatement of control valve noise; Fisher governor co.," by E. E. Allen. GAS 45:53-56, November, 1969.

"Abolition of milieu-induced hyperlipemia in the rat by electrolytic lesion in the anterior hypothalamus," by M. Friedman, et al. PROC SOC EXP BIOL MED 131:288-293, May, 1969.

"Absolute thresholds of human hearing," by E. R. Hermann, et al. AMER INDUSTR HYG ASS J 28:13-20, January-February, 1967.

"Accuracy consideration in fan sound measurement," by P. K. Baade. ASHRAE J 9:94-102, January, 1967.

"Acoustic taruma in frogmen," by G. Borasi, et al. ANN LARING 67: 36-43, January-February, 1968.

"Acoustic cancellation," MECH ENG 92:51, February, 1970.

"Acoustic conditions on harbour tug boats," by C. Szczepanski. BULL INST MAR MED GDANSK 21:215-222, 1970.

"Acoustic damage caused by noise as a cochlear anatomopathological manifestation and the prophylactic problem," by L. Ambrosio, et al. FOLIA MED 51:765-769, October, 1968.

"Acoustic environments of the F-111A aircraft during ground runup. AMRL-TR-68-14," by J. N. Cole, et al. US AIR FORCE AEROSPACE MED RES LAB 1-50, May, 1968.

"Acoustic integrity, flexibility important in building's design; Bell telephone laboratories new development center," HEATING-PIPING 38:304-305, January, 1966.

"Acoustic-lining concepts and materials for engine ducts," by R. A. Mangiarotty. ACOUSTICAL SOC AM J 48:783-794 pt 3, September, 1970.

"Acoustic lining reduces jet noise," ENGINEERING 207:596, April 18, 1969.

"Acoustic microtrauma due to dental drills. Contribution to the technic of early diagnosis of acoustic trauma," by W. Niemeyer. ARCH KLIN OHREN NASEN KEHLKOPFHEILKD 196:227-231, 1970.

"Acoustic noise and vibration of rotating electric machines," by A. J. Ellison, et al. INST E E PROC 115:1633-1640, November, 1968; Discussion 117:127-129, January, 1970.

"Acoustic-noise measurements on nominally identical small electrical machines," by A. J. Ellison, et al. INST E E PROC 117:555-560, March, 1970.

"Acoustic power measuring device," by C. J. Moore, et al. J SCI INSTR series 2 volume 1:659-661, June, 1968.

"Acoustic problems in Sao Paulo," by S. Marone. RESEN CLIN CIENT 38:173-182, July-August, 1969.

"Acoustic resonances and multiple pure tone noise in turbomachinery inlets," by F. F. Ehrich. J ENG POWER 91:253-262, October, 1969.

"Acoustic shelters meet needs as noise pollution grows," by N. P. Chironis. PRODUCT ENG 41:160-161, April 27, 1970.

"Acoustic trauma," by G. Birnmeyer. ARCH KLIN EXP OHR NAS KEHLKOPFHEILK 194:521-526, December 22, 1969.

"Acoustic trauma. Clinical presentation," by D. L. Chadwick. PROC ROY SOC MED 59:957-966, October, 1966.

"Acoustic trauma," by A. Salmivalli. SOTILASLAAK AIKAK 41:79-86, 1966.

"Acoustic trauma. 5 years of experience among flight and ground personnel of Varig airlines," by R. M. Neves Pinto. HOSPITAL (Rio) 75: 959-978, March, 1969.

"Acoustic trauma. Therapeutic trial of PC 63-14 (cogitum)," by P. J. Orsini, et al. ANN OTOLARYNG 86:209-212, March, 1969.

"Acoustic trauma caused by dental turbine drills," by W. Lorenz, et al. DTSCH ZAHN MUND KIEFERHEILKD 54:343-351, June, 1970.

"Acoustic trauma caused by diesel engine noise," by K. Ogata, et al. OTOLARYNGOLOGY 38:279-288, March, 1966.

"Acoustic trauma caused by shooting at shooting ranges. Attempts at prevention," by A. Rigaud. REV CORPS SANTE ARMEES 9:39-60, February, 1968.

"Acoustic trauma due to firearms in the army," by P. Pazat, et al. REV CORPS SANTE ARMEES 9:213-230, April, 1968.

"Acoustic trauma due to percussion wave detonation caused by supersonic airplanes," by O. Kleinsasser. HNO 13:170-175, June, 1965.

"Acoustic trauma from rock and roll," HIGH FIDELITY 17:38+, November, 1967.

"Acoustic trauma from rock-and-roll music," by C. P. Lebo, et al. CALIF MED 107:378-380, November, 1967.

"Acoustic trauma in depth miners," by A. Goubert, et al. J FRANC OTORHINOLARYNG 18:133-135, February, 1969.

"Acoustic trauma in the ground personnel of Varig," by R. M. Pinto, et al. HOSPITAL (Rio) 67:351-354, February, 1965.

"Acoustic trauma in the guinea pig. I. Electrophysiology and histology," by H. A. Beagley. ACTA OTOLARYNG 60:437-451, November, 1965.

--II. Electron microscopy including the morphology of cell junctions in the organ of Corti," by H. A. Beagley. ACTA OTOLARYNG 60:479-495, December, 1965.

"Acoustic trauma in regular army personnel. Clinical audiologic study," by A. Salmivalli. ACTA OTOLARYNG Suppl 222:1-85. 1967.

"Acoustic trauma in scientific and technical workers," by W. Sulkowski, et al. MED PRACY 17:515-518, 1966.

"Acoustic trauma in singers," by A. Profazio, et al. OTORINO-LARINGOL ITAL 37:337-346, August, 1969.

"Acoustic trauma in the sports hunter," by G. D. Taylor, et al. LARYNGOSCOPE 76:863-879, May, 1966.

"Acoustic trauma in the textile industry," by R. Avellaneda. ACTA OTORINOLARING IBER AMER 17:55-59, 1966.

"Acoustic trauma in welders exposed to strong noise in mechanical-technological workshops," by A. Brusin, et al. MED PREGL 22:545-549, 1969.

"Acoustical casing cuts engine room gear noise," SAFETY MAINT 136: 47-48, October, 1968.

"Acoustical control (within the office)," by D. Anderson. AMD MGT 27:66, March, 1966.

"Acoustical enclosures muffle plant noises," by S. Wasserman, et al. PLANT ENG 19:112-115, January, 1965.

"Acoustical glass," GLASS IND 46:229-230, April, 1965.

"Acoustical hazards of children's 'toys'," by D. C. Hodge, et al. J ACOUST SOC AMER 40:911, October, 1966.

"Acoustical locks set quiet mood for visitors to Bell telephone exhibit at N.Y. world's fair," AIR COND HEAT & REFRIG N 106:39, September 13, 1965.

"Acoustical measurements by time delay spectometry," by R. C. Heyser. AUDIO ENG SOC J 15:370-382, October, 1967.

"Acoustical measurements in New York's Philharmonic Hall," by

R. M. Schroeder. BELL LAB REC 43:38-45, February, 1965.

"Acoustical panels control mechanical noise pollution," SAFETY MAINT 136:39-40, July, 1968.

"Acoustical properties of carpets and drapes," by J. W. Simons, et al. HOSPITALS 43:125-127, July 16, 1969.

"Acoustical theory being forced upon fluid power engineers," by G. W. Kamperman. PRODUCT ENG 40:54+, December 15, 1969.

"Acoustics, noise and building. A review," by P. H. Parkin, et al. TOWN PLAN R 40:84-85, April, 1969.

"Action of noise on the cerebral and peripheral rheogram and on the electroencephalogram," by M. Fusco, et al. FOLIA MED 48:88-98, February, 1965.

"Acoustics of Northrop memorial auditorium," by B. S. Ramakrishna, et al. ACOUSTICAL SOC AM J 47:951-960, April, 1970.

"Activity of certain enzymes of intermediate metabolism and behavior of serum proteins and their fractions in workers exposed to strong acoustic and vibrational stimuli," by J. Gregorczyk. ACTA PHYSIOL POL 17:107-118, January- February, 1966.

"Acute hearing loss: etiology, diagnosis and therapy," by H. Krichel-dorff. Z AERZTL FORTBILD 59:562-564, May 15, 1965.

"Adaptation and loudness decrement; a reconsideration," by J. W. Petty, et al. ACOUSTICAL SOC AM J 47:1074-1082 pt 2, April, 1970.

"Adaptation behavior of Corti's organ following ephedrine and acoustic stimulation," by G. Stange, et al. ARCH OHR NAS KEHLKOPFHEILK 184:483-495, 1965.

"The adaptation factor and auditory rest in the appearance of occupational deafness," by L. Teodorescu, et al. OTORINOLARINGOLOGIE 10:109-121, April-June, 1965.

"Adaptive optimum detection: synchronous-recurrent transients," by L. W. Nolte. J ACOUST SOC AMER 44:224-239, July, 1968.

"Adaptive threshold detection of *M*-ary signals in statistically undefined noise," by J. B. Millard, et al. IEEE TRANS COM TECH 14:601-610, October, 1966.

"Additivity of masking," by D. M. Green. J ACOUST SOC AMER 41: 1517-1525, June, 1967.

"Add-on quieting for gas turbine," by W. J. Pietrucha. POWER 114:54-55, May, 1970.

"Adverse conditions found in the use of ultra high-speed equipment," by P. L. Terranova. NEW YORK DENT J 33:143-148, March, 1967.

"Aerodynamics, noise, and the sonic boom," by W. R. Sears. AIAA J 7:577-586, April, 1969.

"Aero-engine noise research laboratory," ENGINEER 224:6-8, July 7, 1967.

"Aesthetic nuisance: an emerging cause of action," NYU L REV 45: 1075, November, 1970.

"Age differences in the control of acquired fear by tone," by J. P. Frieman, et al. CANAD J PSYCHOL 23:237-244, August, 1969.

"Age factor in the response of the albino rat to emotional and muscular stresses," by W. F. Geber, et al. GROWTH 30:87-97, March, 1966.

"Age in the occurrence and evolution of acoustic trauma," by M. Lutovac. SRP ARH CELOK LEK 97:869-874, September, 1969.

"Aggravation of deafness caused by acoustic trauma after retreating from the noisy environment," by P. Pialoux. PROBL ACTUELS OTORHINOLARYNGOL 1-6, 1968.

"Air chambers, are they really cheaper?" by G. Flegel. DOM ENG 206: 70-71, September, 1965.

"Air conditioning and acoustics," by W. H. Schneider. AIR COND HEAT & VEN 63:56-59, August, 1966.

"Air conditioning for all," ENGINEERING 207:413, March 14, 1969.

"Air moving and conditioning association, inc. has already adopted 10-12 watt sound power reference level," AIR COND HEAT & REFRIG N 105:2, August 30, 1965.

"Air pollution by noise," by J. T. Hart. LANCET 1:998, May 9, 1970.

"Air pollution control; industrial noise control," by L. L. Beranek. CHEM ENG 77:227-230, April 27, 1970.

"Aircraft; costly big noise," STATIST 190:1586, December 30, 1966.

"Aircraft noise," by W. Knop. ZBL ARBEITSMED 19:273-275, September, 1969.

"Aircraft noise and development control-the policy for Gatwick airport," by E. Sibert. TOWN PLAN INST J 55:149-152, April, 1969.

"Aircraft noise, can it be cut?" by S. M. Levin. SPACE/AERONAUTICS 46:65-75, August, 1966.

"Aircraft noise; fugitive factor in land use planning," by D. C. McGrath, Jr. AM SOC C E PROC 95 (UP 1 no 6520):73-80, April, 1969.

"Aircraft noise law: a technical perspective," by J. J. Alekshun, Jr. AM BAR ASSN J 55:740-745, August, 1969.

"Aircraft noise; mitigating the nuisance," by E. J. Richards. ASTRONAUTICS & AERONAUTICS 5:34-43, January, 1967; Same. Discussion ASTRONAUTICS & AERONAUTICS 5:43-45, January, 1967.

"Aircraft noise monitor," ENGINEER 219:195, January 22, 1965.

"Aircraft noise; no more shrugging by the industry," SPACE/AERONAUTICS 47:18-20, June, 1967.

"Aircraft noise study set for city centers, suburbs," AVIATION W 90: 27, January 27, 1969.

"Aircraft noise symposium; proceedings," ACOUSTICAL SO AM J 48: 779-842, pt 3, September, 1970.

"Aircraft Noise: A taking of private property without just compensation," by P. W. Fleming, SOUTH CAROLINA L REV 18:593-608, November 4, 1966.

"Airplane cockpit noise levels and pilot hearing sensitivity," by K. J. Kronoveter, et al. ARCH ENVIRON HEALTH 20:495-499, April, 1970.

"Airport noise and the urban dweller: a proposed solution," by C. M. Haar. APPRAISAL J 36:551-558, October 8, 1968; REAL ESTATE APPRAISER 34:21-25, September-October, 1968.

"The airport noise problem and airport zoning (United States)," by E. Seago. MD LAW R 28:120-135, Spring, 1968.

"All those noises that assail us," LIFE 62:4, January 27, 1967.

"Allergy--Acoustic trauma. Review of the medical literature," by E. Mann. ZAHNAERZTL WELT 66:322-324, May 10, 1965.

"Alleviation of aircraft noise," by N. E. Golovin. ASTRONAUTICS & AERONAUTICS 5:71-75, January, 1967.

"Alteration of the masking sound level during and after acoustic stimulation," by G. Muller. ARCH KLIN EXP OHR NAS KEHLKOPFHEILK 195:323-330, 1970.

"American Conference of Governmental Industrial Hygienists' proposed threshold limit value for noise," by H. H. Jones. AMER INDUSTR HYG ASS J 29:537-540, November-December, 1968.

"Amplitude discrimination in noise," by G. B. Henning, et al. J ACOUST SOC AMER 41:1365-1366, May, 1967.

"Amplitude discrimination in noise, pedestal experiments, and additivity of masking," by G. B. Henning. J ACOUST SOC AMER 45:426-435, February, 1969.

"Amplitude distribution of axon membrane noise voltage," by A. A. Verveen, et al. ACTA PHYSIOL PHARMACOL NEERL 15:353-379, August, 1969.

"Amplitude quantization; a new, more general approach," by M. Vinokur. IEEE PROC 57:246-247, February, 1969.

"Analysing effect of noise on sleep," ENGINEER 226:233, August 16, 1968.

"Analysis of the complex electric response of the guinea pig cochlea to click," by C. Vesely, et al. SBORN VED PRAC LEK FAK KARLOV UNIV 10:Suppl:453-460, 1967.

"Analyzer detects malfunctions by examing jet engine noise," by A. J. Kasak, et al. SAE J 76:59-61, July, 1968.

"Anesthesia and 'white noise'," by V. K. Sipko. STOMATOLOGIIA 45:91-92, January-February, 1966.

"Animal experiment studies of the readaptation behavior of acoustic organ," by J. Theissing, et al. Z LARYNG RHINOL OTOL 47:64-70, January, 1968.

"Annoyance reactions to traffic noise in Italy and Sweden," by E. Jonsson, et al. ARCH ENVIRON HEALTH 19:692-699, November, 1969.

"Another tool for hearing conservation--an improved protector," by C. Zenz, et al. AMER INDUSTR HYG ASS J 26:187-188, March-April, 1965.

"Anticoagulation and gunshot concussion in eardrum bleeding," by N. Sonkin. RHODE ISLAND MED J 49:243-244, April, 1966.

"Antinoise campaign in Honolulu," ACOUSTICAL SOC AM J 45:520-521, February, 1969.

"Anti-noise report issued," FLEET OWNER 63:142, December, 1968.

"Appearance combined with soundproofing in a transformer enclosure," ENGINEER 226:232, August 16, 1968.

"Application of an amelioration technic of the signal-noise relationship to the measurement of the visual evoked potentials in the adult rabbit," by P. Magnien, et al. C R ACAD SCI (D) 266:929-932, February 26, 1968.

"Application of integrated circuits to industrial control systems with high-noise environments," by A. Wavre. IEEE TRANS IND & GEN APPLICATIONS 5:278-281, May, 1969.

"Application of the Mossbauer method to ear vibrations," by P. Gilad, et al. ACOUSTICAL SOC AM J 41:1232-1236, May, 1967.

"Application of sound-rated fans and ventilators," by H. R. Bohanon. ASHRAE J 9:73-77, August, 1967.

"Appraisal of Apollo launch noise," by B. O. French. AEROSPACE MED 38:719-722, July, 1967.

"Approach to the objective diagnosis of hoarseness," by N. Isshiki, et al. FOLIA PHONIAT 18:393-400, 1966.

"Apropos of the acoustic accident," by B. Kecht. Z LARYNG RHINOL OTOL 43:280-293, May, 1964.

"Apropos of 2 cases of myoclonic petit mal precipitated by noise in children," by A. Lerique-Koechlin, et al. REV NEUROL 113:269, September, 1965.

"Are academic libraries too noisy?" by J. A. McCrossan. AM LIB 1:396, April, 1970.

"Are we hooked on noise?" by W. Zinsser. LIFE 67:12, October 31, 1969.

"Are you prepared for noise control?" by W. D. Huskonen. FOUNDRY 97:64-67, April, 1969.

"Are you prepared for noise laws?" by F. Haluska. STEEL 164:41+, February 17, 1969.

"ARI issues fancoil unit sound rating," AIR COND HEAT & REFRIG N 108:1+, May 16, 1966.

"ARI issues sound standard for room induction units," AIR COND HEAT & REFRIG N 114:1+, May 6, 1968.

"ARI seeks to educate A-E's, GSA on reduction equipment noise," AIR COND HEAT & REFRIG N 121:28, December 7, 1970.

"ARI's new sound ordinance calls for A-scale readings," AIR COND HEAT & REFRIG N 108:1+, June 27, 1966.

"Army quiet helicopter effort aimed at reduction in losses," AVIATION W 91:33, September 15, 1969.

"Array gain for the case of directional noise," by B. F. Cron, et al. ACOUSTICAL SOC AM J 41:864-867, April, 1967.

"Art and technique of noise and vibration control," by H. C. Carter. DOM ENG 205:68-74, May, 1965.

"Aspects of nervous fatigue in automated systems as a function of age of the operators," by R. Elias, et al. FIZIOL NORM PAT 13:447-454, September-October, 1967.

"Assualt on the ear; city dweller," NEWSWEEK 67:70, April 4, 1966.

"Assessment and control of noise," by D. S. Gordon. INST E E PROC 113:775-776, May, 1966.

"Assessment of aircraft noise disturbance," by C. G. Van Niekerk, et al. AERONAUTICAL J 73:383-396, May, 1969.

"Assessment of the efficacy of periodic medical examinations of persons working in noisy shops of textile mills," by A. A. Tatarskaia, et al. GIG TR PROF ZABOL 9:45-46, July, 1965.

"Assessment of footstep noise through wood-joist and concrete floors," by D. Olynyk, et al. J ACOUST SOC AMER 43:730-733, April, 1968.

"Atrophy of the long process of the incus in a patient with an occupational hearing disorder," by B. M. Gapanavichius. VESTN OTORINO-LARINGOL 32:91-92, November-December, 1970.

"Attenuation of noise and ground vibrations from railways," by P. Grootenhuis. J ENVIRONMENTAL SCI 10:14-19, April, 1967.

"Attenuation provided by fingers, palms, tragi, and V51R ear plugs," by H. H. Holland, Jr. ACOUSTICAL SOC AM J 41:1545, June, 1967.

"Audio frequency analyzer measures auto noise pollution," ELECTRO-TECH 85:24-25, January, 1970.

"Audio measurements course (cont)," by N. H. Crowhurst. AUDIO 50: 28+, December, 1966; 51:36 , January, 1967; 36-38+, February, 1967; 44-46+, March, 1967; 44-45+, April, 1967; 34+, June, 1967; 26-28+, July, 1967.

"Audiogenic seizure susceptibility induced in C57BL-6J mice by prior auditory exposure," by K. R. Henry. SCIENCE 158:938-940, November 17, 1967.

"Audiologic examinations of the masking effect caused by flight noise," by W. Lorenz. MSCHR OHRENHEILK 103:438-444, 1969.

"The audiologic picture and histologic substrate of noise-induced internal ear injury," by W. Lorenz. DEUTSCH GESUNDH 23:2423-2426, December 19, 1968.

"Audiometric configurations associated with blast trauma," by D. I. Teter, et al. LARYNGOSCOPE 80:1122-1132, July, 1970.

"Audiometric findings in a metallurgic industrial plant," by J. E. Fournier, et al. ARCH MAL PROF 28:523-529, June, 1967.

"Audiometric studies in teletype operators," by G. Bocci, et al. BOLL MAL ORECCH 84:190-199, March- April, 1966.

"Audiometric studies of hearing in workers in noisy shops of the Krasnodar factory 'Traktorsel'khozzapchase'," by E. A. Melvnikova. GIG TR PROF ZABOL 10:34-38, July, 1966.

"An audiometric survey of a Canadian armoured regiment. A follow-up to the 1963 report," by G. G. Jamieson. MED SERV J CANADA 23: 1313-1320, December, 1967.

"Audiometric survey of a metallurgy enterprise," by J. E. Fournier, et al. ARCH MAL PROF 31:523-529, June, 1967.

"Audiometry and the medical department's function in management of noise problems," by A. J. Murphy. NAT SAFETY CONGR TRANS 18: 14-17, 1969.

"Auditory and subjective effects of airborne noise from industrial ultrasonic sources," by W. I. Acton, et al. BRIT J INDUSTR MED 24:297-304, October, 1967.

"Auditory discomfort associated with use of the air turbine dental drill," by A. F. Smith, et al. J ROY NAV MED SERV 52:82-83, Summer, 1966.

"Auditory disorders caused by high speed equipment; data on 345 cases," by J. C. Barrancos Mooney. REV ASOC ODONT ARGENT 55: 327-337, August, 1967.

"Auditory effects of acoustic impulses from firearms," by K. D. Kryter, et al. ACTA OTOLARYNG Suppl 211:1-22, 1965.

"Auditory fatigue and predicted permanent hearing defects from rock-and-roll music," by F. L. Dey. NEW ENG J MED 282:467-470, February 26. 1970.

"Auditory figure-background perception in normal children; speech reception threshold," by B. M. Siegenthaler, et al. CHILD DEVELOP 38:1163-1167, December, 1967.

"Auditory hazards of sport guns," by R. R. Coles, et al. LARYNGOSCOPE 76:1728-1731, October, 1966.

"Auditory thresholds during visual stimulation as a function of signal bandwidth," by E. J. Moore, II, et al. J ACOUST SOC AMER 47:659-660, February, 1970.

"Aural trauma caused by electric shock by lightning," by T. Kobayashi. OTOLARYNGOLOGY 40:525-529, July, 1968.

"Auricular barotrauma," by Y. Husson. SEM HOP PARIS 44:2686-2689, October 26, 1968.

"Automobile noise-an effective method for control," U RICHMOND L REV 4:314, Spring, 1970.

"Autonomic reactions to hearing impressions," by G. Lehmann. STUD GEN 18:700-703, 1965.

"Autonomic reactions to low frequency vibrations in man," by R. Coermann, et al. INT Z ANGEW PHYSIOL 21:150-168, September 13, 1965.

"Aversive stimulation as applied to discrimination learning in mentally retarded children," by P. S. Massey, et al. AM J MEN DEFICIENCY 74:269-272, September, 1969.

"Aviation noise--evaluation and possibilities of its elimination," by J. Hilscher. Z GES HYG 15:670-673, September, 1969.

"B. F. Goodrich taps $500 million noise market," RUBBER WORLD 155:32-33, November, 1966.

"Background light, temperature and visual noise in the turtle," by W. R. Muntz, et al. VISION RES 8:787-800, July, 1968.

"The 'bang' of supersonic airplanes," ADM 25:341-351 contd, July-August, 1968.

"Bang on!" ECONOMIST 215:426, April 24, 1965.

"Barotraumatic damage of hearing organ in divers," by Z. Sliskovic. VOJNOSANIT PREGL 26:75-77, February, 1969.

"Basics of hearing protection," ENVIRONMENTAL CONTROL MGT 138:60-62, December, 1969.

"Basics of noise control; sound power and sound pressure," by G. Diorio. HEATING-PIPING 39:194-195, April, 1967.

"Behavior of glucose-6-phosphatase, glycogen and PAS-positive substances in the liver of guinea pigs following chronic exposure to noise," by J. Jonek, et al. Z MIKR ANAT FORSCH 72:256-263, 1965.

"Behavioral and skin potential response correlations in chronic schizophrenic patients," by R. Wyatt, et al. COMPR PSYCHIAT 10:196-200, May, 1969.

"Behind the scenes; rise in hearing impairment attributed to increasing noise pollution," by B. Whyte. AUDIO 53:8+, March, 1969.

"Mr. Benn's parish," by W. A. Shurcliff, et al. NATURE (London) 221:693-694, February 15, 1969.

"Better noise measurement pins down trouble spots," by R. W. Carson. PRODUCT ENG 37:28-30, October 24, 1966.

"Bibliography on transformer noise," INST ELEC & ELECTRONICS ENG TRANS POWER APPARATUS & SYSTEMS 87:372-387, February, 1968.

"'Big-beat' music and acoustic traumas," by H. Kowalczuk. OTOLARYNG POL 21:161-167, 1967.

"Bilateral asymmetry in noise induced hearing loss," by J. E. Watson. ANN OTOL 76:1040-1042, December, 1967.

"Binaural masking of speech by periodically modulated noise," by R. Carhart, et al. J ACOUST SOC AMER 39:1037-1050, June, 1966.

"Binaural summation of thermal noises of equal and unequal power in each ear," by R. J. Irwin. AMER J PSYCHOL 78:57-65, March, 1965.

"Biochemical and hematological reactions to noise in man," by V. Hrubes. ACTIV NERV SUP (Praha) 9:245-248, August, 1967.

"Biochemical observations of the alteration of DPN-diaphorase activity in the organ of Corti after noise exposure," by M. Tateda. J OTORHINOLARYNG SOC JAP 70:1312-1329, July, 1967.

"Bioflavonoid therapy in sensorineural hearing loss: a double-blind study," by J. E. Creston, et al. ARCH OTOLARYNG 82:159-165, August, 1965.

"The biological action of impulse noises (a review of the literature)," by V. N. Morozov, et al. VOENNOMED ZH 8:53-58, August, 1969.

"A bio-physical law describing hearing loss," by E. R. Hermann. INDUSTR MED SURG 34:223-228, March, 1965.

"Block that noise! try earmuffs," PULP & PA 41:58, September 11, 1967.

"The blood system reaction to the occupational effect of vibration and noise," by I. A. Gribova, et al. GIG SANIT 30:34-37, October, 1965.

"Blower whine silenced by tuning inlet and outlet," PRODUCT ENG 37:59+, May 23, 1966.

"Boom! here comes the noise rules explosion," by K. A. Kaufman. IRON AGE 205:19, April 16, 1970.

"Boom nobody wants," NATIONS BSNS 56:76-78, September, 1968.

"Boom problem still clouds SST future," by C. M. Plattner. AVIATION W 86:28-29, January 9, 1967.

"The boom that's brewing a storm; sonic boom generated by Boeing's

SST has stirred up critics who seek to clip the airplane's wings; before craft flies in mid-1970s, airlines and government face some tough decisions," BUS WEEK 64-65+, October 28, 1967.

"Booms," LANCET 2:295-296, August 5, 1967.

"Booms banned; exhausts exempted," NATURE (London) 226:889-890, June 6, 1970.

"Breaking the sound barrier," SR SCHOL 97:15, December 7, 1970.

"Breakthrough in noise measurement," OCCUP HEALTH (London) 22:183, June, 1970.

"Bringing peace to the noisy office," by N. C. Crane. SUPERVISORY MGT 10:42-43, November, 1965.

"Building acoustics," ed. by B. F. Day, et al. Review by C. Pinfold. TOWN PLAN R 41:300-301, July, 1970.

"Building or buying? Here's what to look for," TODAY'S HLTH 43:59, February, 1965.

"The c5 declivity; its interpretation on the basis of universally valid physiological concepts," by E. Lehnhardt. HNO 14:45-52, February, 1966.

"Campaign combats noise problem," MOD HOSP 107:14, September, 1966.

"Can hearing aids damage hearing?" by C. Roberts. ACTA OTOLARYNG 69:123-125, January-February, 1970.

"Can we quiet big tractors?" by B. Coffman. FARM J 93:26F+, August, 1969.

"Can we regain peace and quiet? War on noise sparks some hope," by B. Jackson. FIN POST 64:37-44, February 14, 1970.

"Can you hear the hydraulics system?" by J. S. Noss. MACHINE DESIGN 42:141-145, September 17, 1970.

"Canadian meeting says U.S. accepts too high sound levels," by F. J. Versagi. AIR COND HEAT & REFRIG N 106:1+, November 29, 1965.

"Can't we stop that terrible din?" MACL MAG 83:7, January, 1970.

"Capacity and noise relationships for major hub airports," by R. L. Paullin. IEEE PROC 58:307-313, March, 1970.

"Cardiac hypertrophy due to chronic audiogenic stress in the rat. Rattus norvegicus albinus, and rabbit, Lepus cuniculus," by W. F. Geber, et al. COMP BIOCHEM PHYSIOL 21:273-277, May, 1967.

"Caring for the total patient. Noise in hospitals: its effect on the patient," by P. Haslam. NURS CLIN NORTH AM 5:715-724,, December, 1970.

"Carriers urged to act on social problems," by R. G. O'Lone. AVIATION W 91:53-55, August 11, 1969.

"Case findings in hearing disorders in industry: the 'audiometric car;' 1st results," by J. C. Lafon, et al. J MED LYON 49:1831-1834, November 20, 1968.

"A case of human sacrifice," by P. J. Smith. SPECTATOR 593, May 3, 1968.

"A case of lethal outcome in severe acoustic trauma in the dog," by L. P. Rudenko. ZH VYSSH NERV DEIAT PAVLOV 15:105-108, January-February, 1965.

"The case of the unfortunate construction worker," by S. L. Shapiro. EYE EAR NOSE THROAT MON 49:383-386, August, 1970.

"Catch in the rye; it can make you deaf," by K. Mitchell. MACL MAG 80: 3, August, 1967.

"Central periodicity pitch," by P. C. Nieder, et al. J ACOUST SOC AMER 37:136-138, January, 1965.

"Centrifugal fan sound power level prediction," by G. C. Groff, et al. ASHRAE J 9:71-77, October, 1967.

"Certain aspects of hypacusia caused by acoustic trauma studies with Bekesy's audiometer," by M. Maurizi, et al. ARCH ITAL LARING 76: 131-146, May-June, 1968.

"Certain problems of industrial hygiene in processing plastics by casting under pressure," by A. M. Dzhezhev. GIG SANIT 32:19-22, March, 1967.

"CF6 is tested for noise, flow pattersn; illustrations with text," AVIATION W 91:34-35, October 20, 1969.

"The challenge of hearing protection," by R. Maas. INDUSTR MED SURG 39:124-128, March, 1970.

"Change in the electrical activity of muscles under the effect of noise and vibration," by A. F. Lebedeva. GIG SANIT 33:25-30, March, 1968.

"Changes in acetylcholine concentration in cerebal tissue in rate repeatedly exposed to the action of mechanical vibration," by Z. Brzezinska. ACTA PHYSIOL POL 19:919-926, November-December, 1968.

"Changes in the adrenal glands under the effect of noise," by V. P. Osintseva, et al. GIG SANIT 34:119-122, October, 1969.

"Changes in the auditory pain threshold induced by nitrous oxide," by F. Fruttero, et al. MINERVA OTORINOLARING 16:135-137, July-August, 1966.

"Changes in the bioelectric activity of the brain and in some autonomic and vascular reaction under the influence of noise," by E. A. Drogichina, et al. GIG SANIT 30:29-33, February, 1965.

"Changes in hearing acuity of noise-exposed women," by S. Pell, et al. ARCH OTOLARYNG 83:207-212, March, 1968.

"Changes in the serum level of unsaturated fatty acids in rats repeatedly exposed to noise," by M. Vondrakova. ACTIV NERV SUP (Praha) 7:236-238, August, 1965.

"Changes of the background noise intensity and the bar-pressing response rate," by K. Zielinski. ACTA BIOL EXP 26:43-53, 1966.

"Channel-capacity, intelligibility and immediate memory," by P. M. Rabbitt. QUART J EXP PSYCHOL 20:241-248, August, 1968.

"Characteristics of man's sleep under conditions of continuous protracted effect of broad-band noise of average intensity," by B. I. Miasnikov, et al. IZV AKAD NAUK SSSR 1:89-98, January-February, 1968.

"Choosing ear protectors," SUPERVISORY MANAGEMENT 15:32-35, January, 1970.

"Chords well lost?" by M. E. Drew. NURS MIRROR 131:15, July 31, 1970.

"Christmas cracker or duogastrone symptoms," by C. C. Evans, et al. BRIT MED J 1:120-121, January 11, 1969.

"Chronic otitis media in protection against occupational deafness," by A. Monteiro, et al. HOSPITAL (Rio) 70:1173-1178, November, 1966.

"Chrysler pavilion," by W. R. Farrell, et al. AUDIO 49:28+, April, 1965.

"Cities lend an ear to noise control," BSNS W p 108, January 17, 1970.

"Citizens vs. noise," SCI & CIT 10:31, March, 1968.

"City; a challenge to engineering and political sciences," by P. R. Achenbach. ASHRAE J 11:33-38, March, 1969.

"City noise; designers can restore quiet, at a price," by H. W. Bredin. PRODUCT ENG 39:28-35, November 18, 1968.

"City noise--a sociological and psychological study in a defined environment. Experimental pilot studies on the occurrence of noise nuisances and their correlation with various psychological variables," by O. Arvidsson, et al. NORD HYG T 46:153-188, 1965.

"Click-evoked response patterns of single units in the medial geniculate body of the cat," by L. M. Aitkin, et al. J NEUROPHYSIOL 29:109-123, January, 1966.

"Click-intensity discrimination with and without a background masking noise," by D. H. Raab, et al. J ACOUST SOC AMER 46:965-968, October, 1969.

"The clinical and audiological pattern due to acoustic trauma," by A. Salmivalii. ACTA OTOLARYNG Suppl 224:239+, June 27, 1966.

"Clinical and experimental aspects of the effects of noise on the central nervous system," by F. Angeleri, et al. MED LAVORO 60:759-766, December, 1969.

"Clinical and medico-legal considerations on hypoacusias of metallurgic workers," by M. Ciulla, et al. BOLL SOC MEDICOCHIR CREMONA 19:129-141, January-December, 1965.

"Clinical observations of the hearing loss in perceptive deafness," by I. Kirikae, et al. OTOLARYNGOLOGY 40:599-606, August, 1968.

"Clinical, social and insurance evaluation of professional deafness due to noise," by E. Vensi. ANN LARING 64:337-343, 1965.

"Clinical study of the behavior of auditory adaptation in textile workers," by L. Bertocchi, et al. MINERVA OTORINOLARING 16:106-109, May-June, 1966.

"Clinical trial of EU 4200 in the treatment of sound injuries," by J. Cazaubon, et al. MED TROP 30:403-408, May-June, 1970.

"Clinico-physiologic assessment of the efficacity of certain types of modern antinoise ear caps," by N. Ia. Shalashov. GIG TR PROF ZABOL 14:46-47, May, 1970.

"Coast noise verdict may set precedent," AVIATION W 88:35, January 1, 1968.

"Cochlear hair-cell damage in guinea pigs after exposure to impulse noise," by L. B. Poche, Jr., et al. J ACOUST SOC AMER 46:947-951, October, 1969.

"Cochleo-vestibular and general disorders induced by noise. The role of spas," by P. Molinery, et al. PRESSE THERM CLIMAT 103:5-9, 1966.

"Cockpit noise environment of airline aircraft," ACOUSTICAL SOC AM J 47:449, February, 1970.

"Cockpit noise environment of airline aircraft," by R. B. Stone. AERO-SPACE MED 40:989-993, September, 1969.

"Cockpit noise intensity: fifteen single-engine light aircraft," by J. V.

Tobias. AEROSPACE MED 40:963-966, September, 1969.

"Cockpit noise intensity: fifteen single-engine light aircraft. AM 68-21," by J. V. Tobias. US FED AVIAT AGENCY OFFICE AVIAT MED 1-6, September, 1968.

"Cockpit noise intensity: three aerial application (cropdusting) aircraft," by J. V. Tobias. J SPEECH HEARING RES 11:611-615, September, 1968.

"The code copying capabilities of radiomen under simulated atmospheric noise conditions. Report No. 523," by A. M. Richards. US NAVAL SUBMAR MED CENT 523:1-4, May 15, 1968.

"Cold loads as standard noise sources," by A. Jurkus. IEEE PROC 53: 176-177, February, 1965.

"Combating valve noise in process control systems," by C. B. Schuder, et al. AUTOMATION 16:50-53, September, 1969.

"Combined effects of chronic audio-visual stress and thiouracil administration on the cholesterol-fed rat," by T. A. Anderson, et al. J CELL COMP PHYSIOL 66:141-145, October, 1965.

"Combined effects of I-131 and noise on the cardiac activity in the dog," by T. Mukhamedov. BIULL EKSP BIOL MED 59:43-46, February, 1965.

"Combustion roar of turbulent diffusion flames," by R. D. Giammar, et al. J ENG POWER 92:157-165, April, 1970.

"Coming to terms with traffic noise," ENGINEERING 201:374-375, February 25, 1966.

"Commonly-used terms and definitions for understanding noise regulation," ENVIRONMENTAL CONTROL MGT 139:26-27, January, 1970.

"Communal hygienic studies of the noise problem in Warsaw," by A. Brodniewicz. Z GES HYG 13:760-764, October, 1967.

"Community care today and tomorrow," by A. W. Macara. DISTRICT NURS 9:210+, December, 1966.

"Community noise--the industrial aspect," by K. M. Morse. AMER INDUSTR HYG ASS J 29:368-380, July-August, 1968.

"Community noise ordinances," by A. E. Meling. ASHRAE J 9:40-43+, May, 1967.

"Community noise problems--origin and control," by L. S. Goodfriend. AMER INDUSTR HYG ASS J 30:607-613, November-December, 1969.

"Companies warned: quieter, please!" BSNS W 28-29, July 26, 1969.

"Comparative Bekesy typing with broad and modulated narrow-band noise," by C. T. Grimes, et al. J SPEECH HEARING RES 12:840-846, December, 1969.

"Comparative electrophysiological, histological and biochemical studies on guinea pigs following application of white noise," by E. A. Schnieder, et al. ARCH KLIN EXP OHR NAS KEHIKOPFHEILK 194: 579-583, December 22, 1969.

"Comparative examinations of the visual field in glaucoma and the influence of noise," by E. Ogielska, et al. KLIN OCZNA 36:351-354, 1966.

"Comparative provisions for occupational hearing loss," by M. S. Fox. ARCH OTOLARYNG 81:257-260, March, 1965.

"Comparative study on steady and intermittent exposures to noise. Ammonia content in the brain as an indicator," by K. Matsui, et al. JAP J HYG 23:225-228, June, 1968.

"Comparison between the sound threshold audiogram and a speech audiogram in subjects with hearing impairment due to noise," by G. Fabian. Z GES HYG 14:165-169, March, 1968.

"Comparison of Bekesy threshold for small band and sinus sounds in normal hearing and acoustic trauma," by R. Fischer, et al. ARCH KLIN EXP OHREN NASEN KEHIKOPFHELIKD 196:223-227, 1970.

"Compendium of human responses to the aerospace environment. 9. Sound and noise. NASA CR-1205 (2)," by E. M. Roth, et al. US NASA 9:1-101, November, 1968.

"Compensation claims for loss of hearing: impact of standards," by F. E. Frazier. ARCH ENVIRON HEALTH 10:572-575, April, 1965.

"Compensation claims for loss of hearing; impact of standards," by F. E. Frazier. AMER ASS INDUSTR NURSES J 15:17-19 passim, May, 1967.

"Complex laboratories for measuring vibration and noise," by O. E. Guzeev, et al. GIG SANIT 35:100-101, April, 1970.

"Compressed air: on the quiet," ENGINEERING 206:775, November 22, 1968.

"Computer graphics aid solution to DC-9 cabin noise problem," AVIATION W 87:69+, July 17, 1967.

"Computer polices airport noise," ELECTRONICS 42:183-184, July 7, 1969.

"Computer room noises seen as hearing threat," IND RES 12:24-25, August, 1970.

"Computer units pose hearing threats," AUTOMATION 17:5, August, 1970.

"The concept of susceptibility to hearing loss," by W. D. Ward. J OCCUP MED 7:595-607, December, 1965.

"Concepts of perceived noisiness, their implementation and application," by K. D. Kryter. J ACOUST SOC AMER 43:344-361, February, 1968.

"Concrete breakers must be quieter," ENGINEERING 203:1050, June 30, 1967.

"Conference on acoustic noise and its control, London, January 23-27; with list of authors and titles of contributed papers," ACOUSTICAL SOC AM J 41:1383-1384, May, 1967.

"Conference on noise," NAT PARKS 42:21, August, 1968.

"Congress prepares to sound off on airport noise," by R. Leiser. AM AVIATION 30:81-82, September, 1966.

"Connecticut gears up research program to combat motor vehicle noise pollution on state highways," INSTRUMENTATION TECH 16:18, March, 1969.

"Connecticut may ban noisy vehicles; state police test acoustic radar," by L. Klein. MACHINE DESIGN 42:28-29, April 16, 1970.

"Consequences of noise," by H. Y. Bell. IND ARTS & VOC ED 54:74, March, 1965.

"Conservation of hearing programmes in North America," by O. M. Drew. OCCUP HLTH 20:179+, July- August, 1968.

"Considerations on evaluation of acoustic and vestibular damage in pilots and specialists of military aeronautics," by C. Koch, et al. MINERVA MED 56:3832-3835, November 10, 1965.

"Considerations on the phenomena of adaptation to white noise," by I. De Vincentiis, et al. VALSALVA 42:247-254, October, 1966.

"Considerations on the problem of noise in relation to its influence on some aspects of modern life. I. Behavior of the temporary shift of hearing sensitivity due to high energy level stimulations in persons employed in work with noise," by S. Collatina, et al. CLIN OTORINO-LARING 17:357-370, July-August, 1965.

"Consumers and legislators shout down noisy products," MACHINE DESIGN 42:41, April 30, 1970.

"Continuity effects with alternately sounded tone and noise signals," by L. P. Elfner, et al. MED RES ENGIN 5:22-23, 1966.

"Continuity in alternately sounded tone and noise signals in a free field," by L. F. Elfner. J ACOUST SOC AMER 46:914-917, October, 1969.

"Contractors suggest revisions to proposed Los Angeles noise ordinance," AIR COND HEAT & REFRIG N 121:2, November 16, 1970.

"Contralateral masking and the SISI-test in normal listeners," by B. Blegvad, et al. ACTA OTOLARYNG 63:557-563, June, 1967.

"Contralateral remote masking and the aural reflex," by K. Gjaevenes, et al. J ACOUST SOC AMER 46:918-923, October, 1969.

"Contribution on the problem of acoustic accidents," by G. Kittinger. WIEN MED WSCHR 116:653-655, July 23, 1966.

"Contribution to the evaluation of industrial noise," by J. Mayer. MSCHR OHRENHEILK 101:462-467, 1967.

"Contribution to the investigations on the influence of noise on hearing and the circulatory system in ship-yard employees," by S. Klajman, et al. MED PRACY 16:380-384, 1965.

"Contribution to the study of acoustic stimulation on the blood eosinophil count," by A. Amorelli, et al. ARCH ITAL LARING 73:515-520, November-December, 1965.

"Contribution to the study of the noise caused by street traffic," by G. Sparacio, et al. G IG MED PREV 7:385-394, October-December, 1966.

"Contribution to the study on acoustic trauma and its evolution," by M. Lutovac. VOJNOSANIT PREGL 26:235-236, May, 1969.

"Control of centrifugal fan noise in industrial applications," by F. Oran, et al. AIR COND HEAT & VEN 62:85-96, May, 1965.

"Control of duct generated noise," by V. V. Cerami, et al. AIR COND HEAT & VEN 63:55-64, September, 1965.

"Control of industrial noise through personal protection," by R. R. A. Coles. AM SOC SAFETY ENG J 14:10-15, October, 1969.

"Control of mining noise exposure," by J. H. Botsford. MIN CONG J 53:22-24+, August, 1967.

"Control of neighborhood noise as a hygienic-legal problem," by A. Schubert. Z AERZTL FORTBILD (Jena) 63:1056-1058, October 1, 1969.

"Control of noise," by B. Berger. INST E E PROC 116:444-445, March, 1969.

"Control of noise and hearing conservation," by B. B. Bauer. AUDIO ENG SOC J 17:450+, August, 1969.

"Control of noise from the viewpoint of history. Prehistoric periods, age of the ancient cultures etc.," by H. Wiethaup. ZBL ARBEITSMED 16:120-124, May, 1966.

"The control of vibration and noise," by T. P. Yin. SCI AMER 220: 98-106, January, 1969.

"Controlling foundry noise," by C. H. Borcherding, Jr. FOUNDRY 98: 167-169, June, 1970.

"Controlling noise from rotating electrical machines," ENGINEER 227: 309, February 28, 1969.

"Controlling noise pollution: nascent market," by P. Thomas. IND MKTG 55:37-40, September, 1970.

"Controlling plant noise," by J. M. Handley. MACH 75:81-84, May, 1969.

"Controlling process-plant noise," by J. K. Floyd. MECH ENG 90:23-26, October, 1968.

"Controlling the tonal characteristics of the aerodynamic noise generated by fan rotors," by R. C. Mellin, et al. J BASIC ENG 92:143-154, March, 1970.

"Conveying through a sound barrier," AUTOMATION 16:57, July, 1969.

"Cooperation, understanding make Coral Gables noise law workable," by P. B. Redeker. AIR COND HEAT & REFRIG N 112:1+, December 4, 1967.

"Coping with noise from big motors," by R. L. Nailen. POWER ENG 72:54-57, October, 1968.

"Coral Gables noise code gets grudging approval," by G. Duffy. AIR COND HEAT & REFRIG N 116:1+, April 21, 1969.

"Coral Gables would limit noise from air conditioners," by C. D. Mericle. AIR COND HEAT & REFRIG N 108:1 , May 9, 1966.

"Corona-generated noise in aircraft," by C. E. Cooper. WIRELESS WORLD 72:547-552, November, 1966.

"Correlation of vortex noise data from helicopter main rotors," by S. E. Widnall. J AIRCRAFT 6:279-281, May, 1969.

"Correlations between the noise level and conditioned reflex performance," by K. Hecht, et al. ACTA BIOL MED GERMAN 23:133-143, 1969.

"Costly decibels," by R. O. Fehr. AUDIO ENG SOC J 18:110-111, February, 1970.

"Counting the cost of noise," COUNTRY LIFE 148:444, August 20, 1970.

"Court case on noisy residential unit causes reader to remind all of responsibility," by W. R. Brown. AIR COND HEAT & REFRIG N 105:14, August 9, 1965.

"Criterion for estimating spacecraft shroud acoustic field reductions," by S. M. Kaplan. J ENVIRONMENTAL SCI 12:27-29, February, 1969.

"Crossed centrifugal inhibitions and excitations in the nucleus cochlearis due to long click sequences in guinea pigs," by R. Pfalz. ARCH KLIN EXP OHR NAS KEHLKOPFHEILK 186:9-19, 1966.

"Crusader for quiet," TIME 94:85, December 5, 1969.

"Current methods of evaluation of noise," by J. Calvet, et al. ANN OTOLARYNG 86:113-126, March, 1969.

"Current noise work may add V/STOLs," by W. J. Normyle. AVIATION W 88:123+, June 24, 1968.

"Current prevention of occupational acoustic trauma," by W. Sulkowski. MED PRACY 18:51-59, 1967.

"Cutting noise in data sampling," by T. Kobylarz. ELECTRONICS 41:70-73, May 13, 1968.

"Cutting the noise of an air compressor," ENGINEER 228:34, June 26, 1969.

"Damage caused by noise in weavers and preventive measures to reduce it," by G. Bologni, et al. ANN LARING 66:338-347, 1967.

"Damage due to industrial noise," by H. Guttich. MUNCHEN MED WSCHR 107:1397-1406, July 16, 1965.

"Damage of the ear by thunderbolt," by A. Spirov. VOJNOSANIT PREGL 25:648-651, December, 1968.

"Damage risk criterion and contours based on permanent and temporary hearing loss data," by K. D. Kryter. AMER INDUSTR HYG ASS J 26:34-44, January-February, 1965.

"Damage-risk criterion for the impulsive noise of 'toys'," by K. Gjaevenes. J ACOUST SOC AMER 42:268, July, 1967.

"Dampening properties of hearing protective devices with the view to the degree of inconvenience in their use," by J. Stikar, et al. CESK HYG 10:304-313, June, 1965.

"Damping material cuts noise, vibration (called Deadbeat)," PURCHASING 61:76, December 1, 1966.

"Damping the noise of 6500-12-1 blowers (smoke exhaust fans)," by B. Z. Shushkovskii. GIG TR PROF ZABOL 11:47-48, September, 1967.

"Dance bands can deafen," NEW ZEALAND NURS J 62:14, May, 1969.

"The danger of noise--a threat against our health," SVENSK FARM T 70:775-776, November 20, 1966.

"Dangers of urbanization," DISTRICT NURS 10:125, September, 1967.

"Data for hygenic assessment of impulse noise," by E. Ts. Andreeva-Galanina, et al. GIG SANIT 33:24-29, August, 1968.

"DC-8 stretched jets should produce less noise annoyance at airports," by R. E. Black. SAE J 75:51, June, 1967.

"Deafening music," BRIT MED J 2:127, April 18, 1970.

"Deafness in agricultural tractor drivers," by J. P. Vallee. GAZ MED FRANCE 72:3193-3194, October 10, 1965.

"Deafness in dentists," by C. Wark. J OTOLARYNG SOC AUST 2:89, March, 1967.

"Decade of engineering needed for cleaner, quieter environment," PRODUCT ENG 41:17-18+, January 1, 1970.

"Defense against technology," by N. E. Golovin. ASTRONAUTICS & AERONAUTICS 7:20-23, December, 1969.

"Definition of human requirements in respect to noise, to be used by town planners and builders," J SCI MED LILLE 87:328-329, April, 1969.

"Demand meaningful sound trap ratings!" by V. V. Cerami, AIR COND HEAT & VEN 63:53-54, July, 1966.

"Demonstration of nyctohemeral variations of susceptibility to audiogenic crisis in Swiss albino mice," by C. Poirel. C R SOC BIOL 161:1461-1465, 1967.

"Dentists' hearing: the effect of high speed drill," by B. A. Skurr, et al. AUST DENT J 15:259-260, August, 1970.

"Dentists' hearing: the effect of highspeed drill noise," by V. Bulteau. MED J AUST 2:1111, December 9, 1967.

"Dependence of pharmacologic effects on the noise level, " by K. Hecht, et al. ACTA BIOL MED GERMAN 23:121-132, 1969.

"Design and decisions in machinery sound control," by G. M. Diehl. COMP AIR MAG 75:7-9, August, 1970.

"Design factors and use of ear protection," by C. G. Rice, et al. BRIT J INDUSTR MED 23:194-203, July, 1966.

"Design for quiet," MACHINE DESIGN 39:174-224, September 14, 1967.

"Design for quiet; air-moving systems," by W. F. Walker, et al. MACHINE DESIGN 39:202-208, September 14, 1967.

"Design for quiet; anatomy of noise," by L. L. Beranek, et al. MACHINE DESIGN 39:174-183, September 14, 1967.

"Design for quiet; electrical systems," by J. Campbell. MACHINE DESIGN 39:192-199, September 14, 1967.

"Design for quiet; hydraulic systems," by J. W. Sullivan. MACHINE DESIGN 39:210-215, September 14, 1967.

"Design for quiet; measuring noise," by J. Campbell. MACHINE DESIGN 39:216-224, September 14, 1967.

"Design studies urged at (London) noise conference," AVIATION W 85:67-68, December 12, 1966.

"Detecting and correcting noises in the hospital," by S. Sorenson, et al. HOSPITALS 42:74-80, November 1, 1968.

"Detection of hearing impairment in a large business in the Paris area. Initial results," by P. Housset, et al. ARCH MAL PROF 27:710-713, September, 1966.

"Determination of simple reaction time to visual and acoustic stimuli under conditions of noise interference," by K. Witecki. POL TYG LEK 25:1012-1015, July 6, 1970.

"Determination of sound pressure levels created by noise sources at standard points," by I. K. Razumov. GIG SANIT 34:29-32, November, 1969.

"Detonation-traumatic load during military service," by C. J. Partsch, et al. ARCH KLIN EXP OHR NAS KEHLKOPFHEILK 191:581-586, 1968.

"Detroit studing ordinance to control noisy air conditioners," by G. M. Hanning. AIR COND HEAT & REFRIG N 106:1+, September 20, 1965.

"Development of a personal monitoring instrument for noise," by F. W. Church. AMER INDUSTR HYG ASS J 26:59-63, January-February, 1965.

"The development and results of the study of noise and vibration in railroad transportation," by A. M. Volkov. GIG TR PROF ZABOL 11:58-60, November, 1967.

"Development of engineering practices in jet and compressor noise," by J. B. Large, et al. J AIRCRAFT 6:189-195, May, 1969.

"The development of the startle reflex in the postnatal ontogenesis of the rat," by I. Friedrich, et al. ACTA BIOL MED GERMAN 19:

605-607, 1967.

"DICANNE, a realizable adaptive process," by V. C. Anderson. ACOUSTICAL SOC AM J 45:398-405, February, 1969.

"Dichotic summation of loudness," by B. Scharf. ACOUSTICAL SOC AM J 45:1193-1205, May, 1969.

"Diesel engines don't have to be so noisy," by A. E. W. Austen, et al. SAE J 73:71-75, May, 1965.

"Differential effect of noise on tasks of varying complexity," by D. H. Boggs, et al. J APPL PSYCHOL 52:148-153, April, 1968.

"The differential threshold of sound intensity perception in riveters," by N. Ia. Shalashov. ZH USHN NOS GORL BOLEZ 28:70-75, November-December, 1968.

"Differentiation of the direction of hearing in noise," by J. Laciak. OTOLARYNG POL 22:787-791, 1968.

"The dilute locus, pyridoxine deficiency, and audiogenic seizures in mice," by K. R. Henry, et al. PROC SOC EXP BIOL MED 128:635-638, July, 1968.

"Diminish airport noise? We can't yet measure it," by B. Jackson. FIN POST 62:14, July 20, 1968.

"Direct test of the power function for loudness," by L. E. Marks, et al. SCIENCE 154:1036-1037, November 25, 1966.

"The direction of change versus the absolute level of noise intensity as a cue in the CER situation," by K. Zielinski. ACTA BIOL EXP 25:337-357, 1965.

"Directional audiometry. I. Directional white-noise audiometry," by F. M. Tonning. ACTA OTOLARYNG 69:388-394, June, 1970.

"Discovery and surveillance of occupational acoustic trauma in preventive medicine," by Y. Guerrier, et al. J FRANC OTORHINOLARYNG 14:237-247, May, 1965.

"Discrete frequency noise generation from an axial flow fan blade row," by R. Mani. J BASIC ENG 92:37-43, March, 1970.

"Discussion on deafness," PRACTITIONER 194:691-693, May, 1965.

"Distraction and Stroop Color-Word performance," by B. K. Houston, et al. J EXP PSYCHOL 74:54-56, May, 1967.

"Disturbance by noise from the medical and legal viewpoint," by H. Wiethaup. MED KLIN 60:1218-1220, July 23, 1965.

"Do-it-yourself acoustic structures can be designed, assembled by the purchaser," AIR COND HEAT & REFRIG N 105:12, July 19, 1965.

"Do-it-yourself style: sound attenuation (air diverter roof-mounted to residential air conditioner)," by F. J. Versagi. AIR COND HEAT & REFRIG N 107:24, February 7, 1966.

"Does noise control need the pill?" by C. J. Suchocki. IRON AGE 205: 58-59, April 2, 1970.

"Does rock music damage hearing?" GOOD H 168:208, April, 1969.

"Dofasco's program to save those eardrums," CAN BUS 43:29, June, 1970.

"Don't push the panic button," by E. Watson. OCCUP HLTH NURS 18:10+, October, 1970.

"Doughnuts of foam absorb fan noise," PRODUCT ENG 38:52-53, July 3, 1967.

"Down with decibles!" by O. Schenker-Sprungli. UNESCO COURIER 20:4-7, July, 1967.

"Drilling by ear," by J. Kuletz. AM MACH 110:63-65, July 4, 1966.

"Dust control and sound abatement," by D. Jackson, Jr. COAL AGE 71:66-70+, November, 1966.

"Dust- and soundproof stone-crushing plant protects workers and environment," PIT & QUARRY 62:134, February, 1970.

"Duties of the 'counseling otologist' of the medical labor inspection division of the district council," by H. H. Frey. DEUTSCH GESUNDH 20:687-689, April 15, 1965.

"Dynamic ratings for duct silencers," ENGINEER 223:120, January 20, 1967.

"Dynamic response of middle-ear structures," by H. Fischler, et al. ACOUSTICAL SOC AM J 41:1220-1231, May, 1967.

"Dynamics of glycemic reactions after repeated exposure to noise," by K. Treptow. ACTIV NERV SUP 8:215-216, June, 1966.

"Ear as a measuring instrument," by H. Fletcher. AUDIO ENG SOC J 17:532-534, October, 1969.

"Ear damage from exposure to rock and roll music," by D. M. Lipscomb. ARCH OTOLARYNG 90:545-555, November, 1969.

"An ear defender with peak limited sound transmission--Erdefender," by P. J. Barker. OCCUP HLTH 20:67+, March-April, 1968.

"Ear injuries and industry," by F. Montreuil. UN MED CANADA 95:1299-1306, November, 1966.

"Ear injury," by R. Willis. J OTOLARYNG SOC AUST 2:75-80, March, 1967.

"An ear plug designed to obtund the sound of high speed dental engines," by R. B. Sloane. DENT DIG 72:218-220, May, 1966.

"Ear plugs--or--damaged hearing," by R. R. Coles. J ROY NAV MED SERV 51:19-22, Spring, 1965.

"Ear pollution!" by F. Graham, Jr. AUDUBON 71:34-39, May, 1969.

"Ear protectors," by W. Melnick. OCCUP HEALTH NURS 17:28-31, May, 1969.

"Ear protectors; their usefulness and limitations," by P. L. Michael. ARCH ENVIRON HEALTH 10:612-618, April, 1965.

"Ear stoppies. A defense against snoring," by M. H. Boulware. J FLORIDA MED ASS 57:36, May, 1970.

"Ears need protection from ultrasonics too," SAFETY MAINT 133:41-42, January, 1967.

"East Range Symposium. I. Hearing loss: diagnosis and anatomic considerations," by C. J. Holmberg. J OCCUP MED 7:138-144, April, 1965.

--II. Hearing conservation programs," by M. T. Summar. J OCCUP MED 7:145-146, April, 1965.

--IV. Principles of noise control," by R. Donley. J OCCUP MED 7: 222-226, May, 1965.

--V. Personal protection in noise exposure," by M. T. Summar. J OCCUP MED 7:279-280, June, 1965.

--VI. Criteria for assessing risk of hearing damage," by J. L. Fletcher. J OCCUP MED 7:281-283, June, 1965.

--8. Hearing conservation: preliminary report on a 7-year program," by M. T. Summar. J OCCUP MED 7:334-340, July, 1965.

"Ecological aspects of the affluence and effluents of energetics," by A. B. Campbell. ASHRAE J 11:49-59, March, 1969.

"Economics of aircraft noise suppression," by F. B. Greatrex. AIRCRAFT ENG 38:20+, November, 1966.

"The EEG and the impairment of sleep by traffic noise during the night: a problem of preventive medicine," by H. R. Richter. ELECTROENCEPH CLIN NEUROPHYSIOL 23:291, September, 1967.

"EEG measures of arousal during RFT performance in 'noise'," by R. W. Hayes, et al. PERCEPT MOT SKILLS 31:594, October, 1970.

"Effect of the acoustic reflex on the impedance at the eardrum," by A. S. Feldman, et al. J SPEECH HEARING RES 8:213-222, September, 1965.

"Effect of the acoustic stimulation on energy metabolism of the inner ear, with special reference to glycogen metabolism," by K. Ogawa. J OTOLARYNG JAP 73:335-352, March, 1970.

"The effect of the addition of vitamin C to the diet on the human organism exposed to occupational hazards (in the presence of strong sound stimuli)," by A. M. Margolis, et al. VOP PITAN 25:34-38, March-April, 1966.

"Effect of air-craft noise on gastric function," by C. Y. Kim, et al. YONSEI MED J 9:149-154, 1968.

"Effect of air noise on the human cardiovascular system," by I. A. Sapov, et al. VOEN MED ZH 6:53-54, June, 1970.

"The effect of aircraft noise on the population living in the vicinity of airports," by I. L. Karagodina, et al. GIG SANIT 34:25-30, May, 1969.

"The effect of alcohol and noise on components of a tracking and monitoring task," bt P. Hamilton, et al. BR J PSYCHOL 61:149-156, May, 1970.

"Effect of ambient-noise level on threshold-shift measures of ear-protector attenuation," by R. Waugh. J ACOUST SOC AM 48:597, August, 1970.

"The effect of aviation noise on some indices of protein and vitamin metabolism," by Iu. F. Udalov, et al. VOENNOMED ZH 7:61-64, July, 1966.

"Effect of background conversation and darkness on reaction time in anxious, hallucinating, and severely ill schizophrenics," by A. Raskin. PERCEPT MOTOR SKILLS 25:353-358, October, 1967.

"Effect of the caudate nucleus stimulation on audiogenic seizures in rate," by M. M. Oleshko. FIZIOL ZH 16:443-447, July-August, 1970.

"Effect of cochlear lesions upon audiograms and intensity discrimination in cats," by D. N. Elliott, et al. ANN OTOL 74:386-408, June, 1965.

"Effect of the combined action of noise and vibration on the sensitivity to vibration in adolescents," by A. I. Tsysar'. GIG SANIT 30:30-36, June, 1965.

"Effect of community noises on school children," by V. I. Pal'gov. PEDIAT AKUSH GINEK 5:29-32, September-October, 1966.

"Effect of a competing message on synthetic sentence identification," by C. Speaks, et al. J SPEECH HEARING RES 10:390-395, June, 1967.

"Effect of CR-121 (cleregil) on acoustic trauma," by J. Alavoine. THERAPEUTIQUE 45:526-531, May, 1969.

"The effect of defined noise exposure on the peripheral circulatory system," by A. Fuchs-Schmuck, et al. Z GES HYG 15:651-653, September, 1969.

"Effect of dietary synthetic and natural chelating agents on the zinc-deficiency syndrome in the chick," by F. H. Nielsen, et al. J NUTR 89:35-42, May, 1966.

"The effect of differing noise spectra on the consistency of identification of consonants," by A. C. Busch, et al. LANG SPEECH 10: 194-202, July-September, 1967.

"Effect of duration on amplitude discrimination in noise," by G. B. Henning, et al. J ACOUST SOC AMER 45:1008-1013, April, 1969.

"Effect of experimental impairment of sound conduction in the ear on understanding of speech in noise," by J. Kuzniarz. OTOLARYNG POL 22:531-536, 1968.

"Effect of extremely strong pulsed noise on certain sections of the central and peripheral nervous system," by O. S. Shemiakin. GIG TR PROF ZABOL 14:19-23, November, 1970.

"The effect of hearing aid use on the residual hearing of children with sensorineural deafness," by J. H. Macrae, et al. ANN OTOL 74:408-419, June, 1965.

"Effect of hearing protectors on speech comprehensibility," by Z. Novotny. CESK OTOLARYNG 18:260-264, December, 1969.

"The effect of impact noise with various background noises," by E. P. Orlovskaia. GIG SANIT 32:21-25, September, 1967.

"The effect of impulse noise on man," by A. A. Arkad'evski, et al. GIG SANIT 31:29-33, May, 1966.

"The effect of industrial noise in winding and weaving factories on the arterial pressure of operators," by A. I. Andrukovich. GIG TR PROF ZABOL 9:39-42, December, 1965.

"The effect of industrial noise on the organ of hearing in forging shop workers," by N. Ia. Shalashov. GIG TR PROF ZABOL 9:41-43, July, 1965.

"Effect of intensive noise and neuro-psychic tension on arterial blood pressure levels and frequency of hypertensive disease," by N. N. Shatalov, et al. KLIN MED 48:70-73, March, 1970.

"Effect of the interior noise of a passenger car on efficiency," by R. Schuster. BEITR GERICHTL MED 26:273-277, 1969.

"Effect of isolated mental stimuli on arterial pressure and body weight in Sprague-Dawley rats," by S. Campus, et al. BOLL SOC ITAL BIOL SPER 41:1087-1089, September, 1965.

"Effect of loud impulse noise on the hearing organ of animals," by N. I. Ivanov. VOEN MED ZH 7:24-27, July, 1970.

"Effect of loud industrial noise on hearing," by P. V. Kovalev. VOEN MED ZH 7:22-24, July, 1970.

"Effect of masked noise on the bone conduction threshold," by D. A. Pigulevskii, et al. ZH USHN NOS GORL BOLEX 27:38-43, March-April, 1967.

"Effect of masking noise upon syllable duration in oral and whispered reading," by M. F. Schwartz. J ACOUST SOC AMER 43:169-170, January, 1968.

"The effect of noise and local vibration of permissible levels on the human organism," by N. M. Paran'ko, et al. GIG SANIT 32:25-30, September, 1967.

"The effect of noise impulses on the organism as a function of the frequency of sequence and duration of impulse," by G. A. Suvorov. GIG TR PROF ZABOL 13:4-8, September, 1969.

"Effect of noise in the animal house on seizure susceptibility and growth of mice," by W. B. Iturrian, et al. LAB ANIM CARE 18:557-560, October, 1968.

"Effect of the noise level on the productivity of work," by S. D. Kovrigina, et al. GIG SANIT 30:28-32, April, 1965.

"The effect of the noise of electric machines on certain physiological functions of the organism of machine test personnel," by N. T. Svistunov. GIG TR PROF ZABOL 13:15-19, February, 1969.

"The effect of noise of various levels and spectral composition on several functions of the organism in adolescents," by I. I. Ponomarenko. GIG TR PROF ZABOL 10:32-38, June, 1966.

"Effect of noise on auditive localization of persons with normal hearing disorders of the conductive and receptive type," by J. Laciak. MED PRACY 17:420-434, 1966.

"The effect of noise on basal metabolism," by I. Pinter, et al. ORV HETIL 109:1371-1373, June 23, 1968.

"Effect of noise on the modulation transfer function of the visual channel," by H. Pollehn, et al. J OPT SOC AM 60:842-848, June, 1970.

"An effect of noise on the distribution of attention," by M. M. Woodhead. J APPL PSYCHOL 50:296-299, August, 1966.

"Effect of noise on the functional status of the nervous system in children of preschool age," by Kh. V. Storoshchuk. GIG SANIT 31: 44-48, January, 1966.

"Effect of noise on hearing observed in some small industrial plants," by I. P. Su. J FORMOSA MED ASS 64:171-175, March 28, 1965.

"The effect of noise on the human organism," by J. Kubik, et al. CESK STOMAT 69:339-343, May, 1969.

"The effect of noise on the organism and its control in industry," by R. I. Vorob'ev. FELDSH AKUSH 34:6-7, November, 1969.

"Effect of noise on the perception of sounds in subject with normal and pathological hearing," by S. G. Kristostur'ian. VESTN OTORINOLARING 27:3-9, July-August, 1965.

"Effect of noise on process optimization," by H. S. Heaps, et al. CAN J CHEM ENG 43:319-324, December, 1965.

"Effect of noise on psychological state," by A. Cohen. AM SOC SAFETY ENG J 14:11-15, April, 1969.

"Effect of noise on the reaction time to auditive stimuli," by S. Barbera, et al. BOLL MAL ORECCH 84:202-206, May-June, 1968.

"The effect of noise on the resistance to acute hypoxia," by J. Davidovic, et al. VOJNOSANIT PREGL 22:625-627, October, 1965.

"The effect of noise on synthetic sentence identification," by C. Speaks, et al. J SPEECH HEARING RES 10:859-864, December, 1967.

"Effect of noise on the teaching process," by V. Lejska. CESK PEDIAT 23:357-361, April, 1968.

"Effect of noise and vibration on man," by J. C. Guignard. NATURE 188:533-534.

"The effect of octave noise bands on certain physiological functions of the organism," by S. V. Alekseev. GIG TR PROF ZABOL 12:27-31, June, 1968.

"Effect of oxidoreductive agents and high-energy compounds on the efficiency of the internal ear during acoustic load," by W. Jankowski, et al. OTOLARYNG POL 23:141-144, 1969.

"Effect of physical work and work under conditions of noise and vibration on the human body. I. Behavior of serum alkaline phosphatase, aldolase and lactic dehydrogenase activities," by J. Gregorczyk, et al. ACTA PHYSIOL POL 16:701-708, September-October, 1965.

"The effect of prolonged noise on the immunobiological reactivity of experimental animals," by Kh. V. Storoshchuk, et al. VRACH DELO 4:97-99, April, 1967.

"The effect of prolonged noise on oxidative processes of the rat brain," by N. F. Svadkovskaia, et al. GIG SANIT 32:19-23, July, 1967.

"The effect of pulsating noise and vibration on the body in straightening and hammering out work," by L. N. Shkarinov, et al. GIG SANIT 33:104-106, July, 1968.

"Effect of pulsed masking on selected speech materials," by D. D. Dirks, et al. J ACOUST SOC AMER 46:898-906, October, 1969.

"Effect of pulsed noise on the organism as a function of pulse periodicity," by G. A. Suvorov. GIG TR PROF ZABOL 13:23-27, December, 1969.

"Effect of random noise on plant growth," by C. B. Woodlief. ACOUSTICAL SOC AM J 46:481-482 pt 2, August, 1969.

"Effect of removing the neocortex on the response to repeated sensory stimulation of neurones in the mid-brain," by G. Horn, et al. NATURE (London) 211:754-755, August 13, 1966.

"Effect of resonance hearing protectors in lowering acoustic trauma in the army," by J. Dominik, et al. VOJ ZDRAV LISTY 34:146-149, August, 1965.

"Effect of shock waves from supersonic planes," by J. Calvet, et al. J FRANC OTORHINOLARYNG 18:79-85, February, 1969.

"Effect of short-lasting deafness on the microphonic potential of the guinea pig and an attempt at the pharmacological influencing of this effect," by L. Faltynek. SBORN VED PRAC LEK FAK KARLOV UNIV 8:269-289, 1965.

"The effect of some harmful factors of the confectionery industry on the organism of workers," by N. G. Prokof'eva, et al. GIG TR PROF ZABOL 11:47-49, December, 1967.

"The effect of spatially separated sound sources on speech intelligibility," by D. D. Dirks, et al. J SPEECH HEARING RES 12:5-38, March, 1969.

"Effect of a specific noise on visual and auditory memory span," by S. Dornic. SCAND J PSYCHOL 8:155-160, 1967.

"The effect of speech-type background noise on esophageal speech production," by W. M. Clarke, et al. ANN OTOL 79:653-665, June, 1970.

"The effect of stable high frequency industrial noise on certain physiological functions of adolescents," by I. I. Ponomarenko. GIG SANIT 31:29-33, February, 1966.

"Effect of stimulus duration on localization of direction of noise stimuli," by W. R. Thurlow, et al. J SPEECH & HEARING RES 13: 826-838, December, 1970.

"The effect of stress stimuli on metabolism of laboratory animals.

I. The effect of industrial noise on the behavior of DNA, RNA and soluble proteins in the liver, as well as the relative weight of this organ in the guinea pig," by J. Stanosek, et al. INT ARCH ARBEITS-MED 26:216-223, 1970.

"Effect of supplementary ingestion of vitamins B1 and C on the function of the auditory organ in workers of 'noisy' trades," by Z. F. Nestrugina. ZH USHN NOS GORI BOLEZ 29:91-94, May-June, 1969.

"Effect of temperature on the high-frequency component of the jet noise," by G. Krishnappa. ACOUSTICAL SOC AM J 41:1208-1211, May, 1967.

"The effect of trauma in causing cochlear losses after stapedectomy," by M. M. Paparella. ACTA OTOLARYNG 62:33-43, July, 1966.

"The effect of ultrasonics and high-frequency noise on the blood sugar level," by Z. Z. Ashbel'. GIG TR PROF ZABOL 9:29-33, February, 1965.

"The effect of vibration and noise on protein metabolism of excavator-machinists using several biochemical indices," by A. M. Tambovtseva. GIG SANIT 33:58-61, January, 1968.

"The effect of vitamin B1 and niacin on the course of biochemical processes in organisms exposed to the action of vibration," by D. A. Mikhel'son. VOP PITAN 28:54-58, November-December, 1969.

"Effect on the labyrinth of intense acoustic stimulation," by P. Pialoux, et al. ANN OTOLARYNG 82:610-613, July-August, 1965.

"Effects of achievement motivation and noise conditions on paired-associate learning," by K. Shrable. CALIF J ED RES 19:5-15, January, 1968.

"Effects of air travel on some otorhinolaryngologic diseases," by M. Fayala. TUNISIE MED 44:165-185, May-June, 1966.

"Effects of ambient noise on vigilance performance," by P. H. McCann. FACTORS 11:251-256, June, 1969.

"The effects of ambient noise upon signal detection," by A. Mirabella, et al. HUM FACTORS 9:277-284, June, 1967.

"Effects of auditory stimulation on the performance of brain-injured and familial retardates," by L. S. Schoenfeld. PERCEPT MOTOR SKILLS 31:139-144, August, 1970.

"Effects of delay on the punishing and reinforcing effects of noise onset and termination," by R. C. Bolles, et al. J COMP PHYSIOL PSYCHOL 61:475-477, June, 1966.

"Effects of ear protection on communication," by W. I. Acton. ANN OCCUP HYG 10:423-429, October, 1967.

"Effects of environmental factors on performance; abstract," by F. J. Vilardo. MACHINE DESIGN 41:150+, November 27, 1969.

"Effects of environmental noise," by D. A. Williams, et al. CANAD J PSYCHIAT NURS 10:9-12, October, 1969.

"The effects of environmental noise on pseudo voice after laryngectomy," by S. Drummond. J LARYNG 79:193-202, March, 1965.

"Effects of harmful noise of mechanised workings on workers at the bottom of the mine. Result of a study undertaken at the mines of Houilleres in the Lorraine Bassin," by A. Mas, et al. ARCH MAL PROF 27:815-818, October-November, 1966.

"Effects of high intensity impulse noise and rapid changes in pressure upon stapedectomized monkeys," by J. L. Fletcher, et al. ACTA OTOLARYNG 68:6-13, July-August, 1969.

"Effects of high-speed drill noise and gunfire on dentists' hearing," by W. D. Ward, et al. J AMER DENT ASS 79:1383-1387, December, 1969.

"Effects of illumination and white noise noise on the rate of electrical self-stimulation of the brain in rats," by E. R. Venator, et al. PSYCHOL REP 21:181-184, August, 1967.

"Effects of imaging of signal-to-noise ratio, with varying signal conditions," by S. J. Segal, et al. BRIT J PSYCHOL 60:459-464, November, 1969.

"Effects of intense noise during fetal life upon postnatal adaptability (statistical study of the reactions of babies to aircraft noise)," by Y. Ando, et al. J ACOUST SOC AMER 47:1128-1130, April, 1970.

"Effects of intense noise on processing of cutaneous information of varying complexity," by R. L. Brown, et al. PERCEPT MOTOR SKILLS 20:749-754, June, 1965.

"Effects of intermittent noise on human target detection," by H. D. Warner. HUM FACTORS 11:245-250, June, 1969.

"Effects of low frequency and infrasonic noise on man," by G. C. Mohr, et al. AEROSPACE MED 36:817-824, September, 1965.

"Effects of negative air ions, noise, sex and age on maze learning in rats," by R. A. Terry, et al. INT J BIOMETEOR 13:39-49, June, 1969.

"Effects of noise and difficulty level of input information in auditory, visual and audiovisual information processing," by H. J. Hsia. PERCEPT MOTOR SKILLS 26:99-105, February, 1968.

"Effects of noise in EEG latency changes in an auditory vigilance task," by E. Gulian. ELECTROENCEPH CLIN NEUROPHYSIOL 27: 637, December, 1969.

"Effects of noise measurements in the factories of Katowice Province," by J. Grzesik, et al. MED PRACY 16:489-496, 1965.

"Effects of noise and of signal rate upon vigilance analysed by means of decision theory," by D. E. Broadbent, et al. HUM FACTORS 7: 155-162, April, 1965.

"Effects of noise on arousal level in auditory vigilance," by E. Gulian. ACTA PSYCHOL (Amst) 33:381-393, 1970.

"Effects of noise on commercial v/stol aircraft design and operation," by H. Sternfeld, Jr., et al. J AIRCRAFT 7:220-225, May, 1970.

"Effects of noise on a complex task," by S. J. Samtur. GRAD RES ED 4:63-81, Spring, 1969.

"Effects of noise on cortical evoked potentials in rats," by Z. Chaloupka. ACTIV NERV SUP (Praha) 10:207-208, May, 1968.

"Effects of noise on health," by G. Jansen. GERMAN MED MONTHLY 13:446-448, September, 1968.

"The effects of noise on man," by A. Glorig. JAMA 196:839-842, June 6, 1966.

"Effects of noise on pupil performance," by B. R. Slater. J ED PSYCHOL 59:239-243, August, 1968.

"Effects of noise reduction in a work situation," by D. E. Broadbent, et al. OCCUPATIONAL PSYCHOLOGY Vol. 34, 2:133-140.

"Effects of overpressure on the ear--a review," by F. G. Hirsch. ANN NY ACAD SCI 152:147-162, October 28, 1968.

"Effects of physical work and work under condition of noise and vibration on the human body. II. Behavior of aspartic and alanine aminotransferase, gamma glutamyl transpeptidas, lactic dehydrogenase and L-idotyl dehydrogenase activities," by J. Gregorczyk, et al. ACTA PHYSIOL POL 16:709-714, September-October, 1965.

"Effects of prolonged acoustic stimulation on adrenal glands of the rat," by M. Ameli, et al. CLIN OTORINOLARING 18:211-252, May-June, 1966.

"Effects of random and response contingent noise upon disfluencies of normal speakers," by R. H. Brookshire. J SPEECH HEARING RES 12:126-134, March, 1969.

"Effects of the sonic boom on man," by R. E. Bouille. REV CORPS SANTE ARMEES 7:659-688, October, 1966.

"Effects of sonic boom on people: review and outlook," by H. E. Von Gierke. J ACOUST SOC AMER 39:Suppl:43-50, May, 1966.

"Effects of sonic boom on people: St. Louis, Missouri, 1961-1962," by C. W. Nixon, et al. J ACOUST SOC AMER 39:Suppl:51-58, May, 1966.

"Effects of sonic booms and subsonic jet flyover noise on skeletal muscle tension and a paced tracing task. NASA CR-1522," by J. S. Lukas, et al. US NASA 1-35, February, 1970.

"Effects of too-loud music on human ears. But, mother, rock'n roll has to be loud," by R. R. Rupp, et al. CLIN PEDIAT 8:60-62, February, 1969.

"Effects of traffic noise on health and achievement of high school students of a large city," by G. Karsdorf, et al. Z GES HYG 14:52-54, January, 1968.

"Effects of traffic noise on human life," by O. Guthof, et al. OEFF GESUNDHEITSWESEN 30:1-6, January, 1968.

"Effects of two types of noise on cardiac rhythm of a man held immobile with constant visual attention," by T. Meyer-Schwertz, et al. ARCH SCI PHYSIOL 22:195-228, 1968.

"Effects of unilateral section of the facial nerve in the orientation of the initial motor syndrome of acoustic epilepsy in mice," by J. Requin, et al. C R SOC BIOL 160:1285-1290, 1966.

"Effects of white noise and presentation rate on serial learning in mentally retarded males," by J. R. Haynes. AMER J MENT DEFIC 74:574-577, January, 1970.

"The efficiency of several types of personal noise protectors and their selection as a function of the conditions of use," by L. N. Shkarinov, et al. GIG TR PROF ZABOL 10:38-43, June, 1966.

"Electric motor noise: control of noise at the source," by B. L. Goss. AMER INDUSTR HYG ASS J 31:16-21, January-February, 1970.

"Electrocortical responses to a tonal stimulation and development of the cochlea," by R. Pujol, et al. J PHYSIOL 59:478, 1967.

"Electrodermal and cardiac responses of schizophrenic children to sensory stimuli," by M. E. Bernal, et al. PSYCHOPHYSIOLOGY 7: 155-168, September, 1970.

"Elevator noise; architectural and mechanical considerations," by J. E. Sieffert. ARCH REC 143:199, April, 1968.

"Elimination of Volta-potential noise from plate field mills," by K. Knott. R SCI INSTR 38:602-604, May, 1967.

"Engineering can control industrial noise," by J. H. Botsford. AM SOC SAFETY ENG J 14:7-9, March, 1969.

"Engineering control of furnance noise," by S. H. Judd. AMER INDUSTR

HYG ASS J 30:35-40, January-February, 1969.

"Engineering outline; noise from aircraft," by J. D. Voce. ENGINEERING 204:983-986, December 15, 1967.

"Engineering outline; noise in machines," ENGINEERING 201:1059-1062, June 3, 1966.

"Engineering outline; sound insulation in buildings," by P. H. Parkin. ENGINEERING 205:307-310, February 23, 1968.

"Engineers, contractors seen responsible for controlling outside equipment noise," by J. P. Christoff. AIR CON HEAT & REFRIG N 108: 18, May 9, 1966.

"Engineers search for quiet jet engines," IND RES 11:30, July, 1969.

"Enhancement of speech intelligibility at high noise levels by filtering and clipping," by I. B. Thomas, et al. AUDIO ENG SOC J 16:412-415, October, 1968.

"Environment control at Dayton foundry," by J. C. Miske. FOUNDRY 98: 68-71, May, 1970.

"The environment in relation to otologic disease," by J. Sataloff, et al. ARCH ENVIRON HEALTH 10:403-415, March, 1965.

"Environmental hazards. Community noise and hearing loss," by J. D. Dougherty, et al. NEW ENG J MED 275:759-765, October 6, 1966.

"Environmental insult," PROGRES ARCH 48:25+, February, 1967.

"Environmental noise, hearing acuity, and acceptance criteria," by E. R. Hermann. ARCH ENVIRON HEALTH 18:784-791, May, 1969.

"Environmental noise pollution: a new threat to sanity," by D. F. Anthrop. BUL ATOM SCI 25:11-16, May, 1969.

"Environmental pollution and its control," by M. Eisenbud. BULL NY ACAD MED 45:447-454, May, 1969.

"An epidemiologic approach to in-plant noise problems," by P. R. Ebling, et al. INDUSTR MED SURG 34:508-512, June, 1965.

"Escape from noise," by F. A. Masterson. PSYCHOL REP 24:484-486, April, 1969.

"Establishment of objective criteria reflecting subjective response to roller-bearing noise," by R. F. Lucht, et al. J ACOUST SOC AMER 44:1-4, July, 1968.

"Estimating noisiness of aircraft sounds," by R. W. Young, et al. ACOUSTICAL SOC AM J 45:834-838, April, 1969.

"Ethanol inhibition of audiogenic stress induced cardiac hypertrophy," by W. F. Geber, et al. EXPERIENTIA 23:734-736, September 15, 1967.

"Evaluation and rating of sound," by R. J. Wells. ASHRAE J 9:48-51, April, 1967.

"Evaluation of advances in engine noise technology," by A. L. McPike. AIRCRAFT ENG 42:16+, May, 1970.

"Evaluation of devices for personal hearing protection against noise with special regard to their functional efficiency," by V. Psenickova, et al. PRAC LEK 17:313-317, September, 1965.

"Evaluation of the effect of acoustic stimuli on the rat central nervous system by means of the swimming test in the labyrinth," by J. Grzesik, et al. ACTA PHYSIOL POL 16:379-387, May-June, 1965.

"Evaluation of the effectiveness of measures for lowering the noise of pneumatic rapier looms," by L. I. Maksimova, et al. GIG SANIT 33: 87-89, December, 1968.

"Evaluation of exposures to impulse noise," by K. D. Kryter. ARCH ENVIRON HEALTH 20:624-635, May, 1970.

"Evaluation of noise problems anticipated with future vtol aircraft. AMRL-TR-66-245," by J. N. Cole, et al. US AIR FORCE AERO-SPACE MED RES LAB 1-16, May, 1967.

"Evaluation of the results of various hearing tests for noise deafness," by T. Kawabata. J OTOLARYNGOL JAP 73:1858-1873, December, 1970.

"Evaluation of the risks of hearing impairment due to industrial noise based on exposure parameters," by J. Grzesik. POL TYG LEK 25: 1026-1028, July, 1970.

"Evoked responses to clicks and tones of varying intensity in waking adults," by I. Rapin, et al. ELECTROENCEPH CLIN NEUROPHYSIOL 21:335-344, October, 1966.

"Evolution of the engine noise problem," by F. B. Greatrex, et al. AIRCRAFT ENG 39:6-10, February, 1967.

"Examination of hearing in workers exposed to constant noise," by I. Prvanov. SRPSKI ARH CELOK LEK 93:1041-1046, November, 1965.

"An example of 'engineering psychology': the aircraft noise problem," by K. D. Kryter. AMER PSYCHOL 23:240-244, April, 1968.

"Excessive noise from the viewpoint of the penal law," by H. Wiethaup. MED KLIN 60:1254-1256, July 30, 1965.

"Experience in the United Kingdom on the effects of sonic bangs," by C. H. Warren. J ACOUST SOC AMER 39:Suppl:59-64, May, 1966.

"Experience with acoustic sensory stress. V. Behavior of sodium and potassium in the rat submandibular gland in audiogenic stress," by G. Croce, et al. BOLL MAL ORECCH GOLA NASO 87:300-306, 1969.

"Experience with creating an acoustic complex for studying the effect of industrial noise on the body," by E. Ts. Andreeva-Galanina, et al. GIG SANIT 30:44-47, January, 1965.

"Experience with the dispensary treatment of occupational hypacusia," by K. Huber. CESK OTOLARYNG 15:173-175, June, 1966.

"Experience with glass fiber earplugs," by C. Zenz, et al. AMER INDUSTR HYG ASS J 28:499-500, September-October, 1967.

"Experience with lowering noise of bar glazers in foundries," by L. Jerman, et al. PRAC LEK 17:245-248, August, 1965.

"Experimental application of the word association test in a study of the effect of tractor noise and vibration on the human organism," by V. N. Kozlov, et al. GIG TR PROF ZABOL 13:46-48, December, 1969.

"Experimental data on assessing the effect of continuous whole-body vibration on warm blooded animals," by D. A. Mikhel'son. GIG SANIT 34:129-130, September, 1969.

"An experimental electronic stethoscope for aircraft use. A preliminary report. SAM-TR-67-39," by F. A. Brogan, et al. US AIR FORCE SCH AEROSPACE MED 1-8, May, 1967.

"Experimental investigation of discrete frequency noise generated by unsteady blade forces," by N. J. Lipstein, et al. J BASIC ENG 92: 155-164, March, 1970.

"Experimental method for determination of noise attenuation in air ducts," by C. M. Harman, et al. ASHRAE J 7:43-49, October, 1965.

"Experimental-microscopic study on the problem of localization of industrial noise-conditioned hearing fatigue and of the later resulting permanent acoustic trauma," by H. G. Dieroff, et al. ARCH KLIN EXP OHR NAS KEHLKOPFHEILK 186:1-8, 1966.

"Experimental research on the cochlear function after exposure to intense noise. I. Variations of the microphonic potentials of the intact resected muscles of the middle ear," by W. Mozzo. BOLL SOC ITAL BIOL SPER 44:400-403, March 15, 1968.

"Experimental research on the disturbing effect of airplane noise," by E. Grandjean, et al. INT Z ANGEW PHYSIOL 23:191-202, December 3, 1966.

"Experimental studies of the cochlear potential behaviour after stellatum blockade. II. Cochlear microphonic potential after stellatum blockade to the ear predamaged by noise," by D. Kleinfeldt, et al. ARCH KLIN EXP OHR NAS KEHLKOPFHEILK 190:398-406, 1968.

"Experimental studies on cochlear damage of rabbits following acoustic stimulation," by S. Mizuno. J OTOLARYNGOL JAP 73:1577-1594, October, 1970.

"Experimental studies on the harmfulness of aircraft noise as opposed to industrial noise," by W. Lorenz. MSCHR OHRENHEILK 103:492-498, 1969.

"Experimental studies on the influence of electric and light stimulation upon the susceptibility to acoustic trauma," by T. Nakamura. J OTORHINOLARYNG SOC JAP 69:1439-1454, August, 1966.

"Experimental studies on the influence of impelling noise on various blood circulatory parameters," by A. Fuchs-Schmuck, et al. DTSCH GESUNDHEITSW 25:1951-1954, October 8, 1970.

"Experimental studies on noise susceptibility in impairment of circulation regulating function," by S. Sugiyama. J OTORHINOLARYNG SOC JAP 68:715-731, June, 1965.

"Experimental study of changes in auditory acuity levels by exposing to noise," by S. Funasaka, et al. J OTOLARYNG JAP 72:338-339, February 20, 1969.

"Experimental study of the cochlear injury by sonic stimulation in rabbits," by H. Ouchi, et al. J OTOLARYNG JAP 72:372-373, February 20, 1969.

"Experimental study on sound perception--accuracy of musical scale with sound loading among members of a chorus group," by S. Sakurai, et al. J OTOLARYNGOL JAP 73:1527-1532, September, 1970.

"Experiments on classification of peak and instantaneous value with impulse-rich work noise," by G. Vorwerk, et al. ARCH KLIN EXP OHR NAS KEHLKOPFHEILK 193:259-276, 1969.

"Explosive lesions of the ear," by A. Risavi. VOJNOSANIT PREGL 22:155-161, March, 1965.

"Expo cars get silent treatment," RY AGE 162:25, April 10, 1967.

"Exposure to noise in the textile industry of the U.A.R.," by M. H. Noweir, et al. AMER INDUSTR HYG ASS J 29:541-546, November-December, 1968.

"Extra-auditory effects of noise as a health hazard," by J. R. Anticaglia, et al. AMER INDUSTR HYG ASS J 31:277-281, May-June, 1970.

"FAA anti-noise authority seen clearing House unit," AVIATION W 88:28-29, March 25, 1968.

"FAA fears political tug-of-war on noise abatement programs,"

AVIATION W 85:36-37, December 5, 1966.

"FAA issues aircraft noise-reduction rule," by H. Taylor. AM AVIATION 32:25, January 20, 1969.

"FAA to act on 747 noise requirements," by R. G. O'Lone. AVIATION W 91:35-36, December 8, 1969.

"Facilities and instrumentation for aircraft engine noise studies," by R. E. Gorton. J ENG POWER 89:1-13, January, 1967.

"Fan convector noise," ENGINEERING 207:720-721, May 9, 1969.

"Farm equipment noise exposure levels," by H. H. Jones, et al. AMER INDUSTR HYG ASS J 29:146-151, March-April, 1968.

"Fertility in couples working in noisy factories," by L. Carosi, et al. FOLIA MED 51:264-268, April, 1968.

"A few comments on hearing disorders caused by noise," by T. Takeuchi, et al. J OTOLARYNG JAP 73:Suppl:1002-1003, July, 1970.

"Few questions about sound," by F. J. Versagi. AIR COND HEAT & REFRIG N 114:27, June 17, 1968.

"Field measurement study of the sound levels produced outdoors by residential air-conditioning equipment," by W. E. Blazier, Jr. ASHRAE J 9:35-39, May, 1967.

"Fight against aircraft noise," by E. Jeffs. ENGINEERING 208:108, August 1, 1969.

"The fight against noise," by E. Hubert. AGNES KARLL SCHWEST 21:456, November, 1967.

"Fighting the noise explosion," by H. Lawrence. AUDIO 50:48, June, 1966.

"Findings on traffic noise in the city of Imperia," by I. Murruzzu. NUOVI ANN IG MICROBIOL 18:503-510, November-December, 1967.

"Fitness for work of employees with hearing defects in noisy work places," by Z. Novotny. PRAC LEK 17:63-67, March, 1965

"Flight noise; measurements on airplanes used in competition and hygienic aspects," by H. G. Demus, et al. Z GES HYG 11:1-24, January, 1965.

"Fluctuation of nucleic acid activity in the organ of corti resulting from noise exposure," by R. Nakamura. J OTORHINOLARYNG SOC JAP 70:1818-1827, November, 1967.

"Fluctuation of optic evoked potentials during conditioned avoidance behavior in cats: effects of attention and distraction on primary evoked potentials," by T. Ikeda. FOLIA PSYCHIAT NEUROL JAP 21:19-30, 1967.

"Fluctuation of vitamin-B1 in blood and in the organ of Corti following exposure to noise," by S. Abiko. J OTORHINOLARYNG SOC JAP 69:1117-1133, June, 1966.

"Fluid power engineers take aim at pollution," PRODUCT ENG 41:76, October 12, 1970.

"Focal points in research activities in regional universities, from the viewpoint of hygiene," by N. Saruta. JAP J HYG 23:31-34, April, 1968.

"For effective noise control be practical and plan," by D. L. Person. SAFETY MAINT 136:25-27, December, 1968.

"Forge shop ventilation noise problem overcome," ENGINEER 225: 956, June 21, 1968.

"Foundry noise and hearing in foundrymen," by G. R. Atherley, et al. ANN OCCUP HYG 10:255-261, July, 1967.

"Fractionated sleep. Nocturnal sleep disturbances provoked by noise: electroencephalographic aspects of a preventive medicine problem," by H. R. Richter. REV NEUROL 115:592-595, September, 1966.

"Frequency analysis of the noise of turbocompressors in an oxygen shop," by K. S. Rudakov. GIG SANIT 34:69-71, August, 1969.

"Frequency discrimination following exposure to noise," by J. F. Brandt. J ACOUST SOC AMER 41:448-457, February, 1967.

"Frequency discrimination in noise," by G. B. Henning. ACOUSTICAL

SOC AM J 41:774-777, April, 1967.

"From experience in controlling noise and vibration in industrial plants in Kiev," by I. G. Guslits, et al. GIG TR PROF ZABOL 10: 52-54, June, 1966.

"Functional changes in cerebral blood supply and in hearing acuity occurring under the effect of noise," by L. N. Shkarinov, et al. GIG TR PROF ZABOL 14:23-26, November, 1970.

"Functional changes in the ear produced by high-intensity sound: 5.0-khz stimulation," by G. R. Price. ACOUSTICAL SOC AM J 44: 1541-1545, December, 1968.

"Fundamental audio: loudness and the decibel," by M. Leynard. AUDIO 51:12 , February, 1967.

"Fundamental constraints to sensory discrimination imposed by two kinds of neural noise," by J. L. Stewart. BEHAV SCI 10:271-276, July, 1965.

"Fundamental nature of machine noise," by D. B. Welbourn. ENGINEERING 199:175-176, February 5, 1965.

"Fundamental noise research emphasized; aircraft turbine engines," AVIATION W 92:90, June 22, 1970.

"Fundamentals of fuel oil firing; how to eliminate burner noise," by A. A. Brahams. HEATING-PIPING 37:134-136, April, 1965.

"Further studies on the role of the dentition in the pathogenesis of occupational hearing defects caused by noise (acoustic trauma)," by B. Semczuk, et al. CZAS STOMAT 21:671-675, June, 1968.

"A further study on the temporary effect of industrial noise on the hearing of stapedectomized ears: at 4,000 c.p.s.," by K. Ferris. J LARYNG 81:613-617, June, 1967.

"Gauging a noise's annoyance level," ELECTRONICS 40:202, February 6, 1967.

"General effects of noise," by W. Lorenz. DEUTSCH GESUNDH 23: 2379-2383, December 12, 1968.

"Generating specified shock pulses," by R. O. Brooks. J ENVIRONMENTAL SCI 10:28-33, April, 1967.

"Genetics of audiogenic seizures. I. Relation to brain serotonin and norepinephrine in mice," by K. Schlesinger, et al. LIFE SCI 4:2345-2351, December, 1965.

--II. Effects of pharmacological manipulation of brain serotonin, norepinephrine and gamma-aminobutyric acid," by K. Schlesinger, et al. LIFE SCI 7:437-447, April 1, 1968.

"Get to know the dB better," by G. H. R. O'Donnell. AUDIO 51:44-45, June, 1967.

"Getting used to it," ARCH DESIGN 40:526, October, 1970.

"Give your house a tranquilizer," by M. A. Guitar. AM HOME 72:48+, October, 1969.

"Glossary of noise control terms," ARCH REC 142:204, October, 1967.

"Glycogen in the inner ear after acoustic stimulation. A light and electron microscopic study," by D. Ishii, et al. ACTA OTOLARYNG 67:573-582, June, 1969.

"Going deaf from rock 'n' roll," TIME 92:47, August 9, 1968.

"Government calls for quiet," MOD PLASTICS 46:80-81, December, 1969

"Government plans enforcement of industrial noise standards," AUTOMATION 16:18+, November, 1969.

"Grinders cut noise level by 50 percent," PLASTICS WORLD 23:86, December, 1965.

"Ground runup silencers for Concorde supersonic transport," ACOUSTICAL SOC AM J 41:1558, June, 1967.

"Growing problem of airplane noise, what is being done," GOOD H 168:155-157, February, 1969.

"Guide to airborne, impact and structure borne noise: control in multifamily dwellings," by R. D. Berendt, et al. HEATING-PIPING 41:147, April, 1969.

"Guidelines for designing quieter equipment," by C. H. Allen. MECH ENG 92:29-34, January, 1970.

"Guidelines for noise control specifications for purchasing equipment," IRON & STEEL ENG 47:95-98, May, 1970.

"Guidelines for noise exposure control," AMER INDUSTR HYG ASS J 28:418-424, September-October, 1967.

"Guidelines for noise exposure control," ARCH ENVIRON HEALTH 15:674-678, November, 1967.

"Guidelines for noise exposure control. Intersociety Committee Report-revised, 1970," J OCCUP MED 12:276-281, July, 1970.

"Guiding the air flow cuts down on noise," PRODUCT ENG 39:48, August 12,1968.

"Hardness of hearing due to noise and noise deafness," by S. Mehmke. MED WELT 27:1595-1601, July 8, 1967.

"Harmonic configuration and audiological aspects of acoustic pictures evoked by auditory fatigue. Experimental study with low frequency generator tones," by F. Fruttero. MINERVA OTORINOLARING 17: 83-93, May-June, 1967.

"Hazardous exposure to impulse noise," by R. R. Coles, et al. J ACOUST SOC AMER 43:336-343, February, 1968.

"Hazardous exposure to intermittent and steady-state noise," by K. D. Kryter, et al. J ACOUST SOC AMER 39:451-464, March, 1966.

"Hazardous noise levels in computer labs," J ACOUST SOC AM 48: 860-861, October, 1970.

"Hazards and hurdles in developing standards: a case history-rating the impace-noise resistance of floors," by R. E. Fischer. ARCH REC 147:147-150, May, 1970.

"Hazards from impulse noise," by R. R. Coles, et al. ANN OCCUP HYG

10:381-388, October, 1967.

"Hazards of the arc-air gouging process," by J. T. Sanderson. ANN OCCUP HYG 11:123-133, April, 1968.

"He conquers sound, to relief of people weary of noise," by A. Pease. PRODUCT ENG 41:22-23, January 19, 1970.

"Health problems due to industrial noise. An environmental survey in three large plants," by M. Batawi. J EGYPT PUBLIC HEALTH ASS 40:131-140, 1965.

"Hearing acuity and exposure to patrol aircraft noise," by W. R. Pierson, et al. AEROSPACE MED 40:1099-1101, October, 1969.

"Hearing and noise," by R. Chocholle. ARH HIG RADA 20:47-54, 1969.

"Hearing conservation in industry; an overview," by A. J. Murphy AMER ASS INDUSTR NURSES J 16:15-16, May, 1968.

"Hearing conservation in noise," by S AFR MED J 44:558-561, May 9. 1970.

"Hearing damage following noise from dental turbines," by L. Dunker, et al. DDZ 24:33-35, January, 1970.

"Hearing defects in gunners," by B. Drettner, et al. FORSVARSMEDICIN 1:115-122, July, 1965.

"Hearing disorders caused by noise in loading personnel at a large civilian airport," by G. Pressel, et al. INT ARCH ARBEITSMED 26:231-249, 1970.

"Hearing disorders due to impulsive noise," by J. Kuzniarz, et al. OTOLARYNG POL 22:781-785, 1968.

"Hearing disturbance due to diesel engine noises," by H. Ogata. OTOLARYNGOLOGY 38:279-288. March, 1966.

"Hearing function in women workers under the effect of high-frequency noise in the twisting shop of a synthetic fiber plant," by Z. V. Babaian. ZH USHN NOS GORL BOLEZN 29:31-35, July-August, 1969.

"Hearing hazard from small-bore weapons," by W. I. Acton, et al. J ACOUST SOC AMER 44:817-818, September, 1968.

"Hearing injuries among construction workers in Skaraborg County," NORD MED 83:445-446, April 2, 1970.

"Hearing injury in building workers," by S. Lindqvist. LAKARTIDNINGEN 67:4283-4292, September 16, 1970.

"Hearing loss from exposure to interrupted noise," by J. Sataloff, et al. ARCH ENVIRON HEALTH 18:972-981, June, 1969.

"Hearing loss from noise," by J. L. Konzen, et al. J OCCUP MED 8: 388-389, July, 1966.

"Hearing loss in Canadian Army units," by R. W. Tooley. MED SERV J CANADA 21:173-176, March, 1965.

"Hearing loss in rock-and-roll musicians," by C. Speaks, et al. J OCCUP MED 12:216-219, June, 1970.

"Hearing-loss trend curves and the damage-risk criterion in diesel-engineroom personnel," by J. D. Harris. J ACOUST SOC AMER 37: 444-452, March, 1965.

"Hearing protection in industry," by R. E. Scott. SAFETY MAINT 129: 40-41, January, 1965.

"Hearing protection; keeping up with developments," SAFETY MAINT 137:35-40, February, 1969.

"Hearing protection pitfalls and how to avoid them," by F. P. Haluska. SAFETY MAINT 131:32-34+, January, 1965.

"Hearing tests by means of audiometry using whispered and normal speech for the testing of subjects working in noise," by K. Szymczyk. OTOLARYNG POL 21:277-280, 1967.

"Hearing tests in industry," by M. M. Hipskind. INDUSTR MED SURG 36:393-402, June, 1967.

"The hearing threshold levels of dental practitioners exposed to air turbine drill noise," by W. Taylor, et al. BRIT DENT J 118:206-210, March 2, 1965.

"Heart rate response of anesthetized and unanesthetized dogs to noise near-vacuum decompression," by J. P. Cooke, et al. AEROSPACE MED 37:704-709, July, 1966.

"Helicopter noise," by I. M. Davidson, et al. ROY AERONAUTICAL SOC J 69:325-336, May, 1965.

"Helpless against noise," by K. M. Hartimaier. ZAHNAERZTL MITT 58:382-384, April 16, 1968.

"Hemodynamic reactions during acoustic stimuli," by K. Klein, et al. WIEN KLIN WSCHR 81:705-709, October 8, 1969.

"Hidamets; metals to reduce noise and vibration," by D. Birchon. ENGINEER 222:207-209, August 5, 1966.

"High bypass ratio fan noise research test vehicle," by C. A. Warden. J AIRCRAFT 7:437-441, September, 1970.

"High-intensity noise problems in the Royal Marines," by R. R. Coles, et al. J ROY NAV MED SERV 51:184-192, Summer, 1965.

"High intensity sounds in the recreational environment. Hazard to young ears," by D. M. Lipscomb. CLIN PEDIAT 8:63-68, February, 1969.

"High speed equipment and dentists' health," by R. Von Krammer. J PROSTH DENT 19:46-50, January, 1968.

"Highway noise monitor," MECH ENG 92:52, June, 1970.

"Hirschorn says noise-control standards have industrial impact," AIR COND HEAT & REFRIG N 117:6, August 11, 1969.

"Histochemical investigations of kidneys of guinea pigs following chronic noise influence," by J. Jonek, et al. Z MIKR ANAT FORSCH 73:1-13, 1965.

"Histochemical investigations on the behavior of some enzymes in the adrenal glands in guinea pigs following chronic noise influence," by J. Jonek, et al. Z MIKR ANAT FORSCH 73:174-186, 1965.

"Histochemical studies of the inner ear of guinea pigs subjected to industrial noise," by S. Chodynicki, et al. OTOLARYNG POL 22:831-838, 1968.

"Histologic study of guinea pig cochlea after acoustic trauma," by J. Calvet, et al. REV OTONEUROOPHTALMOL 42:150-152, April, 1970.

"Histological and histochemical aspects of the vaginal cycle of the mature rat during experiments with prolonged acoustic stimulation," by M. De Marini, et al. CLIN OTORINOLARING 18:380-402, July-August, 1966.

"Histological and histochemical changes in the dog brain under the effect of strong sound stimuli," by I. I. Tokarenko. FIZIOL ZH 15:272-279, March-April, 1969.

"Histopathologic changes of the hearing organ in cats caused by noise," by A. Andrevski, et al. GOD ZBORN MED FAK SKOPJE 15:23-35, 1969.

"Homolateral and contralateral masking of subjective tinnitus by broad spectrum noise, narrow spectrum noise and pure tones," by H. Feldmann. ARCH KLIN EXP OHR NAS KEHLKOPFHELIK 194:460-465, December 22, 1969.

"Hospital cooling tower sound problem solved by using stainless fill," AIR COND HEAT & REFRIG N 108:28, May 30, 1966.

"Hospital noises disturb U.S. patients also," MOD HOSP 105:85, December, 1965.

"Hospital quietude," by Maheux. REV INFIRM ASSIST SOC 20:657, July-September, 1970.

"Hospital: silence," SA NURS J 36:5-6, July, 1969.

"Hovercraft noise and its suppression," by E. J. Richards, et al. ROY AERONAUTICAL SOC J 69:387-398, June, 1965.

"How Holmens Bruk ab reduced noise from a new paper machine," by P. Berg, et al. PAPER TR J 154:42-48, November 2, 1970.

"How loud is loud?" by I. B. Berger. SAT R 48:61+, February 27, 1965.

"How much does noise bother apartment dwellers?" by D. R. Prestemon. ARCH REC 143:155-156, February, 1968.

"How noise affects work," by D. E. Broadbent. NEW SOCIETY p.12-14, March 3, 1966.

"How noise control affects you," TEXTILE WORLD 119:43-46, June, 1969.

"How noisy is it?" by H. Cary. IND RES 12:22-23, October, 1970.

"How noisy is a refinery?" by D. A. Tyler. HYDROCARBON PROCESS 48:173-174, July, 1969.

"How Oregon school keeps out airport noise," by J. E. Guerusey. NATIONS SCH 80:65, November, 1967.

"How Ottawa's 'realistic' new traffic noise bylaw operates," FIN POST 64:27, February 14, 1970.

"How quiet is a private room?" by I. D. Snook, Jr. J PRACT NURS 17:33 passim, June, 1967.

"How should civil aviation develop to serve our society best? the AIAA president's forum," ASTRONAUTICS & AERONAUTICS 7:28-55, February, 1969.

"How to control industrial noise," by A. M. Teplitzky. AUTOMATION 17:70-74, March, 1970.

"How to design a quiet school; additions and new structures," AM SCH BD J 156:21-25, October, 1968.

"How to estimate plant noises," by I. Heitner. HYDROCARBON PROCESS 47:67-74, December, 1968.

"How to keep the decibel level low," ADM MGT 28:90-91, May, 1967.

"How to keep down noise levels in computer facilities," by L. L. Boyer, Jr. ARCH REC 145:165-166, May, 1969.

"How to noiseproof a room," MECH ILLUS 65:100-102, March, 1969.

"How to plan a small plant hearing protection program," by R. J. Beaman. SAFETY MAINT 130:36-38, December, 1965.

"How to quiet down a noisy house," BET HOM & GARD 44:120, March, 1966.

"How to save dollars when treating noise," MOD MANUF 2:88-89, October, 1969.

"How to secure privacy of speech in offices," by E. H. Kone. OFFICE 62:14-15+, December, 1965.

"How to select acoustical materials," ARCH REC 138:187-188, July, 1965+.

"How to slash the cost of noise; a practical plan," by J. M. Handley. ENVIRONMENTAL CONTROL & SAFETY MGT 140:58-60+, July, 1970.

"The human controller as an adaptive, low pass filter," by G. H. Robinson. HUM FACTORS 9:141-147, April, 1967.

"Human performance as a function of changes in acoustic noise levels," by R. W. Shoenberger, et al. J ENGIN PSYCHOL 4:108-119, 1965.

"Human response to intense low-frequency noise and vibration," by J. C. Guignard. INST MECH ENG PROC 182 no 3:55-59; Discussion 78-83; Reply 84, 1967-1968.

"Human response to measured sound pressure levels from ultrasonic devices," by C. P. Skillern. AMER INDUSTR HYG ASS J 26:132-136, March-April, 1965.

"Human responses to sonic boom," by C. W. Nixon. AEROSPACE MED 36:399-405, May, 1965.

"Human tolerance to low frequency sound," by B. R. Alford, et al. TRANS AMER ACAD OPHTHAL OTOLARYNG 70:40-47, January-February, 1966.

"Hygienic assessment and establishment of standards for noise in the communications services," by A. P. Mikheev, et al. GIG TR PROF ZABOL 11:14-18, January, 1967.

"Hygienic assessment of noise and vibration on Diesel-powered hydrofoil boats of the "Meteor' and 'Raketa' type," by V. I. Petrov, et al. GIG TR PROF ZABOL 12:23-27, June, 1968.

"Hygienic assessment of the noise factor in the air dissociation workshop of a chemical plant," by M. I. Tsigel'nik, et al. GIG TR PROF ZABOL 13:43-45, October, 1969.

"Hygienic assessment of working conditions at flax mills," by E. A. Krechkovskii. GIG TR PROF ZABOL 13:5-8, December, 1969.

"Hygienic characteristics of noise during testing of automobile motors," by R. A. Medved'. GIG SANIT 30:104-107, March, 1965.

"The hygienic characteristics of noise in agricultural machines," by K. Gruss. PRAC LEK 18:304-308, August, 1966.

"Hygienic characteristics of vibration and noise at work sites of crushing mills in ore enriching plants," by K. P. Antonova. GIG SANIT 34:116-118, November, 1969.

"Hygienic characteristics of working conditions in sugar refineries and ways in which they may be made healthier," by V. V. Paustovskaia, et al. GIG TR PROF ZABOL 14:44-45, May, 1970.

"Hygienic evaluation of airports as noise sources," by I. L. Karafodina, et al. GIG SANIT 31:18-23, July, 1966.

"Hygienic evaluation of manual pneumatic perforators," by N. M. Paran'ko. VRACH DELO 1:94-98, January, 1965.

"Hygienic evaluation of noise in a shop with curved designs," by N. M. Paran'ko, et al. GIG SANIT 31:81-82, May, 1966.

"Hygienic evaluation of the percussion noise of punch-presses," by E. P. Orlovskaia. VRACH DELO 5:89-92, May, 1966.

"Hygienic evaluation of weaving loom machines without shuttles R-105," by E. V. Teterina, et al. GIG TR PROF ZABOL 10:50-51, February, 1966.

"Hygienic importance of occupational factors in plasma-arc metal cutting," by A. V. Il'nitskaia. GIG TR PROF ZABOL 14:14-18, November, 1970.

"Hyperalgesia acustica," by S. L. Shapiro. EYE EAR NOSE THROAT MONTHLY 44:88-90, March, 1965.

"I can't hear the flutes; excessive noise from nonbook learning devices in new Hume library at University of Florida," by L. Cassidy. AM LIB 1:888-889, October, 1970.

"Identification of form in patterns of visual noise," by H. Munsinger, et al. J EXP PSYCHOL 75:81-87, September, 1967.

"Identification of mechanical sources of noise in a Diesel engine; sound emitted from the valve mechanism," by B. J. Fielding, et al. INST MECH ENG PROC 181 no 19:437-446; Discussion 447-450; Reply 451, 1966-1967.

"The identity of the nurse in an industrial hearing conservation program," by A. J. Murphy. OCCUP HLTH NURS 17:32+, May, 1969.

"Impact," by M. Ragon. WORLD HEALTH 19:26+, February-March, 1966.

"Impact noise analyser," ENGINEER 220:433, September 10, 1965.

"Impact-noise rating of various floors," by T. Mariner, et al. ACOUSTICAL SOC AM J 41:206-214, January, 1967.

"Implications of the changing environment to occupational health," by F. D. Yoder. OCCUP HEALTH NURS 18:23-25, July, 1970.

"The importance of certain foodstuffs for maintaining operative work capacity under conditions of intense noise," by Iu. F. Udalov, et al. GIG TR PROF ZABOL 11:18-23, May, 1967.

"Importance of the color of the iris in the evaluation of resistence of hearing to fatigue," by G. Tota, et al. RIV OTONEUROOFTAL 43: 183-192, May-June, 1967.

"The importance of the external ear for hearing in the wind," by H. Feldmann, et al. ARCH KLIN EXP OHR NAS KEHIKOPFHEILK 190: 69-85, 1968.

"Importance of higher frequencies of the speech spectrum for its comprehension in noise," by J. Kuzniarz. OTOLARYNG POL 22:427-435, 1968.

"Importance of the individual factor of muscular exhaustibility of the sound-conducting apparatus in the genesis of the damage caused by continuous acoustic overstimulation," by T. Marullo, et al. CLIN OTORINOLARING 19:265-271, July-August, 1967.

"Improvement of hearing ability by directional information," by M. Ebata, et al. J ACOUST SOC AMER 43:289-297, February, 1968.

"Improving the airport environment; effect of the 1969 FAA regulations on noise," by P. B. Larsen. IA L REV 55:808, April, 1970.

"Impulse duration and temporary threshold shift," by M. Loeb, et al. ACOUSTICAL SOC AM J 44:1524-1528, December, 1968.

"Impulse noise and neurosensory hearing loss. Relationship to small arms fire," by R. J. Keim. CALIF MED 113:16-19, September, 1970.

"Incidence of noise as a cause of nuisance in an industrial city," by M. Braja, et al. IG MOD 60:26-36, January-February, 1967.

"The incidence of snoring as a sleep problem in parents," by E. R. Seller. J ROY COLL GEN PRACT 19:247, April, 1970.

"Inclusion of P32 in the internal organs during the effect of uninterrupted noise of low intensity," by N. F. Svadkovskaia, et al. VRACH DELO 2:105-107, February, 1969.

"Indemnity in occupational deafness," by P. Mounier-Kuhn, et al. J MED LYON 48:1691-1696, November 20, 1967.

"Individual protection against damaging effect of industrial noise," by M. Prazic, et al. LIJECN VJESN 87:409-418, April, 1965.

"Individual reaction to noise," by A. Hedri. PRAXIS 57:1168-1169, August 27, 1968.

"Industrial acoustics to amplify earnings with greater capacity," BARRONS 49:27, November 17, 1969.

"Industrial deafness," LAMP 25:13-14, November, 1968.

"Industrial deafness and the summed evoked potential," by T. G. Heron. S AFR MED J 42:1176-1177, November 9, 1968.

"Industrial hearing loss. Conservation and compensation aspects," by M. S. Fox. OCCUP HEALTH NURS 17:18-24, May, 1969.

"An industrial hygiene study of flame cutting in a granite quarry," by W. A. Burgess, et al. AMER INDUSTR HYG ASS J 30:107-112, March-April, 1969.

"Industrial hygienic remarks on noise at the place of work," by W. Massmann, et al. ZBL ARBEITSMED 16:124-127, May, 1966.

"Industrial medicine and industrial physiology in the tropics," by J. Haas. MED KLIN 63:1001-1004, June, 1968.

"Industrial noise and its control," by J. M. Handley. PLANT ENG 24:56-57, March 19, 1970.

"Industrial noise control is practical," by P. H. Hutton. AMER INDUSTR HYG ASS J 29:499-503, September-October, 1968.

"Industrial noise Medicolegal considerations in prevention," by B. Testa, et al. FOLIA MED 52:311-317, May 5, 1969.

"Industrial noise problems and solutions," by J. H. Botsford. MACHINE DESIGN 42:130+, August 20, 1970.

"Industrial noise; workers lose hearing," CHEM & ENG N 46:22, November 11, 1968.

"Industrial noises--hearing damage. Is prevention possible?" by B. Ingberg. SOCIALMED T 42:379-380, November, 1965.

"Industrial sudden deafness," by S. Kawata, et al. ANN OTOL 76: 895-902, October, 1967.

"Infernal combustion," by A. Brien. NEW STATESMAN 74:318, September 15, 1967.

"Influence of acoustic stimuli on electroretinographic tracings of normal subjects," by G. Maffei, et al. RIV OTONEUROOFTAL 41:503-510, November-December, 1966.

"The influence of age on the origin and progression of noise hearing difficulties," by W. Kup. HNO 14:268-272, September, 1966.

"Influence of experimental vigil on the catecholamine content of rat suprarenal glands," by O. D. Kumanova. UKR BIOKHIM ZH 40:446-448, 1968.

"Influence of fatigue on hearing in the presence of noise," by T. Bystrzanowska. OTOLARYNG POL 20:172-176, 1966.

"Influence of the frequency of interruption of signals on the intelligibility of the spoken language," by R. Lehmann. C R ACAD SCI (D) 261:5653-5656, December 20, 1965.

"Influence of an intensive acoustic stimulation on the cochlear response," by M. Aubry, et al. ACTA OTOLARYNG 60:191-196, September, 1965.

"The influence of a loud acoustic stimulus on the ultra-low frequency acceleration ballistocardiogram in man," by P. J. Pretorius, et al. ACTA CARDIOL 22:238-246, 1967.

"The influence of mechanical noise on the activity rhythms of finches," by M. Lohmann, et al. COMP BIOCHEM PHYSIOL 22:289-296, July, 1967.

"Influence of noise on the cardiovascular system," by C. Gradina, et al. FIZIOL NORM PAT 16:357-367, 1970.

"The influence of noise on emotional states," by R. K. Mason. J PSYCHOSOM RES 13:275-282, September, 1969.

"Influence of noise on hearing function," by D. Filipo, et al. BOLL MAL ORECCH 83:133-145, March-April, 1965.

"Influence of noise on the heart rate and O2 consumption under moderate physical loading," by M. Quaas, et al. INT Z ANGEW PHYSIOL 27: 230-238, 1969.

"The influence of noise on the visual field," by E. Ogielska, et al. ANN OCULIST 198:115-122, February, 1965.

"Influence of repeated 4-hour, intermittent, so-called 'muffled' noise on catecholamine secretion and pulse rate," by W. Hawel, et al. INT Z ANGEW PHYSIOL 24:351-362, 1967.

"Influence of rocket noise upon hearing in guinea pigs," by G. Gonzalez, et al. AEROSPACE MED 41:21-25, January, 1970.

"Influence of stress (electric and audiogenic) on the development of Walker 256 carcinosarcoma in rats," by S. M. Milcu, et al. STUD CERCET ENDOCR 19:131-137, 1968.

"Influence of systemic stresses on the development of an experimental inflammatory reaction. I. Effects of an auditory stress," by L. Thrieblot, et al. J PHYSIOL 57:708-709, September-October, 1965.

"Influence of 3 types of sound stimulation on the blood level of non-esterified fatty acids (FFA) in the sheep," by J. Bost, et al. C R SOC BIOL 160:2340-2343, 1966.

"The influence of various neuroleptic drugs on noise escape response in rats," by C. J. Niemegeers, et al. PSYCHOPHARMACOLOGIA 18:249-259, 1970.

"Influence on auditory acuity of continuous industrial noise reaching 90 plus-or-minus 5 decibels within the range of an octave where it culminates," by P. Martinet. ARCH MAL PROF 30:323-335, June, 1969.

"Influences of research on building design. 2. Helicopters: a noise problem on the doorstep," by W. Allen. BUILDER 1046-1047, June 3 .

"Influences of VARIABLE NOISE ON THE CEREBRAL ULTRASONIC ATTENUATION (CUSA) due to postural change and the function of concentration maintenance (TAF)," by E. Takakuwa, et al. JAP J HYG 23:527-529, February, 1969.

"Infrasound tests human tolerance," by H. M. David. MISS & ROC 17: 31+, October 11, 1965.

"Initial FAA noise proposal hits mainly new aircraft," AVIATION W 90:24, Jamuary 13, 1969.

"Initial notes on content in auditory projective testing," by I. Breger. J PROJ TECH PERS ASSESS 34:125-130, April, 1970.

"Injury to hearing in foundrymen," by G. R. Atherley. OCCUP HEALTH

REV 19:14-16, 1967.

"Inside every fat man," ECONOMIST 226:24+, March 2, 1968.

"Instrumental escape conditioning to a low-intensity noise by rats," by A. K. Myers. J COMP PHYSIOL PSYCHOL 60:82-87, August, 1965.

"Intake noise from axial flow turbochargers and compressors," by I. J. Sharland. INST MECH ENG PROC 3:73-77

"Integrated ceiling system handles air, light and sound," AIR COND HEAT & VEN 65:59-62, April, 1968.

"Integration of aircraft auxiliary power supplies," by J. Wotton. AIRCRAFT ENG 40:22-24+, October, 1968.

"Interaction between the aero engine industry and the growth of air transport," by E. M. Eltis. AIRCRAFT ENG 39:15-24, January, 1967.

"Intreaction of adversive stimuli: summation or inhibition?" by B. A. Campbell. J EXP PSYCHOL 78:181-190, October, 1968.

"The interaction of noise and personality with critical flicker fusion performance," by C. D. Frith. BRIT J PSYCHOL 58:127-131, May, 1967.

"Interactions between synchronous neural responses to paired acoustic signals," by D. C. Teas. J ACOUST SOC AMER 39:1077-1085, June, 1966.

"The intermittent action of high frequency noise in industry," by V. V. Lipovoi. GIG TR PROF ZABOL 13:19-21, February, 1969.

"Intermittent noise and the brain waves, especially alpha-wave," by E. Takakuwa, et al. JAP J HYG 23:370-373, October, 1968.

"Internal noise attenuation of turbine auxiliary power unit must start during design stage," by J. J. Dias. SAE J 76:64-67, January, 1968.

"Internally generated noise from gas turbine engines; measurement and prediction," by M. J. T. Smith, et al. J ENG POWER 89:177-185; Discussion 185-187; Reply 187-190, April, 1967.

"An introduction to noise and its problems," by L. E. Euinton. TRANS SOC OCCUP MED 18:142-155, October, 1968.

"An investigation of the effects of various noise levels as measured by psychological performance and energy expenditure," by D. W. Lehmann, et al. J SCH HEALTH 35:212-214, May, 1965.

"Investigation into noise and its effect on employees carried out in a manufacturing plant," by D. M. Cracknell. OCCUP HEALTH NURSE 20:184-193, July-August, 1968.

"Iroquois plant; horsepower on a raft," MARINE ENG LOG 75:88-89 , October, 1970.

"Is anyone listening," by G. E. Warnaka. IND RES 12:21, October, 1970.

"Is industry facing up to its noise problems?" by R. K. Anderson. SAFETY MAINT 135:43-44, January, 1968.

"Is noise caused by dental turbine drills injurious to hearing?" by J. S. Lumio, et al. MSCHR OHRENHEILK 99:192-199, 1965.

"Is noise harmful to vision?" by R. Grandpierre, et al. PRESSE THERM CLIMAT 102:164-165, 1965.

"Is noise important in hospitals?" by T. W. Hurst. INT J NURS STUD 3:125-135, September, 1966.

"Is the real culprit sound or noise?" by L. deMoll. ADM MGT 29:40+, October, 1968.

"Is that noise necessary?" by S. C. Mott SUPERVISORY MGT 12:16-20, July, 1967.

"Island in a torrent of noise," by J. Clough. TIMES ED SUP 2765:1660, May 17, 1968.

"Isolating equipment for vibration control," by W. E. Whale. HEATING-PIPING 38:122-125, January; 97-100, February; 118-120, March, 1966; Discussion 38:96, June; 94+, July, 1966.

"Isolation of railroad/subway noise and vibration," by L. N. Miller. PROGRES ARCH 46:203-208, April, 1965.

"It's getting noisier," by C. Dreher. NATION 205:238-242, September 18, 1967.

"It's high time the public shouted against the menace of noise," MUNICIPAL J 147, 149, January 15 .

"It's not the noise, it's the annoyance," HOUSE & GARD 127:148-155+, April, 1965.

"It's time to turn down all that noise," by J. M. Mecklin. FORTUNE 80:130-133+, October, 1969.

"Jet age precedent; Nice real estate man sues Air France," TIME 87:67 March 18, 1966.

"Jet engine noise reduced on P & W JT9Ds," AM AVIATION 32:40, September 16, 1968.

"Jet engine oscillations and noises," by S. L. Bragg. ENGINEER 217: 268-269, passim+.

"Jet noise and safety," by R. J. Serling. ACOUSTICAL SOC AM J 45:1574-1575, June, 1969.

"Jet noise and shear flow instability seen from an experimenter's viewpoint," by E. Mollo-Christensen. J AP MECH 34:1-7, March, 1967.

"Jet noise attacked in four ways," MACHINE DESIGN 41:10, March 20, 1969.

"Jet noise, fumes trial could set precedent," AVIATION W 92:44-46, May 11, 1970.

"Jet noise in airport areas: a national solution required (problem of jet transport noise along flight paths near public airports)," MINN LAW R 51:1087-1117, May, 1967.

"Jet noise is getting awful," by R. Sherrill. N Y TIMES MAG 24-25+, January 14, 1968.

"Jet-noise monitor," ELECTRONICS 38:211, January 11, 1965.

"Jet quieted by noise absorbing ducts," MACHINE DESIGN 41:10, May 1, 1969.

"The jet-set and the law: a summary of recent developments in noise law as it relates to airport and aircraft operations in California," by W. S. H. Hood, Jr. PACIFIC LAW JOURNAL 1:581-609, July, 1970.

"Judgments of the relative and absolute acceptability of aircraft noise," by D. E. Bishop. J ACOUST SOC AMER 40:108-122, July, 1966.

"Just a shimmy, no bang-bang; pile driver in Schenectady, N. Y.," AM CITY 80:24, September, 1965.

"Keep the cotton wool handy," by T. Cross. NURS TIMES 61:1524-1525, November, 1965.

"Keep that fan quiet!" by C. J. Trickler. PLANT ENG 21:139-141, June, 1967.

"Keeping aircraft quieter," NATURE (London) 226:4, April 4, 1970.

"Keeping the jet's roar outside the door--it's a difficult and costly process," by B. Lamb. HOUSE & HOME 38:10+, September, 1970.

"Keeping power-station noise within acceptable limits," by R. X. French. IEEE TRANS POWER APPARATUS & SYSTEMS 233-237+.

"Keeping systems quiet," by Y. P. Yerges. DOM ENG 209:46-48+, March, 1967.

"Keeping up with developments; hearing protection," ENVIRONMENTAL CONTROL MGT 139:47-50, February, 1970.

"Keeping up with developments; hearing protection," SAFETY MAINT 133:43-48, February, 1967.

"Keeping up with noise control developments," SAFETY MAINT 129: 42-48, February, 1965+.

"Knotty problem of industrial hearing loss," by L. W. Larson. SAFETY MAINT 135:39-42, May, 1968.

"Know the necessary steps to combat noise," by R. M. Hoover, et al. POWER 113:70-71, October, 1969.

"LA girds for jet-noise showdown," by R. L. Parrish. AM AVIATION 32:23-25, January 6, 1969.

"Laboratory techniques useful in the design of refrigerator compressor valves," by R. Cohen, et al. ASHRAE J 7:106-111, January, 1965.

"Laboratory tests of physiological-psychological reactions to sonic booms," by K. D. Kryter. J ACOUST SOC AMER 39:65-72, May, 1966.

"Laboratory tests of subjective reactions to sonic booms. NASA CR-187," by K. S. Pearsons, et al. US NASA 1-34, March, 1965.

"Landslide noise," by J. D. Cadman, et al. SCIENCE 158:1182-1184, December 1, 1967.

"Late development of acoustic trauma (studies on 500 workers in a noisy environment. II.)," by W. Wagemann. MSCHR UNFALLHEILK 69:23-37, January, 1966.

"Lateralization of hearing loss and vestibular nystagmus in test pilots," by A. Bruner, et al. AEROSPACE MED 41:684-687, June, 1970.

"Lateralization of sounds at the unstimulated ear opposite a noise-adapted ear," by E. C. Carterette, et al. SCIENCE 147:63-65, January 1, 1965.

"Lau starts sound study program," AIR COND HEAT & REFRIG NEWS 111:1+, April, 1967.

"Law of noise," TIME 86:37-38, September 10, 1965.

"LAX studies house insulation as way to decrease jet noise," by F. S. Hunter. AM AVIATION 32:25, September 16, 1968.

"Lead for sound control doesn't impress experts," AIR COND HEAT & REFRIG N 121:30, December 7, 1970.

"Lead sheet for sound absorption; detail sheet," AIR COND HEAT & VEN 65:71-74, April, 1968.

"Lead sound barrier," MECH ENG 90:63, March, 1968.

"Lease and quiet; ways to circumvent construction flaws and restrictions," by J. H. Ingersoll. HOUSE B 109:122-124, August, 1967.

"A legal action for noise deafness," by R. R. Coles. ANN OCCUP HYG 12:223-226, October, 1969.

"Legal aspects of noise control," by F. P. Houston. AUDIO ENG SOC J 17:321+, June, 1969.

"Legal aspects of noise in New York City," by J. J. Allen. NEW YORK LAW JOURNAL 163:1+, June 12, 1970.

"Legal aspects of noise pollution," by H. A. Young. PLANT ENG 23: 66-67, May 29, 1969.

"Lesions of the acoustic organ caused by noise among engine-room crews," by R. Nowak. BULL INST MAR MED GDANSK 17:339-342, 1966.

"Let quiet be public policy; excerpt from The Tyranny of Noise," by R. A. Baron. SAT R 53:66-67, November 7, 1970.

"Lets measure excessive noise," FIN POST 61:23, May 27, 1967.

"Let's take the din out of living," by W. R. Vath. TODAYS HEALTH 43:6-7+, February, 1965.

"Light diesels give promise in mail service," by G. C. Nield. SAE J 74:76-79, July, 1966.

"Lighting, noise and floors in factories," ENGINEERING 201:530, March 18, 1966.

"Living on the firing line," by J. R. Roth. ARCH OTOLARYNG 86: 243-244, September, 1967.

"Location-design control of transportation noise," by A. Cohen. AM SOC C E PROC 93 (UP 4 no 5693):63-86, December, 1967; Discussion by T. E. Parkinson. 94(UP 1 no 6052):95-96, August, 1968; Reply 95(UP 1 no 6487):102-103, April, 1969.

"London noises," LANCET 2:1289, December 13, 1969.

"Long-term industrial hearing conservation results," by M. T. Summar, et al. ARCH OTOLARYNG 82:618-621, December, 1965.

"Long-term observations on the auditory acuity of workers in a noisy environment," by T. Yokoyama, et al. J OTOLARYNG JAP 71:640-672, May, 1968.

"Long-term observations on the effect of noise on drivers of large tanktrucks," by H. Herrmann, et al. ZBL ARBEITSMED 17:73-78, March, 1967.

"Look at lead, a versatile and flexible noise-cutting aid," POWER 113:62-63, August, 1969.

"Lots of rest?" by C. K. Harper. NURS MIRROR 126:22, February 9, 1968.

"Louder, please: noiseless products distracting," TIME 95:92, May 4, 1970.

"Loudness and the decibel," by M. Leynard. AUDIO 51:12+, February, 1967.

"The loudness determination by workers in noisy surroundings. On problems of getting accustomed to noise," by G. Linke, et al. Z LARYNG RHINOL OTOL 47:53-57, January, 1968.

"Loudness, a product of volume times density," by S. S. Stevens, et al. J EXP PSYCHOL 69:503-510, May, 1965.

"Loudness rating of telephone subscribers' sets by subjective and objective methods," by W. D. Cragg. ELEC COM 43 no 1:39-43; no 3:228-232, 1968.

"Loudness recruitment and its measurement with special reference to the loudness discomfort level test and its value in diagnosis," by M. R. Dix. ANN OTOL 77:1131-1151, December, 1968.

"Loudness sensitivity, measurement and trauma caused by impulse noise. I.," by H. Niese. Z LARYNG RHINOL OTOL 44:209-217, April, 1965.

"Loudness sensitivity, measurement and trauma caused by impulse noises. 2.," by H. Weissing. Z LARYNG RHINOL OTOL 44:217-223, April, 1965.

"Machine tool computer resists noise," ELECTRONICS 38:190, October 4, 1965.

"Machinery hazards," by C. J. Moss. ANN OCCUP HYG 12:69-75, April, 1969.

"Machinery noise," AUTOMOBILE ENG 55:149, April, 1965.

"Magnetostriction and transformer noise; abstracts," ULTRASONICS 6:77, April, 1968.

"Magnitude of temporary threshold shift (TTS) in the audiogram and 'response time' of noise-or detonation exposed ears as test of hearing organs endangered by detonation," by F. Pfander. ARCH KLIN EXP OHR NAS KEHLKOPFHEILK 191:586-590, 1968.

"Maine's new dust-free crushed stone plant," by W. E. Trauffer. PIT & QUARRY 63:96-100, August, 1970.

"Man and his cities--impact," AUST NURSES J 64:81+, April, 1966; also in JAMAICAN NURSE 6:17+, August, 1966; NEW ZEALAND NURS J 59:8+, April, 1966; NURS J INDIA 57:100+, April, 1966.

"Man and his cities. Part 1," QUEENSLAND NURSES J 8:38+, June, 1966.

--Part 2.," QUEENSLAND NURSES J 8:5+, July, 1966.

--Part 4.," QUEENSLAND NURSES J 8:5+, September, 1966.

"Man and his Cities. The agression against man," INFIRMIERE 45: 35-37 concl, June, 1967.

"Man and his noises," CANAD MED ASS J 101:109-110, July 26, 1969.

"Masked tonal thresholds in the bottlenosed porpoise," by C. S. Johnson. J ACOUST SOC AMER 44:965-967, October, 1968.

"Masking noise: silence is golden, privacy is pink," by R. Farrell. PROGRES ARCH 48:152-155, November, 1967.

"Masking of speech by aircraft noise," by K. D. Kryter, et al. J ACOUST SOC AMER 39:138-150, January, 1966.

"Masking of speech by continuous noise," by J. Kuzniarz. OTOLARYNG POL 21:401-407, 1967; also in POL MED J 7:1001-1008, 1968.

"Masking of speech by means of impulse noise," by J. Kuzniarz. OTOLARYNG POL 22:421-425, 1968.

"Masking of tones before, during, and after brief silent periods in noise," by L. L. Elliott. J ACOUST SOC AMER 45:1277-1279, May, 1969.

"Math quiets rotating machines," by J. H. Varterasian. SAE J 77:53, October, 1969.

"Means of reducing industrial noise in the mechanical vat shops of glass container factories," by V. P. Goncharenko, et al. GIG TR PROF ZABOL 10:49-52, June, 1966.

"Measurement and reduction of refrigerator noise," by R. J. Sabine. ASHRAE J 7:117-121, January, 1965.

"Measurement and suppression of noise; with special reference to electrical machines, by A. J. King. A review," by K. A. Rose. ROY INST BRIT ARCH J 73:85, February, 1966.

--A review," by H. D. Parbrook. TOWN PLAN R 37:75-76, April, 1966.

"Measurement of acoustic resistance at sound-pressure levels to 171 dB," by P. A. Marino, Jr., et al. ACOUSTICAL SOC AM J 41:1325-1327, May, 1967.

"Measurement of interior noise in passenger cars," by W. Henkel. Z GES HYG 15:225-228, April, 1969.

"Measurements of reaction time in intelligibility tests," by M. H. Hecker, et al. J ACOUST SOC AMER 39:1188-1189, June, 1966.

"Measurements on the impulsive noise from crackers and toy firearms," by K. Gjaevenes. J ACOUST SOC AMER 39:403-404, February, 1966.

"Measures for lowering noise intensity at the Tashkent hydro-electric power stations," by P. E. Popov. GIG SANIT 32:94-95, April, 1967.

"Measuring bearing noise," ENGINEERING 209:358-359, April 10, 1970.

"Measuring fan noise in the lab and in the field," by J. B. Graham. HEATING-PIPING 38:150-157, June, 1966.

"Measuring noise exposure," by S. S. Meyers. INSTRUMENTATION TECH 16:66, December, 1969.

"The mechanism of action of noise on the organism," by E. Ts. Andreeva-Galanina, et al. VESTN AKAD MED NAUK SSSR 24:11-18, 1969.

"Mechanisms of noise generation in a compressor model," by B. T. Hulse, et al. J ENG POWER 89:191-197; Discussion 197-198, April, 1967.

"Medical aspects of industrial noise problem," by M. S. Fox. ENVIRONMENTAL CONTROL MGT 139:22-25+, January, 1970; also in INDUSTR MED SURG 39:241-244, June, 1970; NAT SAFETY CONGR TRANS 18:9-14, 1969.

"Medical audiology," by W. F. Rintelmann. ARCH OTOLARYNG 92: 206-207, August, 1970.

"Medical consequences of environmental home noises," by L. E. Farr. JAMA 202:171-174, October 16, 1967.

"Medical opinion and lawful estimation of defective hearing caused by noise," by G. Kollmorgen. Z AERZTL FORTBILK 64:954-956, September 15, 1970.

"Medical prevention of chronic acoustic trauma," by S. Kubik. PRAC LEK 18:176-179, May, 1966.

"Medico-legal aspects of office hearing evaluations," by C. O. Istre, Jr. et al. ARCH OTOLARYNG 86:645-649, December, 1967.

"Medium speed diesel engine noise," by R. Bertodo, et al. INST MECH ENG PROC 183 no 2:129-138+, 1968-1969.

"Mental-hospital admissions and aircraft noise," by I. Abey-Wickrama, et al. LANCET 2:1275-1277, December 13, 1969+.

"Mental-hospital admissions and aircraft noise," by R. H. Chowns. LANCET 1:467, February 28, 1970.

"Mental hygiene aspects of urbanization and noise," by C. Bitter T ZIEKENVERPL 20:535-537, August 1, 1967.

"The merits of double glazing," by J. Walton. PRACTITIONER 10:Suppl:13-17, March, 1970.

"Metabolism of ammonia under some environmental conditions. 4. Mechanism of increase in the ammonia content of the brain and liver of animals exposed to noise," by C. Sakaguchi. JAP J HYG 21:33-37, April, 1966.

--5. Glutamine synthetase activity in the rat under noise," by C. Sakaguchi. JAP J HYG 21:296-298, October, 1966.

"A method of analyzing individual cortical responses to auditory stimuli," by C. W. Palmer, et al. ELECTROENCEPH CLIN NEUROPHYSIOL 20:204-206, February, 1966.

"A method of determining some parameters of the shock wave of explosions in mines and their hygienic assessment," by B. A. Shaparenko. GIG TR PROF ZABOL 9:9-13, July, 1965.

"Method of estimating the sound power level of fans," by J. B. Graham. ASHRAE J 8:71-74, December, 1966.

"Methodology for the use of primates in the exploration of hazardous environments," by D. N. Farrer. ANN NY ACAD SCI 162:635-645, July 3, 1969.

"Methods of measurement of acoustic noise radiated by an electric machine," by A. J. Ellison, et al. INST E E PROC 116:1419-1431, August, 1969.

"Military noise induced hearing loss: problems in conservation programs," by C. T. Yarington, Jr. LARYNGOSCOPE 78:685-692, April, 1968.

"Minimizing danger in handling systems; noise control, the new necessity," MOD MATERIALS HANDLING 25:65, January, 1970.

"Minimum sonic boom shock strenghts and overpressures," by R. Seebass. NATURE (London) 221:651-653, February 15, 1969.

"Mintech and the makers tackle aero-engine noise," ENGINEERING

204:18-19, July 7, 1967.

"Moderate acousti stimuli: the interrelation of subjective importance and certain physiological changes," by G. R. Atherley, et al. ERGONOMICS 13:536-545, September, 1970.

"Modern-day rock-and-roll music and damage-risk criteria," by J. M. Flugrath. J ACOUST SOC AMER 45:704-711, March, 1969.

"The modern industrial society, its illnesses and the maintaining of its health. 3. The effect of noise on the organ of hearing," by W. Wagemann. HIPPOKRATES 37:138-142, February 28, 1966.

"Modern protection of hearing," by H. Gronemann. ZBL ARBEITSMED 16:371-372, December, 1966.

"Modern sound-stage construction," by D. J. Bloomberg, et al. SMPTE J 75:25-28, January, 1966.

"Monitoring, activation, and disinhibition: effects of white noise masking on spoken thought," by P. S. Holzman, et al. J ABNORM PSYCHOL 75:227-241, June, 1970.

"Morphologic changes in the central structures of the auditory analyzer under the prolonged effects of noise," by A. B. Strakhov, et al. BIULL EKSP BIOL MED 69:95-97, June, 1970.

"Morphologic changes in the hearing organ of experimental animals under the effect of high frequency vibration and noise," by I. P. Enin. VESTN OTORINOLARING 27:25-29, January-February, 1965.

"Morphological and histochemical characteristics of the rat endometrium after prolonged acoustic stimulation," by M. Ameli, et al. CLIN OTORINOLARING 18:354-379, July-August, 1966.

"Morphological changes in the hypothalamus in autonomic disorders caused by strong and auditory stimulus," by G. N. Krivitskaia, et al. ZH NEVROPAT PSIKHIAT KORSAKOV 66:1177-1183, 1966.

"The morphology of the ganglion spirale cochleae," by B. Kellerhals. ACTA OTOLARYNG 226:Suppl:1-78, 1967.

"Muffler quiets air conditioner," AIR COND HEAT & REFRIG N 105: 1+, July 5, 1965.

"Muffling the clamor of urban construction," BSNS W 168-169, December 14, 1968.

"Muffling noise may amplify productivity," STEEL 164:72d+, June 2, 1969.

"Multistep valve design cuts throttling noise," by H. D. Baumann. INSTRUMENTATION TECH 16:92-94, October, 1969.

"Music and noise," by J. C. Waterhouse. MUS & MUS 14:28-29+, January, 1966.

"Music as a source of acoustic trauma," by C. P. Lebo, et al. AUDIO ENG SOC J 17:535-538, October, 1969; also in LARYNGOSCOPE 78: 1211-1218, July, 1968.

"NASA acoustically treated nacelle program," by J. G. Lowry. ACOUSTICAL SOC AM J 48:780-782 pt 3, September, 1970.

"NASA begins major engine noise project," by M. L. Yaffee. AVIATION W 87:38-39+, August 21, 1967.

"NASA seeks quiet aircraft engine design," by W. S. Beller. TECH W 20:20-21, June 19, 1967.

"NASA to launch quiet engine program," by H. Taylor. AM AVIATION 32:23, August 19, 1968.

"NASA's quiet engine program focuses antinoise effort," AVIATION W 92:88-89, June 22, 1970.

"Nacelles cited in jets noise," by L. M. Cafiero. ELECTRONIC N 11:24, September 12, 1966.

"National gypsum (co) offers record in push for acoustical tile," AVD AGE 36:94, December 6, 1965.

"Nearfield infrasonic noise generated by three turbojet aircraft during ground runup operations. AMRL-TR-65-132," by R. T. England, et al. US AIR FORCE AEROSPACE MED RES LAB 1-16, August, 1965.

"Need for inplant control," by P. S. Cowen. FOUNDRY 98:70-71, May, 1970.

"Neuronal convergence of noxious, acoustic, and visual stimuli in the visual cortex of the cat," by K. Murata, et al. J NEUROPHYSIOL 28:1223-1239, November, 1965.

"New aircraft noise requirements detailed," AVIATION W 91:35, November 17, 1969.

"New association of acoustical and insulating material manufacturing," by J. E. Nolan. OFFICE 70:35-36, September, 1969.

"New augered pile technic reduces construction noise," by W. J. Duchaine. MOD HOSP 104:164, March, 1965.

"New BISRA oxy-fuel burner test facility," METALLURGIA 74:28, July, 1966.

"New concept; fan/silencer package," by C. J. Trickler. AIR COND HEAT & VEN 62:76-80, December, 1965.

"A new concept of damage risk criterion," by W. G. Noble. AM OCCUP HYG 13:69-75, January, 1970.

"New data on acute acoustic trauma (explosion trauma) as based on new measurement technics," by J. Mayer. MSCHR OHRENHEILK 101:305-312, 1967.

"New dental turbines and the problem of disposition to acoustic trauma. 2," by L. Dunker, et al. DDZ 23:257-260, June, 1969.

"New engine silencer can be retracted," PRODUCT ENG 38:57, September 25, 1967.

"New fan law for sound," by R. Parker. ASHRAE J 9:83-85, October, 1967.

"New ideas for noise control at home," by A. Lees. POP SCI 197: 94-96, September, 1970.

"A new individual noise dosimeter," by S. Lagerholm, et al. ACTA OTOLARYNG 224:Suppl:234+, June 27, 1966.

"New industrial acoustics facility for dynamic rating of duct silencers," ACOUSTICAL SOC AM J 41:869, April, 1967.

"New medical standards for noise," by E. I. Denisov. GIG TR PROF ZABOL 14:47, May, 1970.

"A new method for rating noise exposures," by J. H. Botsford. AMER INDUSTR HYG ASS J 28:431-446, September-October, 1967.

"New method of noise analysis for high velocity air distribution systems," by R. H. Dean, et al. HEATING-PIPING 40:132-137, January, 1968.

"New noise regulations under the Walsh-Healey Act," by F. A. Van Atta. J OCCUP MED 12:27-29, February, 1970.

"New rating method for duct silencers," by D. B. Callaway, et al. HEATING-PIPING 38:88-95, December, 1966; Discussion 39:79-80+, Reply 88+, April, 1967.

"New silencers quiet turbine exhaust noise," by L. S. Wirt. SAE J 74:88-89, February, 1966.

"New steel laminates can banish noise," PRODUCT ENG 39:100-101, January 15,1968.

"Nixon facing tough choice on air noise," by H. Taylor. ELECTRONIC N 14:18, March 10, 1969.

"Nobody loves an airport," by M. M. Berger. SO CALIF L REV 43:631, Fall, 1970.

"Noise," ADM 25:245-262, May-June, 1968.

"Noise," ELECTRONIC N 12:1+, November 20, 1967.

"Noise," J ROY COLL GEN PRACT 17:135-136, March, 1969.

"Noise," THER NOTES 74:114+, September-October, 1967.

"Noise," by L. L. Beranek. SCI AM 215:66-74+, December, 1966; Discussion 216:8, April, 1967.

"Noise," by P. Grognot. BULL INST NAT SANTE 22:927-939, September-October, 1967.

"Noise," by A. Hustin. ACTA OTORHINOLARYNGOL BELG 24:215-316, 1970.

"Noise," by A. van Meirhaeghe. T ZIEKENVERPL 19:234-237, April 1, 1966.

"Noise abatement by barriers," by M. Rettinger. PROGRES ARCH 46: 168-169, August, 1965.

"Noise abatement in refineries," by R. C. Ewing. OIL & GAS J 67:83-88, October 13, 1969.

"Noise abatement in textile mills," by P. H. R. Waldron. AM DYESTUFF REP 58:17-19, July 28, 1969; also in MOD TEXTILES MAG 50:49-50, July, 1969.

"Noise abatement in the Zetor-Super tractor prototype," by I. Seress. CESK HYG 10:81-85, March, 1965.

"Noise abatement sprayed on pipelines," PIPELINE & GAS J 197: 80, September, 1970.

"Noise abaters," by H. Lawrence. AUDIO 51:82, May, 1967.

"Noise, air pollution study to be defined," PRODUCT ENG 39:29, April, 1968.

"Noise and adrenal function," by H. Sakamoto, et al. JAP J CLIN MED 28:1621-1625, May, 1970.

"Noise annoyance and its assessment," by E. J. Richards. ENGINEERING 9:Suppl:362-364.

"Noise and blood circulation," by R. Heinecker. DEUTSCH MED WSCHR 90:1107-1109, June 11, 1965.

"Noise and the conservation of hearing," by S. S. Keys. TRANS ASS INDUSTR MED OFFICERS 15:12-17, January, 1965.

"Noise and the conservation of hearing," by M. Robinson. RHODE ISLAND MED J 53:146-149, March, 1970.

"Noise and the ear," by L. Stein. PLANT ENG 23:54-55, May 1, 1969.

"Noise: Economic aspects of choice," by C. D. Foster, et al. URBAN STUDIES 7:123-135, June, 1970.

"Noise and emotional stress," by F. I. Catlin. J CHRONIC DIS 18:509-518, June, 1965.

"Noise and health," by G. Lehmann. UNESCO COURIER 20:26-27+, July, 1967.

"Noise and hearing," NURS MIRROR 131:49-51 contd, October 23, 1970+.

"Noise and hearing," by R. F. Balas. ROCKY MOUNTAIN MED J 66:53-54, July, 1969.

"Noise and hearing ability: incidence of hearing defects induced by noise in Finland," by J. S. Lumio. IND MED 34:404-406, May, 1965.

"Noise and hearing in industry," by N. Williams. CANAD J PUBLIC HEALTH 58:514-517, November, 1967.

"A noise and hearing survey of earth-moving equipment operators," by P. LaBenz, et al. AMER INDUSTR HYG ASS J 28:117-128, March-April, 1967.

"Noise and its control in process plants," by B. G. Lacey. CHEM ENG 76:74-84, June 16, 1969.

"Noise and its effects," BRIT MED J 5462:605, September 11, 1965.

"Noise and profits," by L. L. Beranek. WIRE & WIRE PROD 45:106-110, October, 1970.

"Noise and the public health," by C. R. Bragdon. SCI & CIT 10:183-184, September, 1968.

"Noise and urban man," by R. A. Baron. AMER J PUBLIC HEALTH 58: 2060-2066, November, 1968.

"Noise and vibration are analyzed for designers," PRODUCT ENG 41:43, February 16, 1970.

"Noise and vibration; conference, Glasgow, Scotland, April 4-6," ULTRASONICS 6:191, July, 1968.

"Noise and vibration exposure criteria," by H. E. von Gierke. ARCH ARCHIVES ENVIRONMENTAL HEALTH 11:327-329, September, 1965.

"Noise and vibration in moulding shops of iron concrete plants," by I. N. Ivatsevich, et al. GIG SANIT 33:105-107, January, 1968.

"Noise and vibration in plants of prefabricated concrete structures," by Iu. K. Aleksandrovskii, et al. GIG SANIT 30:113-115, August, 1965.

"Noise and vibration on vessels and their standardization," by L. Ia. Skratova. BULL INST MAR MED GDANSK 17:151-154, 1966.

"Noise and your nerves," LIFE AND HLTH 84:6+, May, 1969.

"Noise around us; report of Federal council for science and technology's task force on the problem of environmental noise," SCI N 94:541, November 30, 1968.

"Noise as cause of disease," by G. Jansen. DEUTSCH MED WSCHR 92:2325-2328, December 15, 1967.

"Noise as an environmental distrubance," by O. Kitamura. NAIKA 21:903-906, May, 1968.

"Noise as an environmental factor in industry," by W. Burns. TRANS ASS INDUSTR MED OFFICERS 15:2-11, January, 1965.

"Noise as a health hazard at work, in the community, and in the home," by H. H. Jones, et al. PUBLIC HEALTH REP 83:533-536, July, 1968.

"Noise as an industrial hazard," by J. H. Botsford. WATER & SEWAGE WORKS 116:194-196, November 28, 1969.

"Noise as a pollutant," by L. K. Smith. CAN J PUBLIC HEALTH 61:475-480, November-December, 1970.

"Noise at work," by A. W. Gardner. NURS MIRROR 129:32-33, August 15, 1969.

"Noise at the work place and its significance for human health," by G. Jansen. THERAPIEWOCHE 15:665-667, July, 1965.

"Noise: the audible pollutant; cities could do much to lessen noise levels," by R. A. Baron. NATON'S CITIES 7:28+, September, 1969.

"Noise begins to count; business brief," ECONOMIST 230:66-67, March 22, 1969.

"Noise bill draws opposing views," AM AVIATION 32:21, June 24, 1968.

"Noise bill limitation sought by airlines," AVIATION W 88:41, June 24, 1968.

"Noise-breaks--a possible measure for the prevention of noise-induced deafness," by E. Lehnhardt, et al. INT ARCH GEWERBEPATH 25: 65-74, 1968.

"Noise: builders, users pass buck," by C. J. Suchocki. IRON AGE 204:67-69, July 31, 1969.

"Noise campaigners (growing resentment against aircraft noise has become a force manufacturers must reckon with)," ECONOMIST 221: 490, October 29, 1966.

"Noise can be costly now," by L. F. Yerges. HEATING-PIPING 42: 66-69, February, 1970.

"Noise-caused changes of fine motoricity and sensations of annoyance dependent on certain personality dimensions," by G. Jansen, et al. Z EXP ANGEW PSYCHOL 12:594-613, 1965.

"Noise characteristics in the baby compartment of incubators. Their analysis and relationship to environmental sound pressure levels," by F. L. Seleny, et al. AMER J DIS CHILD 117:445-450, April, 1969.

"Noise: city dwellers, teenagers face deafness from noise," CONG Q W REPT 28:1035-1038, April 17, 1970.

"Noise complaints dwindle quickly as jet service begins at National," AVIATION W 84:50, May 2, 1966.

"Noise-conditioning the air conditioner," by J. W. Sullivan. IEEE TRANS IND & GEN APPLICATIONS 4:527-534, September, 1968.

"Noise conditioning shows results in office operation," FIN POST 62:0-23, May 25, 1968.

"Noise conditions in normal school classrooms," by D. A. Sanders. EXCEPTIONAL CHILD 31:344-353, March, 1965.

"Noise considerations in the application and installation of outdoor air-conditioning equipment," by W. S. Bayless. ASHRAE J 9:52-56, April, 1967.

"Noise control," by R. D. Ford. ANN OCCUP HYG 10:415-422, October, 1967.

"Noise control," by A. C. Raes. ARCH BELG MED SOC 24:316-320, May, 1966.

"Noise control: AFS congress; abstracts of papers," FOUNDRY 93: 164+, July, 1965.

"Noise control also a medical question," by H. Wiethaup. THER GEGENW 107:1504-1506 passim, November, 1968.

"Noise control by acoustic interference; abstract," by I. Y. R. Chen, et al. MACHINE DESIGN 42:92+, October 29, 1970.

"Noise control for electric motors," by S. H. Judd, et al. AMER INDUSTR HYG ASS J 30:588-595, November-December, 1969.

"Noise control in air-handling systems," by C. N. Rink. AM SCH BD J 147:27-28, December, 1968.

"Noise control in air systems," by L. F. Yerges. HEATING-PIPING 41:144-148, March, 1969.

"Noise control in architecture: more engineering than art," ARCH REC 142:193-204, October, 1967.

"Noise control in chemical processing," by W. V. Richings. CHEM & PROCESS ENG 48:77-80, October; 66-68, November, 1967; 49:77-79+, January, 1968.

"Noise control in drop-forges," by H. Wiethaup. ZBL ARBEITSMED 16:227-228, August, 1966.

"Noise control of emergency power generating equipment," by P. J. Torpey. AMER INDUSTR HYG ASS J 30:596-606, November-December, 1969.

"Noise control on the local level," by H. M. Fredrikson. ARCH ENVIRON HEALTH 20:651-654, May, 1970.

"Noise control program is quiet success," by D. R. Carlson. MOD HOSP 105:82-85, December, 1965.

"Noise control: three approaches," by P. Vandervoort, II. HOSP TOP 44:65-66, October, 1966.

"Noise control: three approaches," by M. De Porres. HOSP TOP 44: 66-67, October, 1966.

"Noise control: three approaches," by M. Jeffries. HOSP TOP 44:67 passim, October, 1966.

"Noise control: traditional remedies and a proposal for federal action," by J. M. Kramon. HARV J LEGIS 7:533, May, 1970.

"Noise control: universal and international exhibition of 1967, Montreal," SMPTE J 76:574-577, June, 1967.

"Noise controlled at design stage," by J. E. Seebold. OIL & GAS J 68:56-58, December 21, 1970.

"Noise damage," by G. M. Conger. APPRAISAL J 36:253-254, April, 1968.

"Noise-damage criterion using A weighting levels," by D. M. Mercer. J ACOUST SOC AMER 43:636-637, March, 1968.

"Noise design is a contractor problem," by W. A. Tedesco. AIR COND HEAT & REFRIG N 121:8+, December 14, 1970.

"Noise--the disease of the twentieth century," by B. Semczuk. POL TYG LEK 23:853-855, June 3, 1968.

"Noise disorders caused by a concrete plant," by H. Wiethaup. ZBL ARBEITSMED 16:128-130, May, 1966.

"Noise disturbance due to infants' home," by H. Wiethaup. MED KLIN 62:393, March 10, 1967.

"Noise disturbances by a hydrotherapeutic section of a hospital," by H. Wiethaup. MED KLIN 62:612-613, April 14, 1967.

"The noise dosimeter for measuring personal noise exposure," by S. Lagerholm, et al ACTA OTOLARYNG 263:Suppl:139-144, 1969.

"Noise effects on health, productivity, and well-being," by A. Cohen. TRANS NY ACAD SCI 30:910-918, May, 1968.

"Noise environs and helmet performance for the P-1127 V-STOL aircraft. AMRL-TR-70," by H. C. Sommer, et al. US AIR FORCE AEROSPACE MED RES LAB 1-22, December, 1968.

"Noise: excerpt from The Tyranny of Noise," by R. A. Baron. VOGUE 156:150-151+, November 1, 1970.

"Noise exposure control: guidelines," AMER ASS INDUSTR NURSES J 16:17-21, May, 1968.

"Noise exposures in pulp and paper production," by W. A. Ook, et al. AMER INDUSTR HYG ASS J 30:484-486, September-October, 1969.

"The noise factor and its hygienic evaluation in the vocational training of students in the 9-11th grade," by E. A. Timokhina. GIG SANIT 30:46-50, February, 1965.

"Noise factor reduced for high-pressure eductor," by C. A. Mangold, et al. AM SOC SAFETY ENG J 14:16-17, September, 1969.

"Noise figure measurement," by C. N. G. Matthews. WIRELESS WORLD 73:393-394, August, 1967; Discussion 73:451, 506-507, 554-555, September-November, 1967.

"Noise--the fourth pollution," HLTH SERV WORLD 3:16+, July-August, 1968.

"The noise from the high speed turbine as a cause of hypacusis," by G. Girardi, et al. RASS INT STOMAT PRAT 17:405-415, November-December, 1966.

"Noise from the viewpoint of public health, labor hygiene and economics," by A. Gotze, Jr. ORV HETIL 109:2153-2155, September 29, 1968.

"Noise; the gathering crisis; address, October 8, 1969," by M. O. Hatfield. VITAL SPEECHES 36:130-133, December 15, 1969.

"Noise generation in roots type blowers," by R. N. Arnold, et al. INST MECH ENG PROC 178:202-208+, pt 3.

"Noise gets a slogan; buck-scrambling begins," by C. J. Suchocki. IRON AGE 204:104, November 20, 1969.

"Noise governs transformer selection at UCLA health sciences complex in Los Angeles," by K. K. Leithold. ELEC CONSTR & MAINT 67:106-107, July, 1968.

"Noise, a hazard in the operating room," by D. Kane. AORN J 7:78-80, January, 1968.

"Noise hazard--warning bells," by B. Hoad. TIMES R IND & TECH 5:18-22, August, 1967.

"Noise hazards in iron works," by V. Blaha, et al. PRAC LEK 17:95-101, April, 1965.

"Noise: how much more can we take?" by S. Blum. MCCALLS 94:48-49+, January, 1967.

"Noise, how to reduce the noise hazard, Dow Chemical," by R. D. Sias. MECH ENG 88:26-29, October, 1966.

"Noise in air-moving systems," by R. J. Kenny. MACHINE DESIGN 40: 138-150, September 26, 1968.

"Noise in computer rooms and in mechanical data processing central stations," by B. Kvasnicka. PRAC LEK 17:112-115, April, 1965.

"Noise in evoked cerebral potentials," by J. Woods, et al. ELECTROENCEPH CLIN NEUROPHYSIOL 26:633, June, 1969.

"Noise in the external environment," by H. Creighton. ROY INST BRIT ARCH J 73:465-470, October, 1966.

"Noise in high-speed motors," by B. Brozek. MACHINE DESIGN 42: 123-127, March 5, 1970.

"Noise in the hospital," by W. Schweisheimer. SCHWEST REV 4:29-31, October, 1966.

"Noise in the hospital. One of the main complaints of hospitalized patients," by W. Schweisheimer. MED KLIN 60:816-817, May 14, 1965.

"Noise in hospitals: Its effect on the patient," by P. Haslam. NURS CLIN N AM 5:715+, December, 1970.

"Noise in hospitals--a problem the purchasing agent can help eliminate," by I. H. Hunt. CANAD HOSP 45:80-81, October, 1968.

"Noise in industry," by S. J. Evans, et al. CHEM INDUSTR 9:275-281, March 2, 1968.

"Noise in industry," by M. M. Mackay. VIRGINIA MED MONTHLY 94: 288-292, May, 1967.

"Noise in the iron and steel industry," by J. A. Adam, et al. IRON & STEEL INST J 205:701-713, July, 1967.

"Noise in the laboratory," BR MED J 3:662, September 19, 1970.

"Noise in mass-transit systems," by V. Salmon. STANFORD RESEARCH INSTITUTE J 2-7, September, 1967.

"Noise in the news," by P. L. Michael. AMER INDUSTR HYG ASS J 26:615-618, November-December, 1965.

"Noise in the news," by L. K. Smith. CANAD J PUBLIC HEALTH 60: 299-306, August, 1969.

"Noise in oil-hydraulic systems," ENGINEERING 207:647-648, April 25, 1969.

"Noise in Polish ships," by A. Vent, et al. BULL INST MAR MED GDANSK 17:157-162, 1966.

"Noise in students residences; an investigation," by D. C. R. Porter. ARCHITECTS J 141:971-979, September 21, 1965.

"Noise--an increasing military problem," by R. H. Meyer. MILIT MED 133:550-556, July, 1968.

"Noise-induced health problems in Sao Paulo," by S. Marone. RESEN CLIN CIENT 38:223-234, September-October, 1969.

"Noise induced hearing loss," by P. S. Rummerfield, et al. OCCUP HEALTH NURS 17:23-29 passim, November, 1969.

"Noise-induced hearing loss. Exposures to steady-state noise," by A. Cohen, et al. ARCH ENVIRON HEALTH 20:614-623, May, 1970.

"Noise-induced hearing loss and pop music," by S. Hickling. NEW ZEAL MED J 71:94-96, February, 1970.

"Noise-induced hearing loss and rock and roll music," by W. F. Rintelmann, et al. ARCH OTOLARYNG 88:377-385, October, 1968.

"Noise-induced hearing loss in bench glass-blowers," by J. T. Sanderson, et al. ANN OCCUP HYG 10:135-141, April, 1967.

"Noise induced lesions with special reference to abortions in cattle," by E. Aehnelt. DTSCH TIERAERZTI WOCHENSCHR 77:543-547 contd, October 15, 1970.

"Noise-induced sudden deafness," by T. Tsuiki, et al. OTOLARYNGOLOGY 38:607-614, June, 1966.

"Noise injuries," by B. B. Norill. T NORSK LAEGEFOREN 86:1661-1663, December 1, 1966.

"Noise intermodulation audiometry as a simple method for determination of the intermodulation behavior in normal hearing persons," by M. Hoke, et al. ARCH KLIN EXP OHR NAS KEHLKOPFHEILK 194:482-488, December 22, 1969.

"Noise--the invisible enemy," by A. Striganov. AUST NURSES J 68:20+, January, 1970; also in CATH NURSE 18:24+, September, 1969; COMP AIR MAG 74:10-14, October, 1969; IRISH NURS NEWS 5-6 passim, September-October, 1969; NEW ZEALAND NURS J 62:19+, April, 1969; NURS J INDIA 60:119, April, 1969; QUEENSLAND NURSES J 11:33-35, November, 1969; T ZIEKENVERPL 22:337-339, April 1, 1969.

"Noise is a challenge that must be met," ENGINEERING 101, July 15, 1969.

"Noise is for learning; four types of noise created by junior high school students," by J. Reedy. CLEAR HOUSE 43:154-157, November, 1968.

"Noise is a pollution, too," OCCUP HEALTH NURS 18:26-28, June, 1970.

"Noise is a slow agent of death," by A. Bailey. N Y TIMES MAG 46-47+, November 23, 1969.

"Noise: it hurts!" by B. Ford. SCI DIGEST 68:34-40, October, 1970.

"Noise--it's enough to drive you bats," IND W 166:28-34, March 16, 1970.

"Noise a key in DC-10 pod design," by C. M. Plattner. AVIATION W 90:64-66+, June 16, 1969.

"Noise level control," TIMES ED SUP 2727:324, August 25, 1967.

"The noise level in a childrens hospital and the wake-up threshold in infants," by R. Gadeke, et al. ACTA PAEDIAT SCAND 58:164-170, March, 1969.

"The noise level in installations of the Moscow subway," by P. N. Matveev. GIG TR PROF ZABOL 10:58-61, June, 1966.

"Noise level in processing plants can be controlled in three ways," by F. W. Church. OIL & GAS J 67:64-66, July 21, 1969.

"Noise levels in a clinical chemistry laboratory," by P. D. Griffiths, et al. J CLIN PATHOL 23:445-449, July, 1970.

"Noise levels in industry," by J. Morin. ARCH MAL PROF 28:517-522, June, 1967.

"Noise levels in infant oxygen tents," by R. League, et al. LANCET 2:978, November 7, 1970.

"Noise levels of a newly designed handpiece," by K. R. Cantwell, et al. J PROSTH DENT 15:356-359, March-April, 1965.

"Noise levels of underground mining equipment," by W. M. Ward. OCCUP HEALTH REV 19:16, 1967.

"Noise, a major health problem: ed. by A. Hamliton," by V. Knudson. PARENTS MAG 45:66-68, February, 1970.

"Noise may cause deafness," by E. Podolsky. LIFE AND HLTH 82: 12+, May, 1967.

"Noise measurement," by C. H. G. Mills. AUTOMOBILE ENG 60:111-113, March, 1970.

"Noise measurement in a children's hospital and the awaking noise threshold of infant," by R. Gadeke, et al. MSCHR KINDERHEILK 116:374-375, June, 1968.

"Noise measuring at workshops for autogenous welding and cutting and at a plasma-cutting device," by J. Gabelmann. ZBL ARBEITSMED 19:114-119, April, 1969.

"Noise menace threatens man; hearing and sanity may be affected," by B. J. Culliton. SCI N 90:297-299, October 15, 1966.

"Noise; modern industrial dilemma," by A. Teplitsky. SCI & TECH 36-39, October, 1969.

"Noise monitor tells of impending failure," PRODUCT ENG 38:168+, May 8, 1967.

"Noise: more than a nuisance," U S NEWS 67:40-42, November 10, 1969.

"Noise must be designed out of machines; annual symposium, 6th, Cleveland," PRODUCT ENG 40:24+, October 20, 1969.

"Noise: the new pollution," by J. Maguire. J AMER OSTEOPATH ASS 67:961-967, April, 1968.

"Noise: new pressure on plants," CHEM W 105:17, December 3, 1969.

"Noise, no magic solution," COMP AIR MAG 74:23, January, 1969.

"Noise, noise, noise," IND CAN 71:14, August, 1970.

"The noise nuisance," ROY SOC HEALTH J 89:259-260, November-December, 1969.

"Noise nuisance and noise protection in machine equipment for data processing facilities," by W. Winter. ZBL ARBEITSMED 19:201-204, July, 1969.

"Noise nuisance from motorways; effects on residential areas," by F. J. Langdon. ARCHITECTS J 141:1453-1455, June 23, 1965.

"Noise nuisance may cause redesign of jet engines," MACHINE DESIGN 39:14, July 6, 1967.

"Noise: nuisance or health hazard?" GOOD H 165:199, October, 1967.

"Noise--occupational hazard and public nuisance," INT NURS REV 13:42+, July-August, 1966.

"Noise: An occupational hazard and public nuisance," WHO CHRON 20:191-203, June, 1966.

"Noise an occupational hazard and public nuisance," by A. Bell. WHO PUBLIC HEALTH PAP 30:1-130, 1966.

"Noise of highly turbulent jets at low exhaust speeds," by J. E. F. Williams, et al. AIAA J 3:791-793, April, 1965.

"The noise of the turbines," AN EXP ODONTOESTOMAT 27:253-254, July-August, 1968.

"Noise on the job," by M. C. Brown. AM FEDERATIONIST 74:20-23, May, 1967.

"Noise on transportation trunk lines of the Zakarpatskaia District," by N. F. Grishchenko, et al. GIG SANIT 35:101-103, April, 1970.

"Noise plan threatens older jets," AVIATION W 88:324, March 18, 1968.

"Noise pollutes air," SCI N L 87:389, June 19, 1965.

"Noise: polluting the environment," by B. J. Culliton. SCI N 97:132-133, January 31, 1970.

"Noise pollution," ENG N 184:13, January 8, 1970.

"Noise pollution," JAMA 208:2468, June 30, 1969.

"Noise pollution," JAMA 205:928, September, 23, 1968.

"Noise pollution," by M. A. Brown. RIGHT OF WAY 17:40-42, June, 1970.

"Noise pollution," by A. Cohen. JAMA 206:2523, December 9, 1968.

"Noise pollution efforts now focus on turbines and highways," PRODUCT ENG 40:105, April 7, 1969.

"Noise pollution: a growing menace," by M. Brower. SAT R 50:17-19, May 27, 1967.

"Noise pollution: how many decibels can we take?" by S. Sinclair. CAN BUS 43:22-24, 30+, June, 1970.

"Noise pollution: how much can you afford?" STEEL 164:21-25, January 13, 1969.

"Noise pollution in the molding room; what you should know about it," by A. R. Morse. PLASTICS TECH 15:51-54, July, 1969.

"Noise pollution in the office," by F. S. Burgen. OFFICE 71:120, January, 1970.

"Noise pollution: an introduction to the problem and an outline for future legal research," by J. L. Hildebrand. COLUM L REV 70:652, April, 1970.

"Noise pollution: a new problem," by J. M. Hopkins, et al. HYDROCARBON PROCESS 47:124-126, May, 1968.

"Noise pollution; present and future; editorial," AUDIO ENG SOC J 17:384, June, 1969.

"Noise pollution problem is posed! special report," SAFETY MAINT 137:35-38, May, 1969.

"Noise pollution; symposium," UNESCO COURIER 20:4-31, July, 1967.

"Noise problem," by H. Soule. ASTRONAUTICS & AERONAUTICS 4:6+, May, 1966.

"Noise problem control (checklist of questions)," FACTORY 123:89-90, December, 1965.

"Noise problems connected with the manufacture of nylon and terylene yarn," by J. R. Kerr. PROC ROY SOC MED 60:1121-1126, November 1, 1967.

"Noise problems in aeromedical evacuation operations," by D. C. Gasaway, et al. AEROMED REV 2:1-18, September, 1970.

"Noise problems in industry," by D. C. Murphy. ANN OCCUP HYG 9: 149-163, July, 1966.

"Noise problems of vtol with particular reference to the Dornier DO 31," by M. Flemming, et al. AERONAUTICAL J 73:647-653; Discussion 653-656, August, 1969.

"Noise program saves ears at Union Camp," PULP & PA 40:46, June 6, 1966.

"Noise protection in the industry," by H. Schmidt. MUNCH MED WOCHENSCHR 112:Suppl:52-53, December 25, 1970.

"Noise quenchers already feasible," by B. Barrett. ELECTRONIC N 14:79, March 24, 1969.

"Noise rating, terminology, and usage," by D. H. Ball. AIR COND HEAT & REFRIG N 121:32, October 5; 26, October 26; 17, November 2; 10-12, November 16; 18+, December 14; 15, December 21, 1970.

"Noise reduction," AUTOMOBILE ENG 59:396-412, October, 1969.

"Noise reduction by isolation," by J. Walton. INSTRUMENTS & CONTROL SYSTEMS 39:109-111, February, 1966.

"Noise reduction: engine mountings," AUTOMOBILE ENG 59:404-405, October, 1969.

"Noise reduction could cost millions," by H. Taylor. AM AVIATION 32:24-25, August 5, 1968.

"Noise reduction; engine structural vibrations," AUTOMOBILE ENG 59:396-398, October, 1969.

"Noise reduction in electronic music," by R. M. Dolby. ELEC MUS R 6:33-37, April, 1968.

"Noise reduction in a textile weaving mill," by R. O. Mills. AMER INDUSTR HYG ASS J 30:71-76, January-February, 1969.

"Noise reduction; isolation of the engine and transmission system," AUTOMOBILE ENG 59:402-403, October, 1969+.

"Noise research gets top priority," AIR COND HEAT & REFRIG N 111:1+, May 22, 1967.

"Noise: shut that row," ECONOMIST 222:1122, March 25, 1967.

"Noise specification for industrial plant," by J. B. Erskine. ANN OCCUP HYG 10:407-414, October, 1967.

"Noise-stopping tips from a prefabber help boost rentals for packaged apartments," HOUSE & HOME 31:84, June, 1967.

"Noise study focuses on intakes, exhaust," AVIATION W 86:24, June 19, 1967.

"Noise, the subtle polluter," by L. V. Pace, et al. ENVIRONMENTAL CONTROL & SAFETY MGT 139:30-33+, March, 1970.

"Noise suppression system quiets the jet's blast," SAFETY MAINT 133:45, March, 1967.

"Noise survey helps Japanese plant be a good neighbor," MOD MANUF 3:70-71, April, 1970.

"Noise: a syndrome of modern society," by C. R. Bragdon. SCI & CIT 10:29-37, March, 1968.

"Noise takes toll, say experts," TODAY'S HLTH 45:87, October, 1967.

"Noise transmission through shafts," by W. Sorge. Z GES HYG 13:634-636, August, 1967.

"Noise trauma deafness after stapedectomy with recovery," BULL TR J LARYNG 80:631-633, June, 1966.

"Noise 2000; VI. Internationaler Kongress fur larmbekampfing," by L. Trbuhovie. WEEK 57:429, July, 1970.

"Noise, the underrated health hazard," by S. Golub. RN 32:40-45, May, 1969.

"Noise within a housing area near the Irkutsk airport," by M. I. Nekipelov. GIG SANIT 34:94-96, May, 1969.

"Noise without end?" by J. Barr. NEW SOCIETY 73-74, July 20, 1967.

"Noise, you can get used to it," by J. C. Webster, et al. J ACOUST SOC AMER 45:751-757, March, 1969.

"Noises and hearing," by R. M. Turtur. MINERVA MED 59:Suppl:5-9, June 4, 1968.

"Noises and noises," by J. D. Douglas. CHR TODAY. 13:49-50, October 25, 1968.

"Noises that assail us," by T. J. Clogger, CONTEMPORARY R 211: 113-117, September, 1967.

"Noisiness of noise," by J. D. Webb. ENGINEERING 209:360-362, April 10, 1970.

"The noisiness of tones plus noise. NASA CR-1117," by K. S. Pearsons, et al. US NASA 1-89, August, 1968.

"Noisy air conditioner turns out to be a cooling tower in Hartford case," AIR COND HEAT & REFRIG N 109:1+, September 19, 1966.

"Noise compressors traced to misguided attempt at economy," by A. D. Sullivan. HEATING-PIPING 41:83-85, December, 1969.

"Noisy noise," by R. R. Austin. COMPOSER 32:18-22, Summer, 1969.

"Nonauditory health effects," AM SOC SAFETY ENG J 14:16-17, October, 1969.

"Nontransportational noise control," by R. Donley. ARCH ENVIRON HEALTH 20:644-650, May, 1970.

"Normal and abnormal coordination of movements: a polymyographic approach," by G. Pampiglione. J NEUROL SCI 3:525-538, November-December, 1966.

"Not exactly music to your ears; high sound levels of rock-and-roll music," CONSUMER REP 33:349-350, July, 1968.

"Nucleic acid concentration in cerebral cortex nerve cells upon disruption of higher nervous activity," by I. I. Tokarenko. ZH VYSSH NERV DEIAT 19:692-697, July-August, 1969.

"Nuisance: The law and economics," by P. Burrows. LLOYDS BANK REVIEW 95:36-46, January, 1970.

"The nuisance of noise," by J. L. Burn. THREE BANKS R 18-30, December, 1968.

"The nuisances of traffic in residential areas," by D. M. Winterbottom. TRAFFIC Q 19:384-395, July, 1965.

"Nurse and noise," by P. N. Ghei, et al. NURS J INDIA 60:431 passim, December, 1969.

"Objective studies of the effects of tranquilizing agents," by G. Harrer, et al. ARZNEIM FORSCH 20:921-923, July, 1970.

"Observation of the effect of noise on the general health of workers in heavy industry; attempt at evaluation," by H. Jirkova, et al. PRAC LEK 17:147-148, May, 1965.

"Observations on amplitude modulated acoustical noise signals," by M. Rodenburg, et al. PFLUEGER ARCH 311:197, 198, 1969.

"Observations on hearing loss in higher frequencies with chronic otitis media in excessively noisy environments," by T. Yokoyama, et al. J OTOLARYNG JAP 71:1428-1439, October, 1968.

"Observations upon the relationship of loudness discomfort level and auditory fatigue to sound-pressure level and sensation level," by J. D. Hood. J ACOUST SOC AMER 44:959-964, October, 1968.

"Occlusion of the auditory canal and body-conducted sound," by H. G. Dieroff. Z LARYNG RHINOL OTOL 44:417-426, June, 1965.

"Occupational acoustic trauma in otoscierotic patients (considerations on the clinical and medicolegal evaluation related to 4 personal observations)," by A. Scevola. ARCH ITAL OTOL 78:474-486, July-August, 1967.

"Occupational damages in radio-telegraphers," by D. Petrovic, et al. VOJNOSANIT PREGL 25:123-126, March, 1968.

"Occupational damages of the ear," by E. Lehnhardt. ARCH OHR NAS KEHLKOPFHEILK 185:465-468, 1965.

"Occupational damages of the larynx," by E. Nessel. ARCH OHR NAS KEHLKOPFHEILK 185:474-477, 1965.

"Occupational deafness," by L. P. Sobrinho, et al. REV PAUL MED 70:259-269, June, 1967.

"Occupational deafness. Survey in a business concern in the Paris area," by M. J. Alibert. ANN OTOLARYNG 83:883-887, December, 1966.

"Occupational deafness and visual field," by H. Vynckier. ACTA OTORHINOLARYNG BELG 21:213-222, 1967.

"Occupational deafness in the German Democratic Republic," by H. Zenk. Z GES HYG 11:25-34, January, 1965.

"Occupational deafness in Luxembourg," by E. Faber. BULL SOC SCI MED LUXEMB 102:305-316, November, 1965.

"Occupational deafness in thermoelectric plants with Norberg methane motors," by N. Lo Martire, et al. ATTI ACCAD FISIOCR SIENA 16: 202-212, 1967.

"Occupational deafness in a vocational school," by J. Knops, et al. ARCH BELG MED SOC 24:330-338, May, 1966.

"Occupational deafness in young workers in a noisy environment with ear protection," by Y. Harada. ARCH ITAL OTOL 77:157-165, January-February, 1968.

"Occupational disease potentials in the heavy equipment operator," by F. Ottoboni, et al. ARCH ENVIRON HEALTH 15:317-321, September, 1967.

"Occupational health and safety: the government viewpoint," by F. A. Van Atta. FOUNDRY 98:109-113, February, 1970.

"Occupational hearing loss and high frequency thresholds," JAMA 201: 144, July 10, 1967.

"Occupational hearing loss and high frequency thresholds," by J. Sataloff, et al. ARCH ENVIRON HEALTH 14:832-836, June, 1967.

"Occupational hearing loss in relation to industrial noise exposure," by O. el-Attar. J EGYPT MED ASS 51:183-192, 1968.

"Occupational hearing loss--recent trends and practices," by M. S. B Fox. INDUSTR MED SURG 37:204-208, March, 1968.

"Occupational noise," by J. T. Sanderson. OCCUP HEALTH 18:61-71, March-April, 1966.

"Occupational noise-induced hearing loss," by R. Hinchcliffe. PROC ROY SOC MED 60:1111-1117, November 1, 1967.

"The occurrence of noise disturbances in society. Two questionnaire Studies," by E. Jonsson, et al. NORD HYG T 48:21-34, 1967.

"Offensive noise caused by domestic animals," by H. Wiethaup. MED KLIN 60:1377-1378, August 20, 1965.

"Office noise and employee morale," ADM MGT 26:48-49, March, 1965.

"Old folks at home? life in a Manhattan apartment," by E. G. Smith. ATLAN 220:118-119, December, 1967.

"Ombudsman; not so much what he says (aircraft noise complaints)," ECONOMIST 225:1209, December 23, 1967.

"On alteration of color perception by the exposition to noise," by D. Broschmann. GRAEFE ARCH OPHTHAL 168:250-255, May 13, 1965.

"On changes in lipid metabolism in persons during prolonged action of industrial noise on the central nervous system," by P. S. Khomulo, et al. KARDIOLOGIIA 7:35-38, July, 1967.

"On the character of changes in the enzyme activity in the brain tissue during reflex epilepsy," by A. A. Pokrovskii, et al. ZH VYSSH NERV DEIAT PAVLOV 15:120-127, January-February, 1965.

"On the diagnosis and early detection of noise-induced deafness and on the prevention of acoustic trauma," by G. Fabian. Z GES HYG 14: 508-510, July, 1968.

"On different responses to low intensity noise in man," by J. Havranek, et al. ACTIV NERV SUP (Praha) 7:183, 1965.

"On the drug therapy of non-inflammatory diseases of the inner ear and balancing mechanism," by E. Ziegler. Z ALLGEMEINMED 45:927-929 929, July, 1969.

"On the effect of aviation noises of various intensity and duration," by I. Ia. Borshchevskii, et al. VOENNOMED ZH 2:64-68, February, 1965.

"On the effect of industrial noise on the blood pressure level in workers in machine building plants," by N. N. Pokrovskii. GIG TR PROF ZABOL 10:44-46, December, 1966.

"On the effect of industrial noise on the functional status of the auditory analyzer in adolescents," by L. L. Kovaleva. GIG SANIT 32:52-56, January, 1967.

"On the effect of low frequency ultrasonic waves and high frequency sound waves on the organism of workers," by V. K. Dobroserdov. GIG SANIT 32:17-21, February, 1967.

"On the effect of noise on antibody formation," by N. N. Klemparskaia. GIG TR PROF ZABOL 10:54-56, June, 1966.

"On the effect of noise on the functional status of the ear in weavers," by V. Ia. Kornev. GIG TR PROF ZABOL 12:26-30, October, 1968.

"On the effect of repeated noise stress on rats," by V. Hrubes, et al. ACTA BIOL MED GERMAN 15:592-596, 1965.

"On the effect on the organism of the vibration of pneumatic tools with high rate of rotation," by L. Ia. Tartakovskaia, et al. GIG TR PROF ZABOL 13:16-19, April, 1969.

"On estimating noisiness of aircraft sounds," by R. W. Young, et al. J ACOUST SOC AMER 45:834-838, April, 1969.

"On the evaluation of acute noise trauma," by P. Plath, et al. Z LARYNG RHINOL OTOL 44:754-762, November, 1965.

"On the evaluation of the inverse reactions in the Gelle test with air conduction," by L. Palfalvi. ACTA OTOLARYNG 66:508-514, December, 1968.

"On the extent of hearing damage in workers exposed to noise," by F. Schwetz. MSCHR OHRENHEILK 102:663-668, 1968.

"On the functional interrelationship of certain analyzers in the action of noise-vibration stimuli," by A. F. Lebedeva. GIG TR PROF ZABOL 10:22-28, June, 1966.

"On gauge level frequency measurements in impulse-increased industrial noise," by H. G. Dieroff, et al. Z LARYNG RHINOL OTOL 44:639-648, October, 1965.

"On the hearing impairment of bolt drivers in the building industry," by A. Schurno, et al. Z GES HYG 14:161-165, March, 1968.

"On the hygienic assessment of high-frequency intermittent occupational noise," by A. Z. Mariniako. GIG TR PROF ZABOL 10:18-22, March, 1966.

"On hygienic standardization of medium- and high frequency noise (experimental study)," by N. M. Paran'ko, et al. GIG TR PROF ZABOL 12:48-50, June, 1968.

"On the influence of attitudes to the source on annoyance reactions to noise. An experimental study," by E. Jonsson, et al. NORD HYG T 48:35-45, 1967.

"On the influence of attitudes to the source on annoyance reactions to noise. A field experiment," by R. Cederlof, et al. NORD HYG T 48:46-59, 1967.

"On the influence of continuous noise on the organization of the memory content," by H. Hormann, et al. Z EXP ANGEW PSYCHOL 13:31-38, 1966.

"On the influence of discontinuous noise on the organization of memory contents," by H. Hormann, et al. Z EXP ANGEW PSYCHOL 13:265-273, 1966.

"On the law for protection from construction noise of 9 September, 1965. (1)," by H. Wiethaup. ZBL ARBEITSMED 16:103-105, April, 1966.

"On the legality of anti-noise campaigns," by H. Wiethaup. MED KLIN 60:1095-1099, July 2, 1965.

"On means of protection against aircraft noises," by I. Ia. Borshehevskii. VOENNOMED ZH 3:65-68, March, 1969.

"On measures for controlling noise in textile industry plants," by F. S. Ravinskaia, et al. GIG TR PROF ZABOL 12:54, June, 1968.

"On minimal protective zones for specialized enterprises engaged in everyday services to the population," by V. T. Ivanov. GIG SANIT 33:96-97, June, 1968.

"On noise and vibration exposure criteria," by H. E. von Gierke. ARCH ENVIRON HEALTH 11:327-339, September, 1965.

"On noise annoyance and psychological disposition," by C. R. Johansson. NORD HYG T 47:19-25, 1966.

"On noise evaluation at the working site in relation to hearing defects," by G. Wolff. ZBL ARBEITSMED 17:349-355, November 16, 1967.

"On the noise factor in some industrial plants (light and heavy industry) in Armenia," by Z. V. Babaian. GIG SANIT 30:98-99, June, 1965.

"On the non-specific effect of noise on the human organism. Conclusion," by J. Kubik. CESK HYG 10:553-559, October, 1965.

"On the occurrence of nervous disorders in the residential districts of Berlin. Preliminary report," by G. Feuerhahn, et al. PSYCHIAT NEUROL MED PSYCHOL 20:281-286, August, 1968.

"On the 'paradoxical' character of sound intensification in some forms of neurologic diseases associated with decreased hearing," by A. I. Lopotko. VESTN OTORINOLARING 27:63-68, May-June, 1965.

"On permissible levels of noise in hospitals," by V. I. Pal'gov, et al. VRACH DELO 11:119-124, November, 1965.

"On the pnuematization of the mastoid process in workers due to noise," by D. Kosa, et al. HNO 15:324-325, November, 1967.

"On the possibility of early detection of perceptive hypoacusia in workers exposed to noise," by S. Kossowski, et al. MED PRAC 17: 252-253, 1966.

"On the practice of the measurement of noise and vibrations in factories," by G. Wolff. ZBL ARBEITSMED 15:181-184, August, 1965.

"On the problem of the action of continuous spectrum noise on some physiological functions of the organism," by S. V. Alekseev, et al. GIG TR PROF ZABOL 9:8-11, June, 1965.

"On the problem of assessing the functional status of the vestibular analyzer in persons subjected to the effects of vibration," by N. I. Ponomareva, et al. GIG TR PROF ZABOL 12:31-35, June, 1968.

"On the problem of assessing the hearing and expert opinion on the work capacity of persons exposed to industrial noise," by N. I. Ponomareva, et al. GIG TR PROF ZABOL 11:50-53, October, 1967.

"On the problem of the basis for standard noise levels for adolescents," by E. A. Gel'tishcheva, et al. GIG SANIT 33:34-38, November, 1968.

"On the problem of the effect of general vertical vibration and noise on several indices of protein, fat and carbohydrate metabolism in warm blooded animals," by G. I. Bondarev, et al. GIG TR PROF ZABOL 12:58-59, October, 1968.

"On the problem of the effect of vibration and noise on general morbidity," by S. S. Kangelari, et al. GIG TR PROF ZABOL 10:47-49, June, 1966.

"On the problem of 'noise'," by L. Gottberg. ZBL ARBEITSMED 16: 119-120, May, 1966.

"On the problem of noise-induced lesions of the vestibular apparatus in the evaluation of workers in a noisy environment," by H. G. Dieroff, et al. Z LARYNG RHINOL OTOL 46:746-757, October, 1967.

"On the problem of noise pathology," by E. Ts. Andreeva-Galanina, et al. GIG TR PROF ZABOL 12:3-7, November, 1968.

"On the problem of studying the effect of noise on the organism," by E. Ts. Andreeva-Galanina, et al. GIG SANIT 34:70-75, May, 1969.

"On the problem of visual field reduction due to acoustic trauma," by U. Grohmann. KLIN MBL AUGENHEILK 152:600-602, 1968.

"On problems of individual acoustic protection in commercial air travel," by W. Lorenz. Z GES HYG 14:669-674, September, 1968.

"On the problems of so-called acoustic trauma," by F. Schwetz. Z LARYNG RHINOL OTOL 44:571-577, September, 1965.

"On the prolonged action of permissable parameters of noise on the hearing of workers," by L. I. Maksimova, et al. GIG SANIT 31:11-16, February, 1966.

"On the protection of the flight- and technical personnel from aviation noises," by I. Ia. Borshchevskii. VOENNOMED ZH 6:58-63, June, 1965.

"On the protection of the population in the vicinity of airports from aviation noises," by I. Ia. Borshchevskii. GIG SANIT 31:82-85, September, 1966.

"On the question of occupationally related hearing difficulties in musicians," by M. Flach, et al. Z LARYNG RHINOL OTOL 45:595 605, September, 1966.

"On the separation of useful signal from noise in the neural impulse activity of the cochlear nucleus in cats," by E. A. Radionova. ZH VYSSH NERV DEIAT PAVLOV 15:481-490, May-June, 1965.

"On some hemodynamic changes due to the effects of industrial noise," by N. N. Shatalov. GIG TR PROF ZABOL 9:3-7, June, 1965.

"On standardization of the effect of aircraft noises," by I. Ia.

Borshchevskii, et al. VOENNOMED ZH 10:80-83, October, 1967.

"On standards of industrial noise," by A. P. Pronin. GIG SANIT 30:94-97, November, 1965.

"On the status of meeting noise protection requirements in industrial housing construction," by W. Fasold. Z GES HYG 13:628-631, August, 1967.

"On substantiation of the method of studying higher nervous activity during the effect of noise," by S. V. Alekseev, et al. GIG TR PROF ZABOL 11:35-39, May, 1967.

"On the temporary effect of industrial noise on the hearing at 4,000 c/s of stapedectomized ears," by K. Ferris. J LARYNG 79:881-887, October, 1965.

"On time related differences in the hearing capacity with and without hearing protection in relation to sudden noise of various duration and intensity," by F. Pfander. ARCH OHR NAS KEHLKOPFHEILK 185: 488-510, 1965.

"On the tolerance threshold in acoustic influences," by F. Pfander. HNO 13:27-28, January, 1965.

"On unification of physiological methods in studying the effect of noise on the human organism," by E. Ts. Andreeva-Galanina, et al. GIG TR PROF ZABOL 11:14-18, October, 1967.

"On the use of individual means of protection from industrial noise during the vocational training of textile workers," by M. I. Krasil' shchikov. GIG SANIT 32:40-42, August, 1967.

"On the use of maximal value accumulator for the threshold frequency measurement in intermittent work noise," by H. G. Dieroff, et al. Z LARYNG RHINOL OTOL 47:58-63, January, 1968.

"On the use of sound-protection cotton (Billesholm)," by F. Schwetz, et al. MSCHR OHRENHEILK 103:260-263, 1969.

"On variations of the responses to tests of liminal sensation decay: comparative evaluation," by G. Cervellera, et al. ARCH ITAL LARING 73:Suppl:123-136, 1965.

"On the various degrees of stress of the hearing organ in the metal-processing industry," by H. G. Dieroff, et al. ARCH OHR NAS KEHLKOPFHEILK 185:485-488, 1965.

"An optic and electron microscopic study of the organ of Corti of a mice line having convulsive crises caused by some sound frequencies," by D. Usui, et al. ANN OTOLARYNG (Paris) 87:167-182, March, 1970.

"Optimum reception of M-ary Gaussian signals in Gaussian noise," by T. T. Kadota. BELL SYSTEM TECH J 44:2187-2197, November, 1965.

"Origins of noise," by L. D. Mitchell, et al. MACHINE DESIGN 41:174-178, May 1, 1969.

"Orthogonal tree codes for communication in the presence of white Gaussian noise," by A. J. Viterbi. IEEE TRANS COM TECH 15:238-242, April, 1967.

"Otitis and onset of audiogenic seizures induced with rapid stimulation in mice of a resistant strain," by M. M. Niaussat, et al. C R SOC BIOL 164:57-59, 1970.

"Otology in industrial medicine," by R. L. Watson, Jr. INDUSTR MED SURG 36:731-734, November, 1967.

"Otosclerosis in workers in a noisy environment," by B. Gerth. HNO 14:205-208, July, 1966.

"Ototoxicoses under noise load," by M. Quante, et al. ARCH KLIN EXP OHREN NASEN KEHLKOPFHEILKD 196:233-237, 1970.

"Ouch, that noise! Hearing loss gets compensation," FIN POST 64:4, September 19, 1970.

"Our noise: there is a humming in the land, and a ticking, tapping, buzzing, whining, honking, chugging, pounding, beating, throbbing, roaring, blasting and booming," by F. G. Conn. AM LEGION M 84: 30-35, February, 1968.

"Output, error, equivocation, and recalled information in auditory, visual, and audio-visual information processing with constraint and noise; with a reply by R. E. Jester and rejoinder," by H. J. Hsia.

J COMM 18:325-353, December, 1968.

"Oxygen consumption in the organ of Corti based on observation of the dehydrogenase system following exposure to noise," by T. Takahashi. J OTORHINOLARYNG SOC JAP 70:1702-1715, October, 1967.

"Paced recognition of words masked in white noise," by C. M. Holloway. J ACOUST SOC AMER 47:1617-1618, June, 1970.

"Paper machine and related noise abatement developments," by C. B. Dahl, et al. PAPER TR J 154:44-47, November 30, 1970.

"Parametric studies of the acoustic behavior of duct-lining materials," by J. Atvars, et al. ACOUSTICAL SOC AM J 48:815-825 pt 3, September, 1970.

"Passing the strongly voiced components of noisy speech," by C. M. Holloway. NATURE (London) 226:178-179, April 11, 1970.

"The pathological sensory cell in the cochlea," by H. Engstrom. ACTA OTOLARYNG 63:Suppl:20-26, 1967.

"Pathology characteristic of armored troups," by A. C. Benitte. CONCOURS MED 87:2539-2545, April 10, 1965.

"Pathology of noise in its medico-social aspects," by G. Salvadori. FOLIA MED 49:333-358, May, 1966.

"Pathomorphological changes in the organs of white rats under the prolonged effect of noise," by V. P. Osintseva, et al. GIG TR PROF ZABOL 11:23-27, May, 1967.

"Patterns of hair cell damage after intense auditory stimulation," by C. W. Stockwell, et al. ANN OTOL 78:1144-1168, December, 1969.

"Paying the cost of job aircraft noise," ENGINEERING 199:367, March 19, 1965.

"Perceptive deafness in military aviation technicians caused by noise of F 104 G jet planes," by P. P. Castagliuolo. RIV MED AERO 29: Suppl:361-373, December, 1966.

"Perceptual recognition in the presence of noise by psychiatric patients," by D. W. Stilson, et al. J NERV MENT DIS 142:235-247, March, 1966.

"Perforated stainless may cloak jet noise," STEEL 164:31, February 10, 1969.

"Perforated torus ring quiets 25,000 cfm of compressed air," by S. Butler. PRODUCT ENG 41:73, June 8, 1970.

"Performance and noise testing of oxy-fuel burner," ENGINEER 221: 696, May 6, 1966.

"Performance monitoring technique for arbitrary noise statistics," by G. D. Hingorani, et al. IEEE TRANS COM TECH 16:430-435, June, 1968.

"Performance of cerebral palsied children under conditions of reduced auditory input," by J. Fassler. EXCEPT CHILD 37:201-209, November, 1970.

"Peripheral hearing organ and accidental factors," by T. Shida. J OTOLARYNG JAP 72:653-656, February 20, 1969.

"Personal ear protection," by W. I. Acton. OCCUP HEALTH 22:315-320, October, 1970.

"Personal experiences with a stutter-aid," by W. D. Trotter, et al. J SPEECH & HEARING DIS 32:270-272, August, 1967.

"Personal hearing protection: the occupational health nurse's challenge and opportunity," by R. B. Maas. OCCUP HEALTH NURS 17:25-27, May, 1969.

"Personality and the slope of loudness function," by S. D. Stephens. Q J EXP PSYCHOL 22:9-13, February, 1970.

"The phantom hearing (acouphene) after amputation of auditory field by sound trauma," by P. Pazat, et al. REV OTONEUROOPHTAL 42:81-90, March, 1970.

"Phase cancellation probabilities for several sinusoidal signals in Gaussian noise," by A. R. Cohen. IEEE TRANS COM TECH 15:465-467, June, 1967.

"The physical and psychophysical characteristics of sound and noise," A. C. Raes. ARCH BELG MED SOC 24:305-315, May, 1966.

"Physiologic responses of the albino rat to chronic noise stress," by W. F. Geber, et al. ARCH ENVIRON HEALTH 12:751-754, June, 1966.

"Physiological effects of audible sound; AAAS symposium, December 28-30, 1969," by B. L. Welch. SCIENCE 166:533-534, October 24, 1969.

"Physiological-hygienic assessment of pulsed noise," by E. Ts. Andreeva-Galanina, et al. GIG TR PROF ZABOL 13:8-12, September, 1969.

"A pilot study of hearing loss and social handicap in femal jute weavers," by W. Taylor, et al. PROC ROY SOC MED 60:1117-1121, November 1, 1967.

"Plan before you build; planning a quiet foundry," by G. E. Warnaka. FOUNDRY 98:64-67, May, 1970.

"Plant noise--the disturbing decibel dilemma," FACTORY 125:71-76, November, 1967.

"Plasma lipid responses of rate and rabbits to an auditory stimulus," by M. Friedman, et al. AMER J PHYSIOL 212:1174-1178, May, 1967.

"Plastic foam and the big noise," PLASTICS WORLD 23:74, July, 1965.

"Pollution is an offense to any sense," AM DYESTUFF REP 58:11-12, July 28, 1969.

"Pop music as noise trauma," by E. Fluur. LAKARTIDNINGEN 64:794-796, February 22, 1967.

"Porous steel liner muffles jet whines," PRODUCT ENG 37:82, November 7, 1966.

"Port authority renews warnings on stretched DC-8 noise, weight," by J. W. Carter. AVIATION W 84:40, May 2, 1966.

"Port noise complaint," HARV CIVIL RIGHTS L REV 6:61, December, 1970.

"The possibility of modification of respiration by rhythmic acoustic stimuli," by H. J. Gerhardt, et al. Z LARYNG RHINOL OTOL 46:235-247, April, 1967.

"Potential and development of a v/s.t.o.l. inter-city airliner," by D. H. Jagger, et al. AIRCRAFT ENG 42:6-13, January, 1970.

"Power plants abroad stress noise control," POWER 113:88, March, 1969.

"Practical designs for noise barriers based on lead," by B. Fader. AMER INDUSTR HYG ASS J 27:520-525, November-December, 1966.

"Practical ear enclosure with selectively coupled volume," by A. L. DiMattia. AUDIO ENG SOC J 15:295-298, July, 1967.

"Practical experiences with individual noise protection devices," by E. Glock, et al. Z GES HYG 14:413-418, June, 1968.

"Practical noise control at international airports with special reference to Heathrow," by F. C. Petts. ROY AERONAUTICAL SOC J 70:1051-1059; Discussion 1059-1060, December, 1966.

"Practical problems of partition design," by R. D. Ford, et al. ACOUSTICAL SOC AM J 43:1062-1068, May, 1968.

"Pratt & Whitney's JT9D will be a quiet, high bypass ratio engine," by J. D. Kester, et al. SAE J 76:69-71, June, 1968.

"Predicting hearing loss from noise-induced TTS," by J. C. Nixon, et al. ARCH OTOLARYNG 81:250-256, March, 1965.

"Preferred loudness of recorded music of hospitalized psychiatric patients and hospital employees," by H. L. Bonny. J MUS THERAPY 5:44-52, 1968.

"Preliminary results of the research of noise effect on some vegetative functions," by L. Blazekova. PRAC LEK 18:276-279, August, 1966.

"Pressure could spur noise breakthrough," by H. J. Coleman. AVIATION W 85:49+, December 19, 1966.

"Pressure nozzles on home oil heaters," ELECTRONICS 39:58, November 14, 1966.

"Prevalence of impaired hearing and sound levels at work," by J. H. Botsford. J ACOUST SOC AMER 45:79-82, January, 1969.

"Preventing hearing loss in industry," by V. Hamilton. CANAD NURSE 66:37+, September, 1970.

"Preventive measures for maintaining the health of persons working under noisy conditions," by M. Kvaas, et al. GIG TR PROF ZABOL 10:56-58, June, 1966.

"Prevention of deafness from industrial noise and acoustic trauma," by L. W. Benoay. J AMER OSTEOPATH ASS 68:161-167, October, 1968.

"Prevention of noise in Europe," by N. Simonetti. MINERVA MED 60: Suppl:26-22, March 11, 1969.

"The prevention of noise: a medico-social problem of today," by D. Terzuolo. MINERVA MED 61:3332-3336, August 3, 1970.

"Primer on methods and scales on noise measurement," by W. Rudmose. AM SOC SAFETY ENG J 14:18-26, October, 1969.

"Primer on sound level meters," by B. Katz. AUDIO 53:22-24, July; 42+, August, 1969.

"Privacy: there's too little. Noise: there's too much," by M. Drury. HOUSE B 112:74-76, August, 1970.

"Proactive inhibition, recency, and limited-channel capacity under acoustic stress," by D. Eldredge, et al PERCEPT MOTOR SKILLS 25:85-91, August, 1967.

"The problem of acoustic injury during auricular surgery," by J. Calvet, et al. J FRANC OTORHINOLARYNG 14:807-810, December, 1965.

"The problem of low intensity impulse noises in hospital," by R. Wojtowicz. PRZEGL LEK 25:255-258, 1969.

"The problem of noise in England," by F. Merluzzi. MED LAV 61:181-188, March, 1970.

"The problem of noise in hospitals. Considerations on some studies done in hospitals in Naples," by R. DeCapoa. G IG MED PREV

7:124-132, April-June, 1966.

"The problem of noise in hospitals. I. Phonometric data in various environments and in various conditions," by G. Spaziante. RIV ITAL IG 26:468-511, September-December, 1966.

--II. Results of acoustic spectrum analysis for some noise sources in hospital environments," by G. Spaziante. RIV ITAL IG 29:32-48, January-April, 1970.

"The problem of noise in the Royal Navy and Royal Marines," by R. R. Coles, et al. J LARYNG 79:131-147, February, 1965.

"The problem of noise on board the vessels of the Polish Merchant Marine," by C. Szczepanski. POL TYG LEK 25:760-762, June, 1970.

"The problem of the statistical inference. I. Communication in the presence of noise," by R. F. Wrighton. ACTA GENET 17:178-192, 1967.

--3. Statistical inference as communication in the presence of noise," by R. F. Wrighton. ACTA GENET 18:84-96, 1968.

"Problems associated with measurement of acoustic transients," by G. J. Harbold, et al. AEROSPACE MED 36:767-773, August, 1965.

"Problems in the expert testimony of the occupational disease noise-related hearing disorders and deafness," by H. G. Boenninghaus. Z LARYNG RHINOL OTOL 44:578-582, September, 1965.

"Problems in the field of urban traffic noise control in town buildings," by A. S. Perotskaia, et al. GIG SANIT 35:3-9, August, 1970.

"Problems of adaptation to fatigue and occupational acoustic trauma. Experimental, anatomoclinical and audiometric study," by L. Teodorescu, et al. OTORINOLARINGOLOGIE 15:9-24, January-February, 1970.

"Problems of communication and ear protection in the Royal marines," by M. R. Forrest, et al. J ROY NAV MED SERV 56:162-169, Spring, 1970.

"Problems of deafness due to noise," by H. G. Dieroff. Z GES HYG

11:352-361, May, 1965.

"Problems of industrial hygiene in vulcanizing processes in rubber production," by Z. A. Volkova, et al. GIG SANIT 34:33-40, September, 1969.

"Problems of noise and its effect on hearing in some railroad establishments," by F. Singer, et al. CESK OTOLARYNG 17:13-21, February, 1968.

"Problems of protection from noise in the potassium mining industry," by H. Wolf. Z GES HYG 12:189-195, March, 1966.

"Problems of speech communication in noise and work performance in noise," by K. D. Kryter. AM ASS INDUSTR NURSES J 13:13, April, 1965.

"Production of cleft palate by noise and hunger," by S. Peters, et al. DEUTSCH ZAHNAERZTL Z 23:843-847, August, 1968.

"Products: soft core insulation," PROGRES ARCH 49:52, August, 1968.

"Professional hearing disorder during the influence of strong interrupted noise," by P. S. Kublanova, et al. ZH USHN NOS GORL BOLEZ 27:56-59, September-October, 1967.

"Professional risks in stomatological practice," by S. Bocskay, et al. STOMATOLOGIA 12:455-460, September-October, 1965.

"Professors are skeptical of positive claims made for environmental control (windowless classrooms, negative ions, noise are discussed)," by F. J. Versagi. AIR COND HEAT & REFRIG N 108:42-43, June 6, 1966.

"Progress towards standards for noise and audiometry," by C. N. Davies. ANN OCCUP HYG 10:401-406, October, 1967.

"The progression of hearing loss from industrial noise exposures," by E. J. Schneider, et al. AMER INDUSTR HYG ASS J 31:368-376, May-June, 1970.

"Progressive alterations in cochlear nucleus, inferior colliculus, and medial geniculate responses during acoustic habituation," by M.

Kitzes, et al. EXP NEUROL 25:85-105, September, 1969.

"Prolonged effect of noise of moderate intensity on the functional state of the organism," by O. P. Kozerenko, et al. IZV AKAD NAUK SSSR 4:527-536, July-August, 1967.

"Propeller research gains emphasis," by M. L. Yaffee. AVIATION W 91:56-57+, November 24, 1969.

"Proposed threshold limit value for noise," by H. H. Jones. SAFETY MAINT 137:46-49, January, 1969.

"Protecting hearing at Vulcan mold and iron company," by J. P. Donnelly. FOUNDRY 98:49-51, July, 1970.

"Protecting residual hearing in hearing aid user," by M. Ross, et al. ARCH OTOLARYNG 82:615-617, December, 1965.

"Protection against audiogenic crises in mice by nicotinamide," by A. Lehmann, et al. J PHYSIOL (Paris) 59:Suppl:446, 1967.

"Protection against immission," by H. Wiethaup. ZBL ARBEITSMED 15:237-243, October, 1965.

"Protection from bumps and noise," ENGINEERING 204:446, September 22, 1967.

"Protection of ground personnel against the noise of jet aircraft engines," by J. V. Quercy. ARCH MAL PROF 27:537-541, June, 1966.

"Protective effect of cytochrome C and adenosinotriphosphoric acid in acoustic trauma," by W. Jankowski, et al. OTOLARYNG POL 23:133-140, 1969.

"Protein fractions of serum in animals under the effects of prolonged noise," by N. N. Pushkina, et al. VRACH DELO 9:98-100, September, 1968.

"Proximity damages," by J. L. Sackman. APPRAISAL J 37:177-199, April, 1969.

"Psychic cost of adaptation to an environmental stressor," by D. C. Glass, et al. J PERSONALITY SOC PSYCHOL 12:200-210, July, 1969.

"Psychological and physiological as well as organic damage dur to the effect of noise and high sounds," by K. Jatho. OEFF GESUNDHEIT-SWESEN 29:293-298, July, 1967.

"Psychological and physiological reactions to noise of different subjective valence (TTS and EMG)," by H. Hormann, et al. PSYCHOL FORSCH 33:289-309, 1970.

"The psychological difficulties of control of traumatic hearing disorders due to industrial noise," by D. Hogger. INT ARCH GEWERBEPATH 22:306-314, August 17, 1966.

"Psychological reactions to aricraft noise," by K. D. Kryter. SCIENCE 151:1346-1355, March 18, 1966; Reply by P. K. Holmes. 152:865, May 13, 1966.

"Psychometric studies of subjects exposed to the noise of jet engines," by G. G. Calapaj, et al. MED LAVORO 60:43-52, January, 1969.

"Public health hazards and vertigo," by T. Naito. J OTOLARYNG JAP 72:638-643, February 20, 1969.

"Punishment as a discriminative stimulus and conditioned reinforcer with humans," by T. Ayllon, et al. J EXP ANAL BEHAV 9:411-419, July, 1966.

"Pupil behavior in relation to various sensory stimuli," by N. Orzalesi, et al. BOLL OCULIST 46:284-291, April, 1967.

"Pupillographic changes induced by diverse acoustic stimulations in normal subjects," by B. B. Carenini, et al. BOLL OCULIST 45:75-82, February, 1966.

"Putting correlation to work on engineering test data," by G. T. Roberts. MACHINE DESIGN 42:108-113, January 8, 1970.

"Putting up with transport; society of environmental engineers 10th anniversary symposium. London," ENGINEERING 207:600-602, April 18, 1969.

"Q-Star achieves reduced noise," by C. M. Plattner. AVIATION W 92: 87-88+, April 6, 1970.

"Quantification of the noisiness of approaching and receding sound," by G. Rosinger, et al. ACOUSTICAL SOC AM J 48:843-853 pt 1, October, 1970.

"Quantitative studies on the spiral ganglion of guinea pigs after exposure to noise," by W. Wicke, et al. MONATSSCHR OHRENHEILKD LARYNGORHINOL 104:433-440, 1970.

"Quest for quiet," by B. Swart. FLEET OWNER 60:85-92, February, 1965.

"Quiet foundry," AM MACH 111:116, January 16, 1967.

"Quiet plane sneaks upon ground observers," MACHINE DESIGN 42: 18, April 30, 1970.

"Quiet, please!," by P. Nathan. PARENTS MAG 40:56-57+, October, 1965.

"Quiet, please," by R. B. Newman. ARCH ENVIRON HEALTH 20:561-562, May, 1970.

"Quiet please," by E. Wilson. NM 123:391+, January 27, 1967.

"Quiet, please! Noise pollution," IND DES 17:58-59, July, 1970.

"Quiet, please,' the sign says," by M. E. Hagans. RN 28:92, February, 1965.

"Quiet revolution in design: rooting out noise sources," by A. Hannavy. PRODUCT ENG 37:50-57, October 24, 1966.

"Quiet; San Antonio parade shows advantages sound conditioned homes can offer," AIR COND HEAT & REFRIG N 105:6-7, August 23, 1965.

"Quiet: what you can do to preserve it; excerpts from The Tyranny of Noise," by R. A. Baron. HOUSE & GARD 138:128-129, October, 1970.

"Quietening noisy natural gas," ENGINEERING 204:960-961, December 15, 1967.

"Quieter city is goal of New York City study group," AUTOMOTIVE ENG 78:27-31, August, 1970.

"Quieter diesels at CAV," ENGINEERING 204:872-873, December 1, 1967.

"Quieter transformers," ENGINEERING 204:547, October 6, 1967.

"Quieter turbofans could hit cost snag," by M. L. Yaffee. AVIATION W 91:31+, October 27, 1969.

"Quieting the gas flame," MECH ENG 89:49, July, 1967.

"RFT stability or failure to arouse?" by G. M. Vaught, et al. PERCEPT MOTOR SKILLS 28:378, April, 1969.

"Racket that won't go away," BSNS W 130-132, March 16, 1968.

"Radiated aerodynamics noise effects on boundary-layer transition in supersonic and hyper-sonic wind tunnels," by S. R. Pate, et al. AIAA J 7:450-457, March, 1969.

"Radiated noise from high speed dental handpieces," by H. N. Cooperman, et al. DENT DIG 71:404-407, September, 1965.

"Rail test to evaluate equilibrium in low-level wideband noise. AMRL-TR-66-85," by C. W. Nixon, et al. US AIR FORCE AEROSPACE MED RES LAB 1-16, July, 1966.

"Rapid transit system environment," by W. V. Braktowski. J ENVIRON-MENTAL SCI 10:20-27, April, 1967.

"RB-211 engine given outdoor noise tests," AVIATION W 90:115, April 14, 1969.

"The reaction of the cardiovascular system of working adolescents to sound and vibration stimuli," by A. I. Tsysar'. GIG SANIT 31:33-38, February, 1966.

"Reaction of the hearing organ to sound stimulation in cochlear neuritis," by A. G. Rakhmilevich. VESTN OTORINOLARING 30:28-31, May-June, 1968.

"Reactions of the human nervous and cardiovascular systems to the effects of aviation noises," by V. G. Terent'ev, et al. VOENNOMED ZH 6:55-58, June, 1969.

"Reactions of neurons in cochlear nucleus to acoustic signals of varying duration," by E. A. Radionova. FED PROC 25:389-390, May-June, 1966.

"Reactivity to noise and personality," by S. Dongier. REV MED PSYCHOSOM 9:283-285, October-December, 1967.

"Readjustment reactions of cerebrovascular circulation to chronic noise," by E. Betz. ARCH PHYS THER 17:61-65, January-February, 1965.

"Realistic assessment of the vertiport/community noise problem," by N. Shapiro, et al. J AIRCRAFT 5:407-411, July, 1968.

"Reappraisal of the relationship between noise and human performance by means of a subsidiary task measure," by J. M. Finkelman, et al. J APPL PSYCHOL 54:211-213, June, 1970.

"Recent activity in the noise and hearing field," W. L. Baughn. ARCH ENVIRON HEALTH 12:474-479, April, 1966.

"Recent developments in ear protection," by R. R. Coles. PROC R SOC MED 63:1016-1019, October, 1970.

"Recent noise measurement techniques," by P. L. Tanner. ANN OCCUP HYG 10:375-380, October, 1967.

"Recovery process and the extent of injuries to the organ of Corti (study by the surface specimen technic)," by A. Sugiura. J OTOLARYNGOL JAP 73:1770-1779, November, 1970.

"Recruitment in noise-induced hearing loss," by K. S. Burke, et al. ACTA OTOLARYNG 62:351-361, October-November, 1966.

"Recurrent impact noise from pneumatic hammers," by A. M. Martin, et al. ANN OCCUP HYG 13:59-67, January, 1970.

"Reduce those refinery noise levels," by D. A. Tyler. OIL & GAS J 67:140-141, July 28, 1969.

"Reducing acoustic noise in electrical systems," by D. P. Costa. AUTOMATION 16:82-88, May, 1969.

"Reducing automobile and tractor noise," by V. P. Goncharenko. GIG TR PROF ZABOL 14:46-47, January, 1970.

"Reducing diesel engine noise," by C. Gray. ENGINEERING 209:237, March 6, 1970.

"Reducing machine sound levels," MACH 72:347-348, September, 1965.

"Reducing machinery noise in cement plants," by G. L. Koonsman. IEEE TRANS IND & GEN APPLICATIONS 6:476-479, September, 1970.

"Reducing the noise from diesel engines," ENGINEERING 696, November 18.

"Reducing noise in cams," by R. G. Fenton. MACHINE DESIGN 38: 187-190, April 14, 1966.

"Reducing noise in plant exhaust systems," by L. Klein, et al. PLANT ENG 19:154-157, October, 1965.

"Reduction of aircraft noise measured in several school, motel and residential rooms," by D. E. Bishop. J ACOUST SOC AMER 39:907-913, May, 1966.

"Reduction of compressor noise radiation," by M. V. Lowson. ACOUSTICAL SOC AM J 43:37-50, January, 1968.

"Reduction of noise and vibrations in a hydraulic turbine," by T. Sagawa. J BASIC ENG 91:722-727, December, 1969.

"Reduction of noise from warm air furnace blowers," by A. C. Potter. ASHRAE J 7:112-116, January, 1965.

"Reduction of the noise of railway traffic and of rheostat tests of Diesel locomotives," by E. V. Bobin. GIG SANIT 34:94-97, January, 1969.

"Regeneration of glycogen in the hair cells of the organ of Corti," by S. Chodynicki. FOLIA HISTOCHEM CYTOCHEM 3:211-216, 1965.

"Regional vascular reactions during exposure to vibration and noise, individually and together," by N. M. Paran'ko. GIG SANIT 33:19-24, November, 1968.

"Regulating road traffic noise," ENGINEERING 205:823, May 31, 1968.

"Rehological understanding of origin of inner-ear acoustic fatigue and trauma," by H. Uchiyama, et al. BIORHEOLOGY 6:253, January, 1970.

"The relation between motor activity and risk of death in audiogenic seizure of DBA mice," by I. Lieblich, et al. LIFE SCI 4:2295-2299, December, 1965.

"Relation between sound intensity and amplitude of the auditory evoked response at different stimulus frequencies," by F. Antinoro, et al. ACOUSTICAL SOC AM J 46:1433-1436 pt 2, December, 1969.

"Relation of threshold shift to noise in the human ear," by H. Weissing. ACOUSTICAL SOC AM J 44:610-615, August, 1968.

"Relations of the human vertex potential to acoustic input: loudness and masking," by H. Davis, et al. J ACOUST SOC AMER 43:431-438, March, 1968.

"Relationship between auditory fatigue and noise exposure time," by Y. Katano. J OTORHINOLARYNG SOC JAP 69:1592-1602, September, 1966.

"Relationship between hearing changes and wide range noise intensity and duration of its action," by S. V. Alekseev, et al. GIG TR PROF ZABOL 9:47-49, March, 1965.

"The relationship between the recovery pattern in auditory fatigue and the noise susceptibility," by K. Tsunoda. J OTORHINOLARYNG SOC JAP 69:2088-2096, December, 1966.

"Relationships for temporary threshold shifts produced by three different sources. Rep No. 633," by J. L. Fletcher, et al. US ARMY MED RES LAB 41-45, June 30, 1965.

"Relative annoyance and loudness judgments of various simulated sonic boom waveforms. NASA CR-1192," by L. J. Shepherd, et al. US NASA 1-52, September, 1968.

"Relative aversiveness of noise and shock," by B. A. Campbell, et al. J COMP PHYSIOL PSYCHOL 60:440-442, December, 1965.

"Relative aversiveness of white noise and cold water," by P. J. Woods, et al. J COMP PHYSIOL PSYCHOL 64:493-495, December, 1967.

"Relative effects of figural noise and rotation on the visual perception of form," by E. A. Alluisi, et al. PERCEPT MOT SKILLS 31:547-554, October, 1970

"Reliability of TTS from impulse-noise exposure," by D. C. Hodge, et al. J ACOUST SOC AMER 40:839-846, October, 1966.

"Remarks on the problem of diagnostic criteria of occupational hearing loss resulting from exposure to noise," by J. Borsuk, et al. OTOLARYNG POL 23:273-284, 1969.

"Reminiscences; noise; sound stages," by V. O. Knudsen. AUDIO ENG SOC J 18:436-439, August, 1970.

"Repetitive auditory stimuli and the development of sleep," by B. Tizard. ELECTROENCEPH CLIN NEUROPHYSIOL 20:112-121, February, 1966.

"Report on the effects of noise on the worker. (Bristol Advisory Council on Occupational Health)," J ROY INST PUBLIC HEALTH 31:53-61, March-April, 1968.

"Report on repeated audiometric examinations of industrial workers after having been exposed to harmful noise for a period of 5 years," by S. Podvinec, et al. J FRANC OTORHINOLARYNG 15:53-60, January-February, 1966.

"Research on changes in the auditory capacity in a noisy industrial environment," by J. Morin. ARCH MAL PROF 26:252-256, April-May, 1965.

"Research on ear diseases caused by noise in workers in light metallurgical industry," by E. Manzo, et al. FOLIA MED 51:720-731, September, 1968.

"Research on industrial noise," by M. M. Woodhead. MANAGER 378-380, May.

"Research on regenerated noise in ducts," by A. C. Potter. ASHRAE J 7:52, February, 1965.

"The residue phenomenon in animal experiments," by C. C. Leidbrandt. NEDERL T GENEESK 109:1781, September 18, 1965.

"Resolution of ballistographic records into cardiac and respiratory components," by P. M. Rautaharju, et al. CANAD J PHYSIOL PHARMACOL 44:691-700, September, 1966.

"Response of neurons of the dorsal and posteroventral cochlear nuclei of the cat to acoustic stimuli of long duration," by J. M. Goldberg, et al. J NEUROPHYSIOL 29:72-93, January, 1966.

"Responses of inferior colliculus neurons in the cat to binaural acoustic stimuli having wide-band spectra," by C. D. Geisler, et al. J NEUROPHYSIOL 32:960-974, November, 1969.

"Rest and reminiscence," by L. Moller. NM 123:99+, October 28, 1966.

"Restoration, resolution, and noise," by C. K. Rushforth, et al. OPT SOC AM J 58:539-545, April, 1968.

"Results from the retest of noise-induced temporary threshold shift (NI-TTS)," by T. Yokoyama, et al. J OTORHINOLARYNG SOC JAP 70:1421-1429, August, 1967.

"Results of some audiometric examinations performed in an iron and steel plant," by G. Sparacio, et al. G IG MED PREV 9:90-98, January-March, 1968.

"Results of speech audiometry in acoustic trauma," by E. Vojacek. MSCHR OHRENHEILK 102:152-157, 1968.

"Results of the studies of hearing damages in military tank and infantry personnel," by A. Spirov. VOJNOSANIT PREGL 26:78-80, February, 1969.

"Retention of conditioned noise aversion following medial geniculate lesions in the rat," by M. Lyon. EXP NEUROL 16:1-15, September, 1966.

"Reversible sudden deafness following a brief intensive sound," by M. Abrahamovic. CESK OTOLARYNG 17:200-205, August, 1968.

"A review of hearing damage risk criteria," by W. I. Acton. ANN OCCUP HYG 10:143-153, April, 1967.

"Review of noise in plants of the West Bohemian region," by J. Srutek. J PRAC LEK 17:61-63, March, 1965.

"Review of research and methods for measuring the loudness and noisiness of complex sounds. NASA Contract Rep NASA CR-422," by K. D. Kryter. US NASA 1-57, April, 1966.

"Review of the research into the injurious effect of noise," by S. Kubik. PRAC LEK 18:224-225, June, 1966.

"A review of research on noise, with particular reference to schools," by P. Wrightson. GREATER LONDON RESEARCH QUARTERLY BULLETIN 20-27, September, 1969.

"The risk of occupational deafness in musicians," by M. Flach, et al. GERMAN MED MONTHLY 12:49-54, February, 1967.

"Roaring jet engines can be tested quietly," by V. Singleton, et al. POWER 112:70-71, September, 1968.

"Rock music and hearing," by R. L. Voorhees. POSTGRAD MED 48:108-112, July, 1970.

"Rock physically unsound," SCI DIGEST 63:67-68, June, 1968.

"Rocket blasts and guinea pigs," SCI DIGEST 64:63-64, October, 1968.

"Role of glass in sound control," by L. F. Yerges. AM SCH BD J 147:48-49, November, 1963.

"Role of sound-suppressing curtains in the control of industrial noise," pollution," by I. Singer. WIRE & WIRE PROD 45:111-113, October, 1970.

"Room noise is eliminated by using acoustical sealant," ADHESIVES AGE 11:32, April, 1968.

"Rules lowering noise will increase plant costs," MOD MANUF 2:98-100, September, 1969.

"Safety helmet tunes out noise (blast cleaning)," IRON AGE 205:114N, April 23, 1970.

"San Francisco sets airport noise limit," AVIATION W 93:37, December 21, 1970.

"The sanitary-hygienic characteristics of noise on dry-cargo Diesel ships of 1,200 to 1,800 tons capacity and the efficacy of antinoise measures," by V. I. Petrov, et al. GIG TR PROF ZABOL 12:19-22,

May, 1968.

"Scanning electron microscopy of the organ of Corti," by G. Bredberg, et al. SCIENCE 170:861-863, November 20, 1970.

"Scopolamine effects on go-no go avoidance discriminations; influence of stimulus factors and primacy of training," by N. Rosic, et al. PSYCHOPHARMACOLOGIA 17:203-215, 1970.

"Select an appropriate decibel rating when specifying transformers," ELEC CONSTR & MAINT 68:123-126, May, 1969.

"Selection of strains of rabbits sensitive to an epileptogenic sound stimulus," by F. Horak. PHYSIOL BOHEMOSLOV 14:495-501, 1965.

"Senator Proxmire takes off," NATURE (London) 227:884-885, August 29, 1970.

"The sense of hearing," by C. Stark. STERO R 23:66-74, September, 1969.

"Sensorineural hearing loss associated with firearms," by R. J. Keim. ARCH OTOLARYNG 90:581-584, November, 1969.

"Sensory interaction; perception of loudness during visual stimulation," by R. S. Karlovich. ACOUSTICAL SOC AM J 44:570-575, August, 1968.

"Sensory stimulation and rhesus monkey activity," by W. A. Draper. PERCEPT MOTOR SKILLS 21:319-322, August, 1965.

"A serial study of noise exposure and hearing loss in a group of small and medium size factories," by D. E. Hickish, et al. ANN OCCUP HYG 9:113-133, July, 1966.

"Severe equilibrium disorders following strong sound impression through bilaterally worn earphones," by H. Rudert. ARCH KLIN EXP OHR NAS KEHIKOPFHEILK 188:316-319, 1967.

"Shear noise source terms for a circular jet," by G. Krishnappa. J AP MECH 35:814-815, December, 1968.

"Sheet lead for soundproofing," AIR COND HEAT & VEN 65:46, December, 1968.

"Shhh; retrofitting of jets," FORBES 105:31, March 15, 1970.

"Short-term changes in the threshold of hearing after stimulation with noise and autonomic equilibrium," by R. Tomanek. GIG TR PROF ZABOL 12:14-18, June, 1968.

"Should your plant go underground?" by C. E. Petak. SAFETY MAINT 132:46-49+, September, 1966.

"Shushing refinery noises," CHEM ENG 76:84, March 24, 1969.

"Shut that row," ECONOMIST 222:1122, March 25, 1967.

"Sideline noise new problem the airlines hadn't expected," by B. Jackson. FIN POST 63:12, August 9, 1969.

"Sieve-audiometric studies of workers exposed to noise," by M. Jonsson. DEUTSCH GESUNDH 22:2286-2289, November 30, 1967.

"The significance of binaural hearing for speech communication under the effects of noise," by H. Feldmann. ACTA OTOLARYNG 59:133-139, February-April, 1965.

"Significance of intermission in the strong sound loading test," by T. Shida, et al. J OTOLARYNG JAP 73:Suppl:976-977, July, 1970.

"The significance of noise to health," by W. Klosterkotter. ZBL BAKT 212:336-353, 1970.

"Significance of reports about personal annoyance caused by noise," by O. Arvidsson, et al. NORD HYG T 49:14-20, 1968.

"Silence broken on noise pollution," IRON AGE 205:70-72, April 16, 1970.

"Silence in cars is golden," ENGINEERING 207:88-89, January 17, 1969.

"Silence is golden," BULL NTRDA 56:6+, July-August, 1970.

"Silence pollution," by J. Frye. ELECTR WORLD 79:58-59, March, 1968.

"Silencing airblast circuit-breakers," ENGINEER 225:274-275, February 16, 1968.

"Silencing fan noise in underground garages," by K. A. Traub. AIR COND & VEN 66:47-50, May, 1969.

"Silencing flames," AM GAS ASSN MO 49:15, May, 1967.

"Silencing a gas-turbine generator," ENGINEERING 202:938, November 25, 1966.

"Silencing Impak metric motors," ENGINEER 227:233, February 14, 1969.

"Silencing invisible pollution; anti-noise organizations," by T. Berland. TODAYS HEALTH 48:16-18+, July, 1970.

"Silencing the jets," CHEM & ENG N 47:19, November 24, 1969.

"Silencing of hand-held percussive rock drills for underground operations," by B. H. Weber. CAN MIN & MET BUL 63:163-166, February, 1970.

"Silent conveying," COMP AIR MAG 71:22, May, 1966.

"Silent drills; where they stand now," by J. J. Daly. ROCK PROD 68: 62-63, November, 1965.

"Silent generator," by A. Scott. ENGINEERING 207:904, June 27, 1969.

"Simple guidelines best for noise control," PRODUCT ENG 40:75, June 2, 1969.

"Simple, light muffler with low back pressure cuts noise in light aircraft," by G. A. Alther. SAE J 75:58-60, May, 1967.

"Simple method for identifying acceptable noise exposures," by J. H. Botsford. J ACOUST SOC AMER 42:810-819, October, 1967.

"Simplify your calculations for quiet fan systems," by C. J. Trickler. AIR COND HEAT & VEN 64:69-76, January, 1967.

"Singing flame stills whine in jet engine," PRODUCT ENG 40:22, December 15, 1969.

"A six-year prospective study of the effect of jet-aircraft noise on hearing," by J. J. Knight, et al. J ROY NAV MED SERV 52:92-96, Summer, 1966.

"16 surefire ways to keep your apartment tenants complaining about noise," HOUSE & HOME 36:67-69, July, 1969.

"16,000 miracle cuts out Heathrow jets," by T. Devlin. INNER ED 2864: Suppl:7, April, 10, 1970.

"The Sixth AMA Congress on Environmental Health, Chicago, April 28-29, 1969. Welcoming remarks," by G. D. Dorman. ARCH ENVIRON HEALTH 20:610-611, May, 1970.

"Slight differences in noise stimulation and NI-TTS (noise-induced temporary threshold shift)," by T. Yokoyama, et al. J OTORHINOLARYNG SOC JAP 70:1343-1357, August, 1967.

"Snoring--the listeners' disease," LIFE AND HLTH 82:6+, April, 1967.

"Snoring: theory of compensative resortia," by M. H. Boulware. EYE EAR NOSE THROAT MONTHLY 47:664-668, December, 1968.

"Socially acceptable compressors," by E. C. Hinck. COMP AIR MAG 74:18-21, January, 1969.

"Softly, softly...," NEW ZEALAND NURS J 58:18, July, 1965.

"Softly, softly--anti-noise campaign at Edinburgh Royal Infirmary and associated hospitals," NURS TIMES 61:401, March 19, 1965.

"Solution to the noise control problem," by H. Seelbach. AIR COND HEAT & VEN 64:87-90, April, 1967.

"Solutions to the paper mill noise problem," by R. W. Gray. IND MED 35:257-258, April, 1966.

"Solving the noise problem (abstract)," by E. E. Singer. SUPERVISORY MGT 13:39-42, February, 1968.

"Some aspects of noise in an industrial environment: research performed on a group of hydroelectric power plants," by A. Lacquaniti, et al. MED LAVORO 58:41-59, January, 1967.

"Some cases of acoustic trauma observed in military service," by T. Stern. PRAXIS 58:898-900, July 15, 1969.

"Some clinico-physiological studies of workers subjected to the effect of constant noise," by G. Z. Dumkina. GIG TR PROF ZABOL 10: 23-27, December, 1966.

"Some effects of bone-conducted masking," by Z. G. Schoeny, et al. J SPEECH HEARING RES 8:253-261, September, 1965.

"Some effects of rhythmic distraction upon rhythmic sensori-motor performance," by J. J. Keenan. J EXP PSYCHOL 77:440-446, July, 1968.

"Some harmful effects of noise," by W. Alexander. CANAD MED ASS J 99:27-31, July 6, 1968.

"Some notes on the provision of personal hearing protection for fettlers at an iron foundry," by D. B. Sugden. ANN OCCUP HYG 10:263-268, July, 1967.

"Some particular problems of noise control," by R. E. Fischer. ARCH REC 144:185-192, September, 1968.

"Some physiological factors in noise-induced hearing loss," by M. Lawrence, et al. AMER INDUSTR HYG ASS J 28:425-430, September-October, 1967.

"Some physiological reactions found in persons experimentally exposed to sounds of 500 and 1000 Hz (80 plus or minus 2 db. and 90 plus or minus 2 db.)," by C. Gradina, et al. FIZIOL NORM PAT 14:453-460, September-October, 1968.

"Some problems concerning prevention of inner ear damage caused by noise," by L. Surjan. ORV HETIL 109:1365-1369, June 23, 1968.

"Some problems of the development of an optimal acoustic environment in space ship cabins," by E. M. Iuganov, et al. IZV AKAD NAUK SSSR 1:14-20, January-February, 1966.

"Some problems of improving working conditions in foundries," by V. N. Fomin. GIG TR PROF ZABOL 11:3-8, August, 1967.

"Some spectral features of 'normal' and simulated 'rough' vowels," by F. W. Emanuel, et al. FOLIA PHONIAT 21:401-415, 1969.

"The sonic boom," by R. Caporale. RIV MED AERO 28:199-212, April-June, 1965.

"Sonic boom claims (recommended techniques and procedures for investigating alleged sonic booms by military aircraft in such a manner as to give prompt satisfaction for legitimate damages while protecting the government against spurious claims)," by R. C. Smith. JAG J 23:23-26, July-August, 1968.

"Sonic boom effects on the organ of Corti," by D. A. Majeau-Chargois, et al. LARYNGOSCOPE 80:620-630, April, 1970.

"Sonic boom tests seek forecast data," by C. M. Plattner. AVIATION W 84:55-57, June 20, 1966.

"Sonic boom: what it is...what it will and will not do," TODAY'S HLTH 43:89, May, 1965.

"Sonic booms," by A. B. Lowenfels, et al. SCIENCE 158:313-315, October 20, 1967.

"Soothing silence," by J. C. Burt. LIFE AND HLTH 84:9+, March, 1969.

"Sound absorptive properties of carpeting," by M. J. Kodaras. INTERIORS 128:130-131, June, 1969.

"Sound absorption of draperies," by J. H. Batchelder, et al. ACOUSTICAL SOC AM J 42:573-575, September, 1967.

"Sound and the psyche," MECH ENG 90:40-41, August, 1968.

"Sound and student behavior," by C. E. Schiller, et al. AV INSTR 14: 92, March, 1969.

"The sound around us," IMAGE 9:28+, August, 1967.

"Sound attenuation provided by perforated earmuffs. SAM-TR-68-86," by H. C. Sutherland, Jr., et al. US AIR FORCE SCH AEROSPACE MED 1-4, September, 1968.

"Sound code," by G. M. Diehl. COMP AIR MAG 74:22, January, 1969.

"Sound control in multi-purpose buildings," PARKS & REC 4:46-47, May, 1969.

"Sound control in office buildings," by R. C. Weber. SKYSCRAPER MGT 55:14-15, March; 20-24, April, 1970+.

"Sound drive and paired-associates learning," by H. Levitt. J GEN PSYCHOL 72:343-357, April, 1965.

"Sound effects; massive test vibrator," MECH ENG 87:66, November, 1965.

"Sound generation in subsonic turbomachinery," by C. L. Morfey. J BASIC ENG 92:450-458, September, 1970.

"Sound impulses occurring during drilling processes and their harmful effect on the hearing orgnas," by K. Dietzel, et al. VESTN OTORINOLARINGOL 32:55-57, November-December, 1970.

"Sound insulation and new forms of construction," BUILD RES STA DIGEST 96:1-8, August, 1968.

"Sound insulation of traditional dwellings," BUILD RES STA DIGEST 102:1-7, February; 103:1-8, March, 1969.

"Sound-insulation of windows," by E. Sonntag. Z GES HYG 13:632-633, August, 1967.

"Sound level measurements and frequency analyses in dental working areas," by R. Mayer, et al. DEUTSCH ZAHNAERZTL Z 23:800-805, August, 1968.

"Sound levels and development of audiometric technics in a chemical products firm," by Rety, et al. REV MED SUISSE ROM 88:59-63, February, 1968.

"Sound meter," ENGINEER 223:666, May 5, 1967.

"The sound, the noise, and the fury," by J. S. Chapman. ARCH ENVIRON HEALTH 20:612-613, May, 1970.

"Sound off on noise: Dow," OIL PAINT & DRUG REP 195:7+, May 26, 1969.

"Sound polluted schools," by S. Hammon. SCH MGT 14:14-15, November, 1970.

"Sound pollution; another urban problem," by P. A. Breysse. SCI TEACH 37:29-34, April, 1970.

"Sound of sounds that is New York," by H. C. Schonberg. N Y TIMES MAG 38+, May 23, 1965.

"Sound suppression in supersonic jet tests," ENGINEERING 199:54, January 8, 1965.

"Sound tests show hardboard effective in controlling noise," FOREST IND 96:81, June, 1969.

"Sound; when is it noise?" by J. F. Wiss. PLANT ENG 23:56-57, February 20, 1969.

"Soundproof audiometry room," NT 61:728, May 28, 1965.

"Sound-proof shop office," IND FINISHING 42:106, May 7, 1966.

"Sound-proof unit at West Middlesex Hospital," NURS MIRROR 126: 30-31, February 16, 1968.

"Soundproofing: architect defends house against the jet," HOUSE & HOME 33:8, February, 1968.

"Sound-proofing of buildings," by E. F. Stacy. ROY SOC HEALTH J 88:82-86, March-April, 1968.

"Sound-proofing the surgery," by A. Rathbone. PRACTITIONER 6:19-24, March, 1968.

"Sounds nobody wants," by H. Downs. SCI DIGEST 59:95-97, April, 1966.

"Sounds of music," MED J AUST 1:865-866, April 26, 1969.

"Sounds of progress," SR SCHOL 95:6, October 27, 1969.

"Soundwall soothes substation's neighbors," ELEC WORLD 173:29, March 23, 1970.

"Sources of noise at camping areas and possibilities of their reduction," by E. Lange. Z GES HYG 14:143-145, February, 1968.

"Sources of sound in outdoor air-conditioning equipment," by A. C. Potter. ASHRAE J 9:42-47, April, 1967.

"The specificity of response to stressful stimuli. A comparison of two stressors," by D. Oken, et al. ARCH GEN PSYCHIAT 15:624-634, December, 1966.

"Specifying motors for a quiet plant; absorption, acoustical barriers and distance can quiet airborne motor noise," by R. L. Nailen. CHEM ENG 77:157-161, May 18, 1970.

"Spectral noise levels and roughness severity ratings for normal and simulated rough vowels produced by adult females," by M. A. Lively, et al. J SPEECH HEAR RES 13:503-517, September, 1970.

"Spectrometer performance based upon signal-to-noise toleration," by D. J. Lovell. AM J OPTOM 47:650-656, August, 1970.

"Speech communications as limited by ambient noise," by J. C. Webster. J ACOUST SOC AMER 37:692-699, April, 1965.

"Speech communication in very noisy environments," by C. Cherry, et al. NATURE (London) 214:1164, June 10, 1967.

"Speech intelligibility in a background noise and noise-induced hearing loss," by W. I. Acton. ERGONOMICS 13:546-554, September, 1970.

"Speech intelligibility in presence of ambient noise," by G. Grisanti. VALSALVA 42:348-372, December, 1966.

"Sprayed mineral fiber manufacturers' association formed," AIR COND HEAR & REFRIG N 107:23, March 7, 1966.

"Ssh! ordinances on the books in New York and other cities," NEWS-WEEK 74:52, September 8, 1969.

"SST noise critical to airport compatibility," AVIATION W 92:83-84, January 5, 1970.

"SST: why don't they ban the boom? (caused by supersonic flights; United States)," by A. R. Karr. WALL ST J 175:18, April 30, 1970.

"SST's bag of mischief," by W. D. Lynn. SCIENCE 164:129, April 11, 1969.

"The stability and control of conditioned noise aversion in the tilt cage," by M. Halpern, et al. J EXP ANAL BEHAV 9:357-367, July, 1966.

"Standard procedure for noise computation," SAFETY MAINT 136:37, December, 1968.

"Standardization of high frequency industrial noise for adolescents," by I. I. Ponomarenko. GIG SANIT 33:34-38, August, 1968.

"Standardization of vibration and noise and their combined action on man," by N. M. Paran'ko, et al. VRACH DELO 8:102-106, August, 1969.

"Stapedectomy and air travel," by J. Bastien. ANN OTOLARYNG 83: 69-74, January-February, 1966.

"Startle response of rats after the production of lesions at the junction of the mesencephalon and the diencephalon," by S. G. Carlsson. NATURE (London) 212:1504, December 24, 1966.

"Startling noise and resting refractive state," by N. Roth. BRIT J PHYSIOL OPT 23:223-231, 1966.

"State and evaluation of hearing in persons working under conditions of intense production of noise and vibration," by V. E. Ostapkovich, et al. KLIN MED 48:79-83, March, 1970.

"Static visual noise and the Ansbacher Effect," by G. Stanley. Q J EXP PSYCHOL 22:43-48, February, 1970.

"Statistical data from consultation in occupational deafness," by J. Bonnefoy. J MED LYON 48:1699-1700, November 20, 1967.

"Statistical data on consultation for occupational deafness," by J. Bonnefoy. J MED LYON 49:1835-1836, November 20, 1968.

"Statistical evaluation of hearing losses in military pilots," by G. von Schulthess, et al. ACTA OTOLARYNG 65:137-145, January-February, 1968+.

"Statistical theory of extremal systems with dependent search under influence of multi-plicative and additive plant noise," by N. G. Gorelik. AUTOMATION & REMOTE CONTROL 1077-1083, July, 1967.

"Statistics in otology," by A. Morgon. ACTA OTOLARYNG 63:304-310, February-March, 1967.

"The status of hearing and of arterial pressure in exposure to intense industrial noise," by N. N. Shatalov, et al. GIG TR PROF ZABOL 13:12-15, April, 1969.

"The status of hearing in the chronic action of impulse noise on the organism," by N. I. Ivanov. VESTN OTORINOLARING 29:73-78, March-April, 1967.

"Status of the organ of hearing in workers of the cutting shop of a shoes factory," by A. A. Tatarskaia. GIG TR PROF ZABOL 14:45-46, February, 1970.

"Steels double up for composites," IRON AGE 200:70-72, November 16, 1967.

"Step closer to quiet," by R. J. Hayes. MIN CONG J 55:68-72, April, 1969.

"Steps to measure floor noises," by E. G. Marklew. ENGINEERING 674-675, May 13.

"Stimulation of efferent olivocochlear bundle causes release from low level masking," by P. Nieder, et al. NATURE (London) 227:184-185, July 11, 1970.

"Stol future keyed to noise limitations," AVIATION W 90:46+, February 24, 1969.

"Stol noise standards spur interest in propeller design," AVIATION W 90:87, June 16, 1969.

"Stomatologic research: high speed handpieces and loss of hearing," by H. Gelb. NEW YORK J DENT 35:353-354, December, 1965.

"Stop that din! An ancient medical opinion on the phrase 'why don't you shut up?', " ZAMBIA NURSE 1:7+, December, 1965.

"Stop that noise," by C. Kucyn. MOTOR B 116:46-47 , July, 1965.

"Stop that noise!" by J. R. Pritchard. CAN PERS 17:17-21, May, 1970.

"Stopping all that racket," by K. A. Kaufman. IRON AGE 203:46, February 6, 1969.

"Street noises in Naples," by R. De Capoa, et al. G IG MED PREV 7:20-47, January-March, 1966.

"Stresses in skin panels subjected to random acoustic loading," by B. L. Clarkson. AERONAUTICAL J 72:1000-1010, November, 1968.

"The struggle for quiet," by P. Deeley. OBSERVER 20:11, November, 1966.

"Studies of air pollution and noise in urban Korea," by N. H. Kim, et al. YONSEI MED J 8:40-52, 1967.

"Studies of audiogenic hypertension. I. Preventive effect of ethyl-crotonylurea, maprobamate and pentobarbital on experimental audiogenic hypertension in rate," by F. Gesmundo. BOLL SOC ITAL BIOL SPER 43:647-651, June 15, 1967.

"Studies of children under imposed noise and heat stress," by D. P. Wyon. ERGONOMICS 13:598-612, September, 1970.

"Studies of the effect of acoustic stimuli on the degree of acidification of sweat," by B. Semczuk, et al. OTOLARYNG POL 24:47-52, 1970.

"Studies of the effect of noise on the cardiorespiratory efficiency," by B. Semczuk, et al. POL TYG LEK 25:83-85, January 19, 1970.

"Studies on acoustic stress with the effect of simultaneous horizontal mechanical low-frequency vibrations on biochemical changes in guinea pigs," by J. Gregorczyk, et al MED PRACY 16:124-129, 1965.

"Studies on the acoustic stress action with the simultaneous action of horizontal mechanical vibration of lower frequencies on biochemical changes in guinea pigs. II. Effect on the number of eosinophils in the blood on the ascorbic acid level in adrenals and on the behaviour of the relative organ weight," by A. Lewandowska-Tokarz, et al. MED PRACY 16:278-282, 1965.

"Studies on the action of the pulse after a given amount of industrial noise. Economic research program with the help of sequential analysis. I," by A. Fuchs-Schmuck. INT Z ANGEW PHYSIOL 22: 1-9, March 3, 1966.

"Studies on behaviour of the pulse following definite industrial noise. Economical experimental design, using sequential analysis. II," by A. Fuchs-Schmuck. INT Z ANGEW PHYSIOL 23:345-353, March 7, 1967.

"Studies on the dependency of the intraocular pressure from noise," by H. P. Vick. KLIN MBL AUGENHEILK 153:356-360, 1968.

"Studies on the determination of the true auditory threshhold under the influence of experimental exposure to mine noise," by A. Dobrowolski. OTOLARYNG POL 21:43-48, 1967.

"Studies on the effect of acoustic stimuli on rats," by O. Ribari, et al. ACTA CHIR ACAD SCI HUNG 11:97-106, 1970.

"Studies on the effect of acoustic and ultraacoustic fields on biochemical processes. IX. Effect on some blood components in workers under noise conditions," by S. Jozkiewicz, et al. ACTA PHYSIOL POL 16:727-737, September-October, 1965.

"Studies on the effect of acoustic stress with simultaneous action of horizontal mechanical vibration of lower frequencies on the biochemical changes in guinea pigs. V. Effect on total lipids and phospholipids level and on total cholesterol and its esters content in guinea pigs blood serum," by J. Stanosek, et al. MED PRACY 16: 434-437 contd, 1965.

"Studies on the effects of acoustic stimuli on guinea pigs-experimental studies on the effects of long-term and repeated acoustic stimuli on the internal ear and pituitary--adrenal cortex system," by K. Sakashita. J OTORHINOLARYNG SOC JAP 70:1666-1701, October, 1967.

"Studies on the function of concentration maintenance (TAF). 7. CPT-swing degree and TAF under exposure to noise," by E. Takakuwa, et al. JAP J HYG 20:359-363, February, 1966.

"Studies on the functional and perception efficiency in industrial workers working in noise," by B. Rogacka-Trawinska. POL TYG LEK 25:1023-1025, July 6, 1970.

"Studies on the influence of acoustic stimuli on respiratory movements," by B. Semczuk. POL MED J 7:1090-1096, 1968.

"Studies on late development of deafness following skull injuries," by W. Wagemann, et al. HNO 13:256-264, September, 1965.

"Studies on noise hazards for commercial aviators," by W. Lorenz, et al. Z GES HYG 13:10-17, January, 1967.

"Studies on phosphorus metabolism in rabbit brain under some environmental factors. I. Changes in the concentrations of inorganic, organic and total phosphorus," by K. Nakao. JAP J HYG 21:38-43, April, 1966.

"Studies on the polycystic ovaries of rats under continuous auditory stress," by K. B. Singh, et al. AM J OBSTET GYNECOL 108;557-564, October 15, 1970.

"Studies on the problem of hearing damage caused by the use of high speed dental turbines," by O. Arentsschild, et al. ZAHNAERZTL RUNDSCH 75:217-223, June 8, 1966.

"Studies on the relation between cerebral ammonia content and intensity of noise exposure," by K. Matsui, et al JAP J HYG 22:478-480, October, 1967.

"Studies on the role of the state of dentition in the physiopathology of the auditory organ. IV. Studies on the effect of dentition on the etiology of acoustic trauma," by B. Semczuk. ANN UNIV CURIE SKLODOWSKA 22:173-178, 1967.

"Studies on thiamine metabolism in the brain and liver under various environmental conditions," by K. Horio. JAP J HYG 22:487-495, October, 1967.

"Studies on turbine noise in the hearing and ultrasonic range. 1," by L. Dunker, et al. DDZ 23:211-218, May, 1969.

"Studies on the variations of monaomineoxidase and acetylcholinesterase activities in the brain under physical environmental conditions. 2. Influences of different physical environmental conditions on monoamineoxidase and acetylcholinesterase activities," by T. Kojima. JAP J HYG 21:20-26, April, 1966.

"A study of the acoustic reflex in infantrymen," by M. H. Hecker, et al. ACTA OTOLARYNG 207:Suppl:1-16, 1965.

"Study of acoustical treatments for jet-engine nacelles," by A. H. Marsh. ACOUSTICAL SOC AM J 43:1137-1156, May, 1968.

"Study of cochlear reactions in workers handling pneumatic hammers," by G. Jullien, et al. ARCH MAL PROF 28:297-300, January-February, 1967.

"Study of noise and hearing in jute weaving," by W. Taylor, et al. J ACOUST SOC AMER 38:113-120, July, 1965.

"A study of noise and its relationship to patient discomfort in the recovery room," by B. B. Minckley. NURS RES 17:247-250, May-June, 1968.

"Study of noise deafness," by T. Kawabata, et al. J OTOLARYNG JAP 72:396-397, February 20, 1969.

"A study of recruitment phenomenon of audition and its mechanism," by A. Watanabe. J OTORHINOLARYNG SOC JAP 68:1391-1403, November, 1965.

"Study of the residual noise of average evoked potentials," by D. Arnal, et al. ELECTROENCEPH CLIN NEUROPHYSIOL 27:315-321, September, 1969.

"Study of sound levels in houses," by J. J. Mize, et al. J HOME ECON 58:41-45, January, 1966.

"Study of tinnitus induced temporarily by noise," by G. R. Atherley, et al. J ACOUST SOC AMER 44:1503-1506, December, 1968.

"Study on acoustic trauma in a vocational school," by P. Van de Calseyde, et al. ACTA OTORHINOLARYNG BELG 22:664-666, 1968.

"A study on the effect of repeated noise in rats," by V. Hrubes, et al. ACTIV NERV SUP (Praha) 7:165-167, 1965.

"Subjective basis for aircraft noise limitation," by D. W. Robinson. ROY AERONAUTICAL SOC J 71:396-400, June, 1967.

"Subjective complaints in noise-deafness. (Studies on 500 workers in noisy environment. I.)," by W. Wagemann. MSCHR UNFALLHEILK 68:546-553, December, 1965.

"Subjective matching of anxiety to intensities of white noise," by R. Sullivan. J ABNORM PSYCHOL 74:646-659, December, 1969.

"Subjective response to synthesized flight noise signatures of several types of V-STOL aircraft. NASA CR-1118," by E. G. Hinterkeuser, et al. US NASA 1-92, August, 1968.

"Supersonic blues," MED J AUST 1:1142-1143, May 31, 1969.

"Supersonic boom," LIFE 67:51-52, November 7, 1969.

"Supersonic 'boom'. Its nature and its effects," by J. Lavernhe, et al. PRESSE MED 74:1973-1975, September 17, 1966.

"Supersonic booms. A current problem," by J. Causse, et al. ANN OTOLARYNG 85:419-423, July-August, 1968.

"Is supersonic transport worth the noise?" NATURE (London) 227:873-874, August 29, 1970.

"Supplying gas quietly," ENGINEERING 204:130, July 28, 1967.

"A survey of engine room noise in the Royal New Zealand Navy," by N. Roydhouse. NEW ZEAL MED J 67:133-140, January, 1968.

"A survey of hearing conservation programs in representative aerospace industries. I. Prevalence of programs and monitoring audiometry," by W. F. Rintelmann, et al. AMER INDUSTR HYG ASS J 28:372-380, July-August, 1967.

"Survey of industrial acoustic conditions," by J. Morin. ARCH MAL PROF 31:517-522, June, 1967.

"Survey of noise measurement methods," by J. R. Ranz. MACHINE DESIGN 38:199-206, November 10, 1966.

"Symposium: environmental protection. Introduction. N. A. Rockefeller; Water quality control: a modern approach to state regulation. F. E. Maloney, R. C. Ausness; Federal enforcement under the refuse act of 1899. J. T. B. Tripp, R. M. Hall; Comments; environmental control in New York City. Introduction. Environmental problems of an expanding population. Problems of solid waste disposal. Noise pollution. Abandoned cars-the facts. Aesthetic considerations in land use planning," ALBANY L REV 35:23, 1970.

"Symposium: residuals and environmental quality management. Environmental quality management. B. T. Bower, W. Spofford; Governmental responsibility for waste management in urban regions. R. T. Anderson;

The community noise problem: factors affecting its management. C. Bragdon; Pesticide residues and environmental economics. W. F. Edwards, M. Langham, J. C. Headley; Environmental litigation-where the action is? F. P. Grad, L. Rockett," NATURAL RESOURCES J 10:655, October, 1970.

"System keeps eye on noise," ELECTRONICS 41:197-199, June 10, 1968.

"System's analyst's view of noise and urban planning," by M. Wachs, et al. AM SOC C E PROC 96(UP 2 no 7639):147-159, October, 1970.

"Systems approach key to reducing plant noise," PRODUCT ENG 40: 102, December 1, 1969.

"Taking the squeal out of brakes," by J. E. Gieck. SAE J 73:83, September, 1965.

"Tape noise reduction center," by B. Whyte. AUDIO 54:10+, May, 1970.

"Teletypewriter used near phones and dictators," OFFICE 68:32, August, 1968.

"Telltale hearts," by K. M. Farrell. AJN 67:1239+, June, 1967.

"Temporal effects in simultaneous masking and loudness," by E. Zwicker J ACOUST SOC AMER 38:132-141, July, 1965.

"Temporary and permanent hearing loss. A ten-year follow-up," by J. Sataloff, et al. ARCH ENVIRON HEALTH 10:67-70, January, 1965.

"Temporary changes of the auditory system due to exposure to noise for one or two days," by J. H. Mills, et al. J ACOUST SOC AM 48:524-530, August, 1970.

"Temporary changes of the auditory threshold after the exposure to noise at different frequencies," by S. Kubik, et al. PRAC LEK 17:240-243, August, 1965.

"The temporary effects of 125 c.p.s. octave-band noise on stapedectomized ears," by K. Ferris. J LARYNG 80:579-582, June, 1966.

"Temporary hearing loss due to vibration and noise," by K. Yamamura, et al. JAP J HYG 25:472-478, December, 1970.

"Temporary shifts in auditory thresholds of chinchilla after exposure to noise," by E. N. Peters. J ACOUST SOC AMER 37:831-833, May, 1965.

"Temporary threshold shift and damage-risk criteria for intermittent noise exposures," by W. D. Ward. J ACOUST SOC AM 48:561-574, August, 1970.

"Temporary threshold shift from impulse noise," by J.G. Walker. ANN OCCUP HYG 13:51-58, January, 1970.

"Temporary threshold shift in rock-and-roll musicians," by J. Jerger, et al. J SPEECH HEARING RES 13:221-224, March, 1970.

"Temporary threshold shift produced by exposure to high-frequency noise noise," by P. E. Smith, Jr. AMER INDUSTR HYG ASS J 28:447-451, September-October, 1967.

"Temporary threshold shifts in hearing from exposure to combined impact-steady-state noise conditions," by A. Cohen, et al. J ACOUST SOC AMER 40:1371-1380, December, 1966.

"10 years of hearing conservation in the Royal Air Force," by G. Jacobs, et al. NEDERL MILIT GENEESK T 18:212-218, July, 1965.

"Teratogenic effects of audiogenic stress in albino mice," by C. O. Ward, et al. J PHARM SCI 59:1661-1662, November, 1970.

"A test for susceptibility to noise-induced hearing loss," by P. E. Smith, Jr. AMER INDUSTR HYG ASS J 30:245-250, May-June, 1969.

"Test your knowledge; noise; questions and answers," by L. Ibbotson. WIRELESS WORLD 74:320+, September, 1968.

"Testing and evaluating hearing protectors," by F. P. Beguin. AM ASS INDUSTR NURSES J 13:11+, April, 1965.

"Testing sonic booms," by R. Clark. NATURE (London) 215:1122-1123, September 9, 1967.

"That noise you hear may be pollution," BSNS W 40-41, April 22, 1967.

"There's profit in noise control," IRON AGE 195:90-91, January 28, 1965.

"Thermal efficiency, silencing and practicability of gas-fired industrial pulsating combustors," by D. Reay. COMBUSTION 41:16-24, January, 1970.

"These guidelines suggested to limit noise pollution (digest of *Rational approach to legislative control of noise*)," by G. J. Thiessen. FIN POST 64:44, February 14, 1970.

"Third London airport: choice cannot be based on cost alone says RIBA," RIBA J 77:224-225, May, 1970.

"This blight of noise," by L. H. Nahum. CONN MED 33:252-255, April, 1968.

"Thousands of tiny holes quiet noise from jumbo jet engines," PRODUCT ENG 41:96, January 1, 1970.

"3 recommendation inquiries to the Board of Health," by R. J. Kruisinga. T ZIEKENVERPL 22:220, March 1, 1969.

"Three-shaft jetliner engine makes only half as much noise," PRODUCT ENG 39:62-63, November 4, 1968.

"Three-sided squeeze on noise," SAFETY MAINT 137:41-42, April, 1969.

"Threshold ultrasonic dosages for structural changes in the mammalian brain," by F. J. Fry, et al. J ACOUST SOC AM 48:Suppl 2:1413+, December, 1970.

"Time perception for helicopter vibration and noise patterns," by G. R. Hawkes, et al. J PSYCHOL 76:71-77, September, 1970.

"Time-rated acoustical ceiling assemblies," ARCH REC 137:197-198, February, 1965.

"Time to consider aircraft noise," ENGINEERING 262:961, December 2, 1966.

"Timidity and metabolic elimination patterns in audiogenic-seizure susceptible and resistant female rats," by A. S. Weltman, et al. EXPERIENTIA 22:627-629, September 15, 1966.

"Tolerable limit of loudness: its clinical and physiological significance," by J. D. Hood, et al. J ACOUST SOC AMER 40:47-53, July, 1966.

"Tomorrow's markets: noise abatement: can marketers ignore the din?" SALES MGT 103:45-46+, November 10, 1969.

"Tonal patterns of cochlear impairment following intense stimulation with pure tones," by F. Suga, et al. LARYNGOSCOPE 77:784-805, May, 1967.

"Tone and speech hearing in acoustic trauma," by F. Schwetz. MSCHR OHRENHEILK 103:105-110, 1969.

"Tone height shifts under noise conditions," by H. G. Dieroff, et al. FOLIA PHONIAT 18:247-255, 1966.

"Tort Liability in Light Civilian Aircraft Operation," NEW YORK L FORUM 8, 3:545-593, Fall, 1967.

"Total and free thiamine and thiamine diphosphate content of brains from rats with audiogenic epilepsy," by A. I. Goshev. VOP MED KHIM 15:581-583, November-December, 1969.

"Toward more uniform noise ratings," by R. Warren. AIR COND HEAT & VEN 66:40-44, January, 1969; Discussion 66:4+, March; 7+, June, 1969.

"Towards a criterion for impulse noise in industry," by R. R. Coles, et al. ANN OCCUP HYG 13:43-50, January, 1970.

"Traffic noise," by B. H. Sexton. TRAFFIC O 23:427-439, July, 1969.

"Traffic noise draws R & D attention," by W. J. Kalb. IRON AGE 201: 21, June 20, 1968.

"Traffic noise from roundabouts," by M. A. McCormick, et al. ARCHITECTS J 146:861-862, October 4, 1967.

"Traffic noise: technique for assessing motorway noise and protecting buildings against it," by J. Simmons. ARCHITECTS J 144:239-244, July 27, 1966.

"Transformer noise," ENGINEER 224:399, September 29, 1967.

"Transport noise reduction," by M. L. Yaffee. AVIATION W 89:39+, November 25; 61+, December 2, 1968.

"Transportation noise sources," by R. C. Potter. AUDIO ENG SOC J 18:119-127, April, 1970.

"Transportation vehicle noise control: application and acceptability," by P. A. Franken. ARCHIVES OF ENVIRONMENTAL HEALTH 20:636-643, May, 1970.

"Traumatic deafness from nailing gun," by J. Delatour. J FRANC OTORHINOLARYNG 15:701-703, October, 1966.

"Treatment of patients with suddenly occuring perception deafness with ATP," by L. Fal'tinek. VESTN OTORINOLARING 27:60-62, May-June, 1965.

"Treatment with vascular aim of recent acoustic traumas, especially after stapedectomy," by J. Causse, et al. REV LARYNG 87:360-368, May-June, 1966.

"A trial application of the labyrinth test in a water basin for studies on the combined action of phrenotropic drugs and strong acoustic stimuli in white rats," by J. Grzesik, et al. ACTA PHYSIOL POL 17:327-333, March-April, 1966.

"Trials and measurement results of a plastic ear mould as individual protection from noise," by C. J. Partsch, et al. HNO 14:107-109, April, 1966.

"Triumph over traffic; New York's court of appeals declares noise a compensable injury," TIME 92:74, July 12, 1968.

"The Tullio phenomenon and a possibility for its treatment," by G. Lange. ARCH KLIN EXP OHR NAS KEHLKOPFHEILK 187:643-649, 1966.

"Tune in on your employees' hearing problems," MOD MANUF 3:88, October, 1970.

"Turbine noise problem licked," GAS 44:60, September, 1968.

"Turbofan-engine noise suppression," by R. E. Pendley, et al. J AIRCRAFT 5:215-220, May, 1968.

"Turbulence as a producer of noise in proportional fluid amplifiers," by D. W. Prosser, et al. J AP MECH 33:728-734, December, 1966.

"2 LA suburbs plan noise ordinances," AIR COND HEAT & REFRIG N 121:29, October 5, 1970.

"U.K., industry join in anti-noise research," by H. J. Coleman. AVIATION W 87:61+, August 14, 1967.

"Underexpanded jet noise reduction using radial flow impingement," by D. S. Dosanjh, et al. AIAA J 7:458-464, March, 1969.

"Unilateral inhibition of sound-induced convulsions in mice," by R. L. Collins. SCIENCE 167:1010-1011, February 13, 1970.

"Unresolved noise problems," ENVIRONMENTAL CONTROL MGT 138:47, October, 1969.

"Unusual Tullio phenomena," by S. K. Kacker, et al. J LARYNG 84:155-166, February, 1970.

"Unwanted noise," TRANS SOC OCCUP MED 18:125, October, 1968.

"Unwanted noise, a by-product of galloping technology, threatens civilized life," by W. H. Ferry. AAUW J 63:129-132, March, 1970.

"The uproar of our cities: electroencephalographic aspects," by H. R. Richter. CONFIN NEUROL 25:215-223, 1965.

"Urban noise control (focuses on efforts in New York city)," COLUMBIA J LAW & SOCIAL PROBLEMS 4:105-119, March, 1968.

"U.S. mapping program to alleviate noise," by E. J. Bulban. AVIATION W 85:45+, October 24, 1966.

"U.S. Public Health Service field work on the industrial noise hearing loss problem," by A. Cohen. OCCUP HEALTH REV 17:3-10 passim, 1965.

"Use of probe microphones to measure sound pressures in the ear," by B. M. Johnstone, et al. ACOUSTICAL SOC AM J 46:1404-1405, December, 1969.

"Use of sensation level in measurements of loudness and of temporary threshold shift; discussion," ACOUSTICAL SOC AM J 41:714-715, March, 1967.

"Validation of the single-impulse correction factor of the CHABA impulse-noise damage-risk criterion," by D. C. Hodge, et al. J ACOUST SOC AM 48:Suppl 2:1429+, December, 1970.

"Variable air volume air conditioning; air distribution," by S. Daryanani, et al. AIR COND HEAT & VEN 63:69-72, March, 1966.

"Variables that influence sound pressures generated in the ear canal by and audio-metric earphone," by N. P. Erber. ACOUSTICAL SOC AM J 44:555-562, August, 1968.

"Variations of amylasuria and amylasemia under the effect of sound and vibratory 'stress' in industry," by S. Miclesco-Groholsky, et al. ARCH MAL PROF 31:277-282, June, 1970.

"Variations of amylasuria under the effect of sound and vibration injuries (preliminary note)," by S. Miclesco-Groholsky. ANN OTOLARYNG 86: 251-258, April-May, 1969.

"Vehicle noise meter meets new law," ENGINEER 225:774, May 17, 1968.

"Vibration and noise characteristics of an aircraft-type gas turbine used in a marine propulsion system," by R. E. Harper. NAVAL ENG J 81: 103-110, December, 1969.

"Vibration and structure-borne noise control," by L. L. Eberhart. ASH-RAE J 8:54-60, May, 1966.

"The virtues of softness," by A. Schreiner. CLAIRER 6:41-43, November 7, 1967.

"V/stol gets big lift in expanding aviation work," by W. Andrews. AERO-SPACE TECH 21:88+, November 20, 1967.

"Waffled cone cuts throttle-valve noise," MACHINE DESIGN 42:150, February 19, 1970.

"Wall consturction affects noise level inside structure," by F. J. Versagi. AIR COND HEAT & REFRIG N 110:22-23, April 24, 1967.

"Walsh-Healy act changes spur design-out of noise to protect workers' hearing," by L. L. Beranek. POWER 114:52-53, May, 1970.

"Walsh-Healy, the first nine months; an evaluation of the noise control considerations," ENVIRONMENTAL CONTROL MGT 139:33-34, January, 1970.

"Walsh-Healy noise regulations cannot be ignored! interview with A. A. Fletcher," SAFETY MAINT 138:34-39, August, 1969.

"Walsh-Healy revisited; Northrop reviews hearing loss prevention policies and procedures," by J. L. Morrow. ENVIRONMENTAL CONTROL & SAFETY MGT 139:32-34, June, 1970.

"War on noise; some small victories," ECONOMIST 214:1308, March 20, 1965.

"Watch those published sound power levels," by F. J. Versagi. AIR COND HEAT & REFRIG N 105:13, August 23, 1965.

"Water and alchol consumption of mice sensitive or refractory to audiogenic crises," by A. Duveau, et al. C R SOC BIOL 160:791-794, 1966.

"Water hammer stops; join the anti-noise arsenal," by R. Zell. DOM ENG 206:72-73, September, 1965.

"Ways to reduce plant noises," by B. G. Golden. HYDROCARBON PROCESS 47:75-78, December, 1968.

"We live amid acoustic torture and it isn't doing us any good," FIN POST 63:1-2, June 28, 1969.

"Wearing ear protectors in metal factories," by A. Lefort. ARCH MAL PROF 30:357-360, June, 1969.

"The well-tuned ear," by N. Calder. NEW STATESMAN 472, April 3, 1970.

"We're poisoning ourselves with noise," by J. Stewart. READ DIGEST 96:137-190+, February, 1970; also in TURK HEMSIRE DERG 20:16-18, July-September, 1970.

"What belongs in acoustical specifications," by R. Farrell. ARCH REC 138:227-230+, September; 203-206, November, 1965.

"What can be done about plant noise," by H. S. Sharp. SUPERVISORY MGT 10:51-53, February, 1965.

"What do you know about silencing aluminum spars," by D. B. Hoisington. YACHTING 125:83+, April, 1969.

"What ever became of the big noise about the quiet home?" HOUSE & HOME 30:88-91, December, 1966.

"What goes into a hearing protection program?" SAFETY MAINT 132: 37-40, July, 1966.

"What is a sound-level meter?" by F. Van Veen. SAFETY MAINT 130: 45, August, 1965.

"What price silence?" by H. H. Lett. CAN BANK 77:22-25, July-August, 1970.

"What to do about plant noise," by B. G. Golden. CHEM ENG PROG 65: 47-52, April, 1969.

"What you should know about noise control," MOD MATERIALS HANDLING 25:43-45, April, 1970.

"When noise annoys: Time essay," TIME 88:24-25, August 19, 1966.

"When noise is a problem, design it out instead of in," by A. R. Gardner. PRODUCT ENG 37:86-91, April 11, 1966.

"When standards are set, it will be thumbs down on noise," by K. Mansfield. PRODUCT ENG 40:56, May 5, 1969.

"Where noise regulated," FIN POST 64:44, February 14, 1970.

"White House science unit seeks government attack on jet noise," AVIATION W 84:44, March 28, 1966.

"White noise?" by D. Flooke. CRAWDADDY 11:17-18, September-October, 1967.

"White noise and the delta function," by J. H. Park, Jr. IEEE PROC 56:114-115, January, 1968.

"Who is going to do what about your noise problem?" by F. X. Worden. SAFETY MAINT 132:42-44, December, 1966.

"Why life near airports will get quieter," BSNS W 44-45, February 7, 1970.

"Why 92 dB A?" by V. H. Hill. AMER INDUSTR HYG ASS J 31:189-197, March-April, 1970.

"Why, where and when of noise measurement," by A. D. Woewucki, Jr. SAFETY MAINT 134:42-44, December, 1967.

"Will you reduce noise on your designs for the 1970's?" by G. L. Charson. HYDRAULICS & PNEUMATICS 22:85-89, November, 1969.

"The work of adolescents in environments exposed to noise," by M. Fisarova. CESK NEUROL 29:396-401, November, 1966.

"Work prohibition for expectant and breast feeding mothers in harmful noise," MUNCHEN MED WSCHR 31:Suppl:6, August 2, 1968.

"You can do something about noise," by L. J. Cormack. POWER 111: 79-81, May, 1967.

"Your world is getting noisier day by day--and no apparent way to stop it," by B. Jackson. FIN POST 61:23, May 27, 1967.

SUBJECT INDEX

ATP:TREATMENT

"Treatment of patients with suddenly occuring perception deafness with ATP," by L. Fal'tinek. VESTN OTORINOLARING 27:60-62, May-June, 1965.

ACOUSTIC ACCIDENT

"Apropos of the acoustic accident," by B. Kecht. Z LARYNG RHINOL OTOL 43:280-293, May, 1964.

ACOUSTICS:BRAZIL

"Acoustic problems in Sao Paulo," by S. Marone. RESEN CLIN CIENT 38:173-182 continued, July-August, 1969.

ADENOSINOTRIPHOSPHORIC ACID

"Protective effect of cytochrome C and adenosinotriphosphoric acid in acoustic trauma," by W. Jankowski, et al. OTOLARYNG POL 23:133-140, 1969.

ADOLESCENTS

see: Youth

AGE

"Age in the occurrence and evolution of acoustic trauma," by M. Lutovac. SRP ARH CELOK LEK 97:869-874, September, 1969.

"Aspects of nervous fatigue in automated systems as a function of age of the operators," by R. Elias, et al. FIZIOL NORM PAT 13:447-454, September-October, 1967.

"The influence of age on the origin and progression of noise hearing difficulties," by W. Kup. HNO 14:268-272, September, 1966.

AIRCRAFT DESIGN

"Design studies urged at (London) noise conference," AVIATION W 85:67-68, December 12, 1966.

"Effects of noise on commercial v/stol aircraft design and operation," by H. Sternfeld, Jr., et al. J AIRCRAFT 7:220-225, May, 1970.

AIRCRAFT NOISE

"Acoustic environments of the F-111A aircraft during ground runup. AMRL-TR-68-14," by J. N. Cole, et al. US AIR FORCE AERO-SPACE MED RES LAB 1-50, May, 1968.

"Acoustic lining reduces jet noise," ENGINEERING 207:596, April 18, 1969.

"Acoustic trauma due to percussion wave detonation caused by supersonic airplanes," by O. Kleinsasser. HNO 13:170-175, June, 1965.

"Aerodynamics, noise, and the sonic boom," by W. R. Sears. AIAA J 7:577-586, April, 1969.

"Aero-engine noise research laboratory," ENGINEER 224:6-8, July 7, 1967.

"Aircraft; costly big noise," STATIST 190:1586, December 30, 1966.

"Aircraft noise," by W. Knop. ZBL ARBEITSMED 19:273-275, September, 1969.

"Aircraft noise and development control--the policy for Gatwick airport," by E. Sibert. TOWN PLAN INST J 55:149-152, April, 1969.

"Aircraft noise, can it be cut?" by S. M. Levin. SPACE/AERONAUTICS 46:65-75, August, 1966.

"Aircraft noise; fugitive factor in land use planning," by D. C. McGrath, Jr. AM SOC C E PROC 95(UP 1 no 6520):73-80, April, 1969.

"Aircraft noise law: a technical perspective," by J. J. Alekshun, Jr.

AM BAR ASSN J 55:740-745, August, 1969.

"Aircraft noise; mitigating the nuisance," by E. J. Richards. ASTRONAUTICS & AERONAUTICS 5:34-43, January, 1967; Discussion 5:43-45, January, 1967.

"Aircraft noise monitor," ENGINEER 219:195, January 22, 1965.

"Aircraft noise; no more shrugging by the industry," SPACE/AERONAUTICS 47:18-20, June, 1967.

"Aircraft noise study set for city centers, suburbs," AVIATION W 90:27, January 27, 1969.

"Aircraft noise symposium; proceedings," ACOUSTICAL SO AM J 48:779-842, September, 1970, pt 3.

"Aircraft noise: a taking of private property without just compensation," by P. W. Fleming. SOUTH CAROLINA L REV 4,18: 593-608, 1966.

"Airplane cockpit noise levels and pilot hearing sensitivity," by K. J. Kronoveter, et al. ARCH ENVIRON HEALTH 20:495-499, April, 1970.

"Alleviation of aircraft noise," by N. E. Golovin. ASTRONAUTICS & AERONAUTICS 5:71-75, January, 1967.

"Analyzer detects malfunctions by examining jet engine noise," by A. J. Kasak, et al. SAE J 76: 59-61, July, 1968.

"Assessment of aircraft noise disturbance," by C. G. Van Niekerk, et al. AERONAUTICAL J 73:383-396, May, 1969.

"Audiologic examinations of the masking effect caused by flight noise," by W. Lorenz. MSCHR OHRENHEILK 103:438-444, 1969.

"Aviation noise--evaluation and possibilities of its elimination," by J. Hilscher. Z GES HYG 15:670-673, September, 1969.

"Bang on!" ECONOMIST 215:426, April 24, 1965.

AIRCRAFT NOISE

"CF6 is tested for noise, flow patterns; illustrations with text," AVIATION W 91:34-35, October 20, 1969.

"Cockpit noise environment of airline aircraft," ACOUSTICAL SOC AM J 47:449, February, 1970.

"Cockpit noise environment of airline aircraft," by R. B. Stone. AEROSPACE MED 40:989-993, September, 1969.

"Cockpit noise intensity: fifteen single-engine light aircraft," by J. V. Tobias. AEROSPACE MED 40:963-966, September, 1969.

"Cockpit noise intensity: fifteen single-engine light aircraft. AM 68-21," by J. V. Tobias. US FED AVIAT AGENCY OFFICE AVIAT MED 1-6, September, 1968.

"Cockpit noise intensity: three aerial application (cropdusting) aircraft," by J. V. Tobias. J SPEECH HEARING RES 11:611-615, September, 1968.

"Compendium of human responses to the aerospace environment. 9. Sound and noise. NASA CR-1205 (2)," by E. M. Roth, et al. US NASA 9:1.101, November, 1968.

"Computer graphics aid solution to DC-9 cabin noise problem," AVIATION W 87:69+, July 17, 1967.

"Corona-generated noise in aircraft," by C. E. Cooper. WIRELESS WORLD 72:547-552, November, 1966.

"Current noise work may aid V/STOLs," by W. J. Normyle. AVIATION W 88:123+, June 24, 1968.

"DC-8 stretched jets should produce less noise annoyance at airports," by R. E. Black. SAE J 75:51-55, June, 1967.

"Development of engineering practices in jet and compressor noise," by J. B. Large, et al. J AIRCRAFT 6:189-195, May, 1969.

"Economics of aircraft noise suppression," by F. B. Greatrex. AIRCRAFT ENG 38:20+, November, 1966.

AIRCRAFT NOISE

"Effect of air-craft noise on gastric function," by C. Y. Kim, et al. YONSEI MED J 9:149-154, 1968.

"The effect of aircraft noise on the population living in the vicinity of airports," by I. L. Karagodina, et al. GIG SANIT 34:25-30, May, 1969.

"The effect of aviation noise on some indices of protein and vitamin metabolism," by Iu. F. Udalov, et al. VOENNOMED ZH 7:61-64, July, 1966.

"Effect of temperature on the high-frequency component of the jet noise," by G. Krishnappa. ACOUSTICAL SOC AM J 41:1208-1211, May, 1967.

"Effects of air travel on some otorhinolaryngologic diseases," by M. Fayala. TUNISIE MED 44:165-185, May-June, 1966.

"Effects of intense noise during fetal life upon postnatal adaptability (statistical study of the reactions of babies to aircraft noise)," by Y. Ando, et al. J ACOUST SOC AMER 47:1128-1130, April, 1970.

"Engineering outline; noise from aircraft," by J. D. Voce. ENGINEERING 204:983-986, December 15, 1967.

"Engineers search for quiet jet engines," IND RES 11:30, July, 1969.

"Estimating noisiness of aircraft sounds," by R. W. Young, et al. ACOUSTICAL SOC AM J 45:834-838, April, 1969.

"Evaluation of advances in engine noise technology," by A. L. McPike. AIRCRAFT ENG 42:16+, May, 1970.

"Evaluation of noise problems anticipated with future vtol aircraft. AMRL-TR-66-245," by J. N. Cole, et al. US AIR FORCE AEROSPACE MED RES LAB 1-16, May, 1967.

"Evolution of the engine noise problem," by F. B. Greatrex, et al. AIRCRAFT ENG 39:6-10, February, 1967.

AIRCRAFT NOISE

"An example of 'engineering psychology': the aircraft noise problem," by K. D. Kryter. AMER PSYCHOL 23:240-244, April, 1968.

"An experimental electronic stethoscope for aircraft use. A preliminary report. SAM-TR-67-39," by F. A. Brogan, et al. US AIR FORCE SCH AEROSPACE MED 1-8, May, 1967.

"Experimental research on the disturbing effect of airplane noise," by E. Grandjean, et al. INT Z ANGEW PHYSIOL 23:191-202, December 3, 1966.

"Experimental studies on the harmfulness of aircraft noise as opposed to industrial noise," by W. Lorenz. MSCHR OHRENHEILK 103:492-498, 1969.

"FAA anti-noise authority seen clearing house unit," AVIATION W 88:28-29, March 25, 1968.

"FAA fears political tug-of-war on noise abatement programs," AVIATION W 85:36-37, December 5, 1966.

"FAA issues aircraft noise-reduction rule," by H. Taylor. AM AVIATION 32:25, January 20, 1969.

"FAA to act on 747 noise requirements," by R. G. O'Lone. AVIATION W 91:35-36, December 8, 1969.

"Facilities and instrumentation for aircraft engine noise studies," by R. E. Gorton. J ENG POWER 89:1-13, January, 1967.

"Fight against aircraft noise," by E. Jeffs. ENGINEERING 208: 108, August 1, 1969.

"Flight noise; measurements on airplanes used in competition and hygienic aspects," by H. G. Demus, et al. Z GES HYG 11:1-24, January, 1965.

"Fundamental noise research emphasized; aircraft turbine engines," AVIATION W 92:90, June 22, 1970.

"Growing problem of airplane noise, what is being done," GOOD H

168:155-157, February, 1969.

"Hearing acuity and exposure to patrol aircraft noise," by W. R. Pierson, et al. AEROSPACE MED 40:1099-1101, October, 1969.

"Hearing disorders caused by noise in loading personnel at a large civilian airport," by G. Pressel, et al. INT ARCH ARBEITSMED 26:231-249, 1970.

"High bypass ratio fan noise research test vehicle," by C. A. Warden. J AIRCRAFT 7:437-441, September, 1970.

"Hovercraft noise and its suppression," by E. J. Richards, et al. ROY AERONAUTICAL SOC J 69:387-398, June, 1965.

"How should civil aviation develop to serve our society best? the AIAA president's forum," ASTRONAUTICS & AERONAUTICS 7:28-55, February, 1969.

"Initial FAA noise proposal hits mainly new aircraft," AVIATION W 90:24, January 13, 1969.

"Integration of aircraft auxiliary power supplies," by J. Wotton. AIRCRAFT ENG 40:22-24+, October, 1968.

"Interaction between the aero engine industry and the growth of air transport," by E. M. Eltis. AIRCRAFT ENG 39:15-24, January, 1967.

"Jet age precedent; Nice real estate man sues Air France," TIME 87:67, March 18, 1966.

"Jet engine noise reduced on P&W JT9Ds," AM AVIATION 32:40, September 16, 1968.

"Jet engine oscillations and noises," by S. L. Bragg. ENGINEER 217:268-269 passim plus.

"Jet noise and safety," by R. J. Serling. ACOUSTICAL SOC AM J 45:1574-1575, June, 1969.

AIRCRAFT NOISE

''Jet noise and shear flow instability seen from an experimenter's viewpoint,'' by E. Mollo-Christensen. J AP MECH 34:1-7, March, 1967.

''Jet noise attacked in four ways,'' MACHINE DESIGN 41:10, March 20, 1969.

''Jet noise, fumes trial could set precedent,'' AVIATION W 92:44-46, May 11, 1970.

''Jet noise in airport areas: a national solution required (problem of jet transport noise along flight paths near public airports),'' MINN LAW R 51:1087-1117, May, 1967.

''Jet noise is getting awful,'' by R. Sherrill. N Y TIMES MAG 24-25+, January 14, 1968.

''Jet-noise monitor,'' ELECTRONICS 38:211, January 11, 1965.

''Jet quieted by noise absorbing ducts,'' MACHINE DESIGN 41:10, May 1, 1969.

''Judgments of the relative and absolute acceptability of aircraft noise,'' by D. E. Bishop. J ACOUST SOC AMER 40:108-122, July, 1966.

''Keeping aircraft quieter,'' NATURE (London) 226:4, April 4, 1970.

''Keeping the jet's roar outside the door - it's a difficult and costly process,'' by B. Lamb. HOUSE & HOME 38:10+, September, 1970.

''LA girds for jet-noise showdown,'' by R. L. Parrish. AM AVIATION 32:23-25, January 6, 1969.

''LAX studies house insulation as way to decrease jet noise,'' by F. A. Hunter. AM AVIATION 32:25, September 16, 1968.

''Masking of speech by aircraft noise,'' by K. D. Kryter, et al. J ACOUST SOC AMER 39:138-150, January, 1966.

''Mental-hospital admissions and aircraft noise,'' by R. H. Dhowns.

LANCET 1:467, February 28, 1970.

"Mental-hospital admissions and aircraft noise," by I. Abey-Wickrama, et al. LANCET 2:1275-1277, December 13, 1969+.

"Mintech and the makers tackle aero-engine noise," ENGINEERING 204:18-19, July 7, 1967.

"Nacelles cited in jets noise," by L. M. Cafiero. ELECTRONIC N 11:24, September 12, 1966.

"Nearfield infrasonic noise generated by three turbojet aircraft during ground runup operations. AMRL-TR-65-132," by R. T. England, et al. US AIR FORCE AEROSPACE MED RES LAB 1-16, August, 1965.

"New aircraft noise requirements detailed," AVIATION W 91:35, November 17, 1969.

"Nixon facing tough choice on air noise," by H. Taylor. ELECTRONIC N 14:18, March 10, 1969.

"Noise campaigners (growing resentment against aircraft noise has become a force manufacturers must reckon with)," ECONOMIST 221:490, October 29, 1966.

"Noise complaints dwindle quickly as jet service begins at National," AVIATION W 84:50, May 2, 1966.

"Noise environs and helmet performance for the P-1127 V-STOL aircraft. AMRL-TR-68-70," by H. C. Sommer, et al. US AIR FORCE AEROSPACE MED RES LAB 1-22, December, 1968.

"Noise a key in DC-10 pod design," by C. M. Plattner. AVIATION W 90:64-66+, June 16, 1969.

"Noise nuisance may cause redesign of jet engines," MACHINE DESIGN 39:14, July 6, 1967.

"Noise of highly turbulent jets at low exhaust speeds," by J. E. F. Williams, et al. AIAA J 3:791-793, April, 1965.

"Noise plan threatens older jets," AVIATION W 88:324, March 18, 1968.

"Noise problems in aeromedical evacuation operations," by D. C. Gasaway, et al. AEROMED REV 2:1-18, September, 1970.

"Noise problems of vtol with particular reference to the Dornier DO 31," by M. Flemming, et al. AERONAUTICAL J 73:647-653; Discussion 653-656, August, 1969.

"Noise reduction could cost millions," by H. Taylor. AM AVIATION 32:24-25, August 5, 1968.

"Noise study focuses on intakes, exhaust," AVIATION W 86:24, June 19, 1967.

"Noise suppression system quiets the jet's blast," SAFETY MAINT 133:45, March, 1967.

"Ombudsman; not so much what he says (aircraft noise complaints)," ECONOMIST 225:1209, December 23, 1967.

"On the effect of aviation noises of various intensity and duration," by I. Ia. Borshchevskii, et al. VOENNOMED ZH 2:64-68, February, 1965.

"On estimating noisiness of aircraft sounds," by R. W. Young, et al. J ACOUST SOC AMER 45:834-838, April, 1969.

"On means of protection against aircraft noises," by I. Ia. Borshchevskii. VOENNOMED ZH 3:65-68, March, 1969.

"On problems of individual acoustic protection in commercial air travel," by W. Lorenz. Z GES HYG 14:669-674, September, 1968.

"On the protection of the flight- and technical personnel from aviation noises," by I. Ia. Borshchevskii. VOENNOMED ZH 6:58-63, June, 1965.

"On the protection of the population in the vicinity of airports from aviation noises," by I. Ia. Borshchevskii. GIG SANIT 31:82-85, September, 1966.

AIRCRAFT NOISE

"On standardization of the effect of aircraft noises," by I. Ia. Borshchevskii, et al. VOENNOMED ZH 10:80-83, October, 1967.

"Paying the cost of job aircraft noise," ENGINEERING 199:367, March 19, 1965.

"Perceptive deafness in military aviation technicians caused by noise of F 104 G jet planes," by P. P. Castagliuolo. RIV MED AERO 29:Suppl:361-373, December, 1966.

"Perforated stainless may cloak jet noise," STEEL 164:31, February 10, 1969.

"Porous steel liner muffles jet whines," PRODUCT ENG 37:82, November 7, 1966.

"Port authority renews warnings on stretched DC-8 noise, weight," by J. W. Carter. AVIATION W 84:40, May 2, 1966.

"Potential and development of a v/s.t.o.l. inter-city airliner," by D. H. Jagger, et al. AIRCRAFT ENG 42:6-13, January, 1970.

"Practical noise control at international airports with special reference to Heathrow," by F. C. Petts. ROY AEROMAUTICAL SOC J 70:1051-1059; Discussion 1059-1060, December, 1966.

"Pratt & Whitney's JT9D will be a quiet, high bypass ratio engine," by J. D. Kester, et al. SAE J 76:69-71, June, 1968.

"Protection of ground personnel against the noise of jet aircraft engines," by J. V. Quercy. ARCH MAL PROF 27:537-541, June, 1966.

"Psychological reactions to aircraft noise," by K. D. Kryter. SCIENCE 151:1346-1355, March 18, 1966; Reply by P. K. Holmes 152:865, May 13, 1966.

"Psychometric studies of subjects exposed to the noise of jet engines," by G. G. Calapaj, et al. MED LAVORO 60:43-52, January, 1969.

AIRCRAFT NOISE

"Quiet plane sneaks upon ground observers," MACHINE DESIGN 42: 18, April 30, 1970.

"Reactions of the human nervous and cardiovascular systems to the effects of aviation noises," by V. G. Terent'ev, et al. VOENNO-MED ZH 6:55-58, June, 1969.

"Realistic assessment of the vertiport/community noise problem," by N. Shapiro, et al. J AIRCRAFT 5:407-411, July, 1968.

"Reduction of aircraft noise measured in several school, motel and residential rooms," by D. E. Bishop. J ACOUST SOC AMER 39:907-913, May, 1966.

"Roaring jet engines can be tested quietly," by V. Singleton, et al. POWER 112:70-71, September, 1968.

"Shear noise source terms for a circularjet ," by G. Krishnappa. J AP MECH 35:814-815, December, 1968.

"Shhh; retrofitting of jets," FORBES 105:31, March 15, 1970.

"Silencing the jets," CHEM & ENG N 47:19, November 24, 1969.

"Simple, light muffler with low back pressure cuts noise in light aircraft," by G. A. Alther. SAE J 75:58-60, May, 1967.

"Singing flame stills whine in jet engine," PRODUCT ENG 40:22, December 15, 1969.

"A six-year prospective study of the effect of jet-aircraft noise on hearing," by J. J. Knight, et al. J ROY NAV MED SERV 52: 92-96, Summer, 1966.

"16,000 miracle cuts out Heathrow jets," by T. Devlin. TIMES ED 2864:Suppl:7, April 10, 1970.

"Sound suppression in supersonic jet tests," ENGINEERING 199: 54, January 8, 1965.

"Stapedectomy and air travel," by J. Bastien. ANN OTOLARYNG

83:69-74, January-February, 1966.

"Stol future keyed to noise limitations," AVIATION W 90:46+, February 24, 1969.

"Stol noise standards spur interest in propeller design," AVIATION W 90:87, June 16, 1969.

"The struggle for quiet," by P. Deeley. OBSERVER 11, November 20, 1966.

"Study of acoustical treatments for jet-engine nacelles," by A. H. Marsh. ACOUSTICAL SOC AM J 43:1137-1156, May, 1968.

"Subjective basis for aircraft noise limitation," by D. W. Robinson. ROY AERONAUTICAL SOC J 71:396-400, June, 1967.

"Subjective response to synthesized flight noise signatures of several types of V-STOL aircraft. NASA CR-1118," by E. G. Hinterkeuser, et al. US NASA 1-92, August, 1968.

"Thousands of tiny holes quiet noise from jumbo jet engines," PRODUCT ENG 41:96, January 1, 1970.

"Three-shaft jetliner engine makes only half as much noise," PRODUCT ENG 39:62-63, November 4, 1968.

"Time to consider aircraft noise," ENGINEERING 262:961, December 2, 1966.

"Tort liability in light civilian aircraft operation," NEW YORK L FORUM 8,3:545-593, Fall, 1967.

"Transport noise reduction," by M. L. Yaffee. AVIATION W 89:39+, November 25; 61+, December 2, 1968.

"Underexpanded jet noise reduction using radial flow impingement," by D. S. Dosanjh, et al. AIAA J 7:458-464, March, 1969.

"White House science unit seeks government attack on jet noise," AVIATION W 84:44, March 28, 1966.

AIRLINES

"Acoustic trauma. 5 years of experience among flight and ground personnel of Varig airlines," by R. M. Neves Pinto. HOSPITAL 75:959-978, March, 1969.

"Acoustic trauma in the ground personnel of Varig," by R. M. Pinto, et al. HOSPITAL 67:351-354, February, 1965.

"Noise bill limitation sought by airlines," AVIATION W 88:41, June 24, 1968.

"Sideline noise new problem the airlines hadn't expected," by B. Jackson. FIN POST 63:12, August 9, 1969.

AIRPLANES

"Carriers urged to act on social problems," by R. G. O'Lone. AVIATION W 91:53-55, August 11, 1969.

"V/stol gets big lift in expanding aviation work," by W. Andrews. AEROSPACE TECH 21:88+, November 20, 1967.

AIRPORTS

"Airport noise and the urban dweller; a proposed solution (address)," by C. M. Haar. REAL ESTATE APPRAISER 34:21-25, September-October, 1968.

"Airport noise and the urban dweller: a proposed solution," by C. M. Haar. APPRAISAL J 36:551-558, October, 1968.

"The airport noise problem and airport zoning (United States)," by E. Seago. MD LAW R 28:120-135, Spring, 1968.

"Capacity and noise relationships for major hub airports," by R.L. Paulin. IEEE PROC 58:307-313, March, 1970.

"Computer polices airport noise," ELECTRONICS 42:183-184, July 7, 1969.

"Congress prepares to sound off on airport noise," by R. Leiser. AM AVIATION 30:81-82, September, 1966.

AIRPORTS

"Diminish airport noise? We can't yet measure it," by B. Jackson. FIN POST 62:14, July 20, 1968.

"The effect of aircraft noise on the population living in the vicinity of airports," by I. L. Karagodina, et al. GIG SANIT 34:25-30, May, 1969.

"How Oregon school keeps out airport noise," by J. E. Guerusey. NATIONS SCH 80:65, November, 1967.

"Hygienic evaluation of airports as noise sources," by I. L. Karagodina, et al. GIG SANIT 31:18-23, July, 1966.

"Improving the airport environment: effect of the 1969 FAA regulations on noise," by P. B. Larsen. IA L REV 55:808, April, 1970.

"Nobody loves an airport," by M. M. Berger. SO CALIF L REV 43: 631, Fall, 1970.

"Noise within a housing area near the Irkutsk airport," by M. I. Nekipelov. GIG SANIT 34:94-96, May, 1969.

"On the protection of the population in the vicinity of airports from aviation noises," by I. Ia. Borshchevskii. GIG SANIT 31:82-85, September, 1966.

"Practical noise control at international airports with special reference to Heathrow," by F. C. Petts. ROY AERONAUTICAL SOC J 70:1051-1059; Discussion 1059-1060, December, 1966.

"San Francisco sets airport noise limit," AVIATION W 93:37, December 21, 1970.

"Third London airport: choice cannot be based on cost alone says RIBA," RIBA J 77:224-225, May, 1970.

"Why life near airports will get quieter," BSNS W 44-45, February 7, 1970.

AIR CONDITIONING

"Air chambers, are they really cheaper?" by G. Flegel. DOM ENG 206:70-71, September, 1965.

"Air conditioning and acoustics," by W. H. Schneider. AIR COND HEAT & VEN 63:56-59, August, 1966.

"Air conditioning for all," ENGINEERING 207:413, March 14, 1969.

"Air moving and conditioning association, inc. has already adopted 10-12 watt sound power reference level," AIR COND HEAT & REFRIG N 105:2, August 30, 1965.

"ARI issues fancoil unit sound rating," AIR COND HEAT & REFRIG N 108:1+, May 16, 1966.

"ARI issues sound standard for room induction units," AIR COND HEAT & REFRIG N 114:1+, May 6, 1968.

"ARI seeks to educate A-E's, GSA on reduction equipment noise," AIR COND HEAT & REFRIG N 121:28, December 7, 1970.

"ARI's new sound ordinance calls for A-scale readings," AIR COND HEAT & REFRIG N 108:1+, June 27, 1966.

"Canadian meeting says U.S. accepts too high sound levels," by F. J. Versagi. AIR COND HEAT & REFRIG N 106:1+, November 29, 1965.

"Control of duct generated noise," by V. V. Cerami, et al. AIR COND HEAT & VEN 63:55-64, September, 1966.

"Coral Gables would limit noise from air conditioners," by C. D. Mericle. AIR COND HEAT & REFRIG N 108:1+, May 9, 1966.

"Court case on noisy residential unit causes reader to remind all of responsibility," by W. R. Brown. AIR COND HEAT & REFRIG N 105:14, August 9, 1965.

"Demand meaningful sound trap ratings!" by V. V. Cerami. AIR COND HEAT & VEN 63:53-54, July, 1966.

AIR CONDITIONING

"Design for quiet; air-moving systems," by W. F. Walker, et al. MACHINE DESIGN 39:202-208, September 14, 1967.

"Detroit studying ordinance to control noisy air conditioners," by G. M. Hanning. AIR COND HEAT & REFRIG N 106:1+, September 20, 1965.

"Do-it-yourself style: sound attenuation (air diverter roof-mounted to residential air conditioner)," by F. J. Versagi. AIR COND HEAT & REFRIG N 107:24, February 7, 1966.

"Engineers, contractors seen responsible for controlling outside equipment noise," by J. P. Christoff. AIR COND HEAT & REFRIG N 108:18, May 9, 1966.

"Experimental method for determination of noise attenuation in air ducts," by C. M. Harman, et al. ASHRAE J 7:43-49, October, 1965.

"Few questions about sound," by F. J. Versagi. AIR COND HEAT & REFRIG N 114:27, June 17, 1968.

"Field measurement study of the sound levels produced outdoors by residential air-conditioning equipment," by W. E. Blazier, Jr. ASHRAE J 9:35-39, May 1967.

"Forge shop ventilation noise problem overcome," ENGINEER 225: 956, June 21, 1968.

"Guiding the air flow cuts down on noise," PRODUCT ENG 39:48, August 12, 1968.

"LAU starts sound study program," AIR COND HEAT & REFRIG N 111:1+, August, 1967.

"Muffler quiets air conditioner," AIR COND HEAT & REFRIG N 105: 1+, July 5, 1965.

"New industrial acoustics facility for dynamic rating of duct silencers," ACOUSTICAL SOC AM J 41:869, April, 1967.

"New method of noise analysis for high velocity air distribution systems," by R. H. Dean, et al. HEATING-PIPING 40:132-137, January, 1968.

"New rating method for duct silencers," by D. B. Callaway, et al. HEATING-PIPING 38:88-95, December 1966; Discussion 39:79-80 ; Reply 88+, April, 1967.

"Noise-conditioning the air conditioner," by J. W. Sullivan. IEEE TRANS IND & GEN APPLICATIONS 4:527-534, September, 1968.

"Noise considerations in the application and installation of outdoor air-conditioning equipment," by W. S. Bayless. ASHRAE J 9:52-56, April, 1967.

"Noise control in air systems," by L. F. Yerges. HEATING-PIPING 41:144-148, March, 1969.

"Noise in air-moving systems," by R. J. Kenny. MACHINE DESIGN 40:138-150, September 26, 1968.

"Noise research gets top priority," AIR COND HEAT & REFRIG N 111:1+, May 22, 1967.

"Noisy air conditioner turns out to be a cooling tower in Hartford case," AIR COND HEAT & REFRIG N 109:1+, September 19, 1966.

"On the evaluation of the inverse reactions in the Gelle test with air conduction," by L. Palfalvi. ACTA OTOLARYNG 66:508-514, December, 1968.

"Research on regenerated noise in ducts," by A. C. Potter. ASHRAE J 7:52, February, 1965.

"Sources of sound in outdoor air-conditioning equipment," by A. C. Potter. ASHRAE J 9:42-47, April, 1967.

"Variable air volume air conditioning; air distribution," by S. Daryanani, et al. AIR COND HEAT & VEN 63:69-72, March, 1966.

"Watch those published sound power levels," by F. J. Versagi. AIR COND HEAT & REFRIG N 105:13, August 23, 1965.

AIR MOVEMENT
see: Air Conditioning

ALANINE AMINOTRANSFERASE
"Effects of physical work and work under condition of noise and vibration on the human body. II. Behavior of aspartic and alanine aminotransferase, gamma glutamyl transpeptidas, lactic dehydrogenase and L-idotyl dehydrogenase activities," by J. Gregorczyk, et al. ACTA PHYSIOL POL 16:709-714, September-October, 1965.

ALCOHOL
"The effect of alcohol and noise on components of a tracking and monitoring task," by P. Hamilton, et al. BR J PSYCHOL 61:149-156, May, 1970.

ALDOLASE
"Effect of physical work and work under conditions of noise and vibration on the human body. I. Behavior of serum alkaline phosphatase, aldolase and lactic dehydrogenase activities," by J. Gregorczyk, et al. ACTA PHYSIOL POL 16:701-708, September-October, 1965.

ALUMINUM SPARS
see: Sailing

AMPLIFIERS
"Turbulence as a producer of noise in proportional fluid amplifiers," by D. W. Prosser, et al. J AP MECH 33:728-734, December, 1966.

AMPLITUDE DISCRIMINATION
"Amplitude discrimination in noise," by G. B. Henning, et al. J ACOUST SOC AMER 41:1365-1366, May, 1967.

"Amplitude discrimination in noise, pedestal experiments, and additivity of masking," by G. B. Henning. J ACOUST SOC AMER 45:426-435, February, 1969.

"Amplitude distribution of axon membrane noise voltage," by A. A. Verveen, et al. ACTA PHYSIOL PHARMACOL NEERL 15:353-379, August, 1969.

AMPLITUDE DISCRIMINATION

"Click-intensity discrimination with and without a background masking noise," by D. H. Raab, et al. J ACOUST SOC AMER 46:965-968, October, 1969.

"Effect of duration on amplitude discrimination in noise," by G. B. Henning, et al. J ACOUST SOC AMER 45:1008-1013, April, 1969.

"Observations on amplitude modulated acoustical noise signals," by M. Rodenburg, et al. PFLUEGER ARCH 311:197-198, 1969.

"Relation between sound intensity and amplitude of the auditory evoked response at different stimulus frequencies," by F. Antinoro, et al. ACOUSTICAL SOC AM J 46:1433-1436, pt 2, December, 1969.

AMYLASURIA

"Variations of amylasuria and amylasemia under the effect of sound and vibratory 'stress' in industry," by S. Miclesco-Groholsky, et al. ARCH MAL PROF 31:277-282, June, 1970.

"Variations of amylasuria under the effect of sound and vibration injuries (preliminary note)," by S. Miclesco-Groholsky. ANN OTOLARYNG 86:251-258, April-May, 1969.

APARTMENT NOISE

see also: Architecture

"How much does noise bother apartment dwellers? reprint," by D. R. Prestemon. ARCH REC 143:155-156, February, 1968.

"Lease and quiet: ways to circumvent construction flaws and restrictions," by J. H. Ingersoll. HOUSE B 109:122-124, August, 1967.

"Noise-stopping tips from a prefabber help boost rentals for packaged apartments," HOUSE & HOME 31:84, June, 1967.

"Old folks at, home? life in a Manhattan apartment," by E. G. Smith. ATLAN 220:118-119, December, 1967.

"16 surefire ways to keep your apartment tenants complaining about noise," HOUSE & HOME 36:67-69, July, 1969.

ARCHITECTURE

"Acoustic integrity, flexibility important in building's design; Bell telephone laboratories new development center," HEATING-PIPING 38:304-305, January, 1966.

"Acoustics, noise and building, by P. H. Parkin, et al. A review," by H. C. Pinfold. TOWN PLAN R 40:84-85, April, 1969.

"Acoustics of Northrop memorial auditorium," by B. S. Ramakrishna, et al. ACOUSTICAL SOC AM J 47:951-960, April, 1970.

"Appearance combined with soundproofing in a transformer enclosure," ENGINEER 226:232, August 16, 1968.

"Are Academic libraries too noisy?" by J. A. McCrossan. AM LIB 1:396, April, 1970.

"Assessment of footstep noise through wood-joist and concrete floors," by D. Olynyk, et al. J ACOUST SOC AMER 43:730-733, April, 1968.

"Bringing peace to the noisy office," by N. C. Crane. SUPERVISORY MGT 10:42-43, November, 1965.

"Building acoustics, edited by B. F. Day, et al. A review," by C. Pinfold. TOWN PLAN R 41:300-301, July, 1970.

"Building or buying? Here's what to look for," TODAY'S HLTH 43: 59, February, 1965.

"Definition of human requirements in respect to noise, to be used by town planners and builders," J SCI MED LILLE 87:328-329, April, 1969.

"Elevator noise; architectural and mechanical considerations," by J. E. Sieffert. ARCH REC 143:199, April, 1968.

"Engineering outline; sound insulation in buildings," by P. H. Parkin. ENGINEERING 205:307-310, February 23, 1968.

"Engineers, contractors seen responsible for controlling outside equipment noise," by J. P. Christoff. AIR COND HEAT & REFRIG N 108:18, May 9, 1966.

"Environmental insult," PROGRES ARCH 48:25+, February, 1967.

"Getting used to it," ARCH DESIGN 40:526, October, 1970.

"Hazards and hurdles in developing standards: a case history-rating the impact-noise resistance of floors," by R. E. Fischer. ARCH REC 147:147-150, May, 1970.

"How to design a quiet school; additions and new structures," AM SCH BD J 156:21-25, October, 1968.

"Hygienic evaluation of noise in a shop with curved designs," by N. M. Paran'ko, et al. GIG SANIT 31:81-82, May, 1966.

"Impact-noise rating of various floors," by T. Mariner, et al. ACOUSTICAL SOC AM J 41:206-214, January, 1967.

"Integrated ceiling system handles air, light and sound," AIR COND HEAT & VEN 65:59-62, April, 1968.

"Lead for sound control doesn't impress experts," AIR COND HEAT & REFRIG N 121:30, December 7, 1970.

"Lease and quiet: ways to circumvent construction flaws and restrictions," by H. H. Ingersoll. HOUSE B 109:122-124, August, 1967.

"Modern sound-stage construction," by D. J. Bloomberg, et al. SMPTE J 75:25-28, January, 1966.

"Muffling the clamor of urban construction," BSNS W 168-169, December 14, 1968.

"New augered pile technic reduces construction noise," by W. J. Duchaine. MOD HOSP 104:164, March, 1965.

"Noise: builders, users pass buck," by C. J. Suchocki. IRON AGE 204:67-69, July 31, 1969.

ARCHITECTURE

"Noise control in architecture: more engineering than art," ARCH REC 142:193-204, October, 1967.

"Noise control: universal and international exhibition of 1967, Montreal," SMPTE J 76:574-577, June, 1967.

"Noise design is a contractor problem," by W. A. Tedesco. AIR COND HEAT & REFRIG N 121:8+, December 14, 1970.

"Noise-stopping tips from a prefabber help boost rentals for packaged apartments," HOUSE & HOME 31:84, June, 1967.

"On the hearing impairment of bolt drivers in the building industry," by A. Schurno, et al. Z GES HYG 14:161-165, March, 1968.

"On the status of meeting noise protection requirements in industrial housing construction," by W. Fasold. Z GES HYG 13:628-631, August, 1967.

"Practical problems of partition design," by R. D. Ford, et al. ACOUSTICAL SOC AM J 43:1062-1068, May, 1968.

"Soundproofing: architect defends house against the jet," HOUSE & HOME 33:8, February, 1968.

"Sound-proofing of buildings," by E. F. Stacy. ROY SOC HEALTH J 88:82-86, March-April, 1968.

"Sound insulation and new forms of construction," BUILD RES STA DIGEST 96:1-8, August, 1968.

"Wall construction affects noise level inside structure," by F. J. Versagi. AIR COND HEAT & REFRIG N 110:22-23, April 24, 1967.

"What belongs in acoustical specifications," by R. Farrell. ARCH REC 138:227-230+, September; 203-206, November, 1965.

"When noise is a problem, design it out instead of in," by A. R. Gardner. PRODUCT ENG 37:86-91, April 11, 1966.

ASPARTIC AMINOTRANSFERASE

"Effects of physical work and work under condition of noise and vibration on the human body. II. Behavior of aspartic and alanine aminotransfersse, gamma glutamyl transpeptidas, lactic dehydrogenase and L-idotyl dehydrogenase activities," by J. Gregorczyk, et al. ACTA PHYSIOL POL 16:709-714, September-October, 1965.

ATTENTION

see also: Learning

"Changes of the background noise intensity and the bar-pressing response rate," by K. Zielinski. ACTA BIOL EXP 26:43-53, 1966.

"Distraction and Stroop Color-Word performance," by B. K. Houston, et al. J EXP PSYCHOL 74:54-56, May, 1967.

"An effect of noise on the distribution of attention," by M. M. Woodhead. J APPL PSYCHOL 50:296-299, August, 1966.

"Effect of a specific noise on visual and auditory memory span," by S. Dornic. SCAND J PSYCHOL 8:155-160, 1967.

"Effects of noise on a complex task," by S. J. Samtur. GRAD RES ED 4:63-81, Spring, 1969.

AUTOMOBILE NOISE

"Audio frequency analyzer measures auto noise pollution," ELECTRO-TECH 85:24-25, January, 1970.

"Automobile noise--an effective method for control," U RICHMOND L REV 4:314, Spring, 1970.

"Connecticut gears up research program to combat motor vehicle noise pollution on state highways," INSTRUMENTATION TECH 16:18, March, 1969.

"Connecticut may ban noisy vehicles; state police test acoustic radar," by S. Klein. MACHINE DESIGN 42:28-29, April 16, 1970.

"Effect of the interior noise of a passenger car on efficiency," by R. Schuster. BEITR GERICHTL MED 26:273-277, 1969.

AUTOMOBILE NOISE

"Expo cars get silent treatment," RY AGE 162:25, April 10, 1967.

"Hygienic characteristics of noise during testing of automobile motors," by R. A. Medved'. GIG SANIT 30:104-107, March, 1965.

"Measurement of interior noise in passenger cars," by W. Henkel. Z GES HYG 15:225-228, April, 1969.

"Noise reduction," AUTOMOBILE ENG 59:396-412, October, 1969.

"Noise reduction: engine mountings," AUTOMOBILE ENG 59:404-405, October, 1969.

"Noise reduction; engine structural vibrations," AUTOMOBILE ENG 59:396-398, October, 1969.

"Noise reduction; isolation of the engine and transmission system," AUTOMOBILE ENG 59:402-403, October, 1969 .

"Reducing automobile and tractor noise," by B. P. Goncharenko. GIG TR PROF ZABOL 14:46-47, January, 1970.

"Silence in cars is golden," ENGINEERING 207:88-89, January 17, 1969.

BAROTRAUMA

"Acoustic trauma in frogmen," by G. Borasi, et al. ANN LARING 67:36-43, January-February, 1968.

"Auricular barotrauma," by Y. Husson. SEM HOP PARIS 44:2686-2689, October 26, 1968.

"Barotraumatic damage of hearing organ in divers," by Z. Sliskovic. VOJNOSANIT PREGL 26:75-77, February, 1969.

BASAL METABOLISM

"The effect of noise on basal metabolism," by I. Pinter, et al. ORV HETIL 109:1371-1373, June 23, 1968.

BEARING NOISE

"Establishment of objective criteria reflecting subjective response to roller-bearing noise," by R. F. Lucht, et al. J ACOUST SOC AMER 44:1-4, July, 1968.

"Measuring bearing noise," ENGINEERING 209:358-359, April 10, 1970.

BEHAVIOR

see also: Attention, Learning, Personality, Psychology, Psychotics

"Adaptation and loudness decrement; a reconsideration," by J. W. Petty, et al. ACOUSTICAL SOC AM J 47:1074-1082, pt 2, April, 1970.

"Effects of ambient noise on vigilance performance," by P. H. McCann. HUMAN FACTORS 11:251-256, June, 1969.

"Effects of delay on the punishing and reinforcing effects of noise onset and termination," by R. C. Bolles, et al. J COMP PHYSIOL PSYCHOL 61:475-477, June, 1966.

"How noise affects work," by D. E. Broadbent. NEW SOCIETY 12-14, March 3, 1966.

"Muffling noise may amplify productivity," STEEL 164:72d+, June 2, 1969.

"Noise effects on health, productivity, and well-being," by A. Cohen. TRANS NY ACAD SCI 30:910-918, May, 1968.

"Normal and abnormal coordination of movements: a polymyographic approach," by G. Pampiglioen. J NEUROL SCI 3:525-538, November-December, 1966.

"On different responses to low intensity noise in man," by J. Havranek, et al. ACTIV NERV SUP 7:183, 1965.

"Punishment as a discriminative stimulus and conditioned reinforcer with humans," by T. Ayllon, et al. J EXP ANAL BEHAV 9:411-419, July, 1966.

BEHAVIOR

"Pupillographic changes induced by diverse acoustic stimulations in normal subjects," by B. B. Carenini, et al. BOLL OCULIST 45:75-82, February, 1968.

BEKESY TYPING

see: Inner Ear

BIBLIOGRAPHY

"Allergy--acoustic trauma. Review of the medical literature," by E. Mann. ZAHNAERZTL WELT 66:322-324, May 10, 1965.

"Bibliography on transformer noise," INST ELEC & ELECTRONICS ENG TRANS POWER APPARATUS & SYSTEMS 87:372-387, February, 1968.

"The biological action of impulse noises (a review of the literature)," by V. N. Morozov, et al. VOENNOMED ZH 8:53-58, August, 1969.

"Effects of overpressure on the ear--a review," by F. G. Hirsch. ANN NY ACAD SCI 152:147-162, October 28, 1968.

"Evaluation of advances in engine noise technology," by A. L. McPike. AIRCRAFT ENG 42:16+, May, 1970.

"Facilities and instrumentation for aircraft engine noise studies," by R. E. Gorton. J ENG POWER 89:1-13, January, 1967.

BIOCHEMICAL PROCESSES

"Biochemical and hematological reactions to noise in man," by V. Hrubes. ACTIV NERV SUP 9:245-248, August, 1967.

"The effect of vibration and noise on protein metabolism of excavator-machinists using several biochemical indices," by A. M. Tambovtseva. GIG SANIT 33:58-61, January, 1968.

"The effect of vitamin B 1 and niacin on the course or biochemical processes in organisms exposed to the action of vibration," by D. A. Mikhel'son. VOP PITAN 28:54-58, November-December, 1969.

BIOELECTRICITY

"Changes in the bioelectric activity of the brain and in some autonomic and vascular reaction under the influence of noise," by E. A. Drogichina, et al. GIG SANIT 30:29-33, February, 1965.

BIOFLAVONOID

"Bioflavonoid therapy in sensorineural hearing loss: a double-blind study," by J. E. Creston, et al. ARCH OTOLARYNG 82:159-165, August, 1965.

BIOLOGY

"Activity of certain enzymes of intermediate metabolism and behavior of serum proteins and their fractions in workers exposed to strong acoustic and vibrational stimuli," by J. Gregorczyk. ACTA PHYSIOL POL 17:107-118, January-February, 1966.

BRAKES

"Taking the squeal out of brakes," by J. E. Gieck. SAE J 73:83, September, 1965.

BURNERS

"Fundamentals of fuel oil firing; how to eliminate burner noise," by A. A. Brahams. HEATING-PIPING 37:134-136, April, 1965.

"Performance and noise testing of oxy-fuel burners," ENGINEER 221:696, May 6, 1966.

"Pressure nozzles on home oil heaters," ELECTRONICS 39:58, November 14, 1966.

CR-121

"Effect of CR-121 (cleregil) on acoustic trauma," by J. Alavoine. THERAPEUTIQUE 45:526-531, May, 1969.

CAMPING AREAS

"Sources of noise at camping areas and possibilities of their reduction," by E. Lange. Z GES HYG 14:143-145, February, 1968.

CAMS

"Reducing noise in cams," by R. G. Fenton. MACHINE DESIGN 38:187-190, April 14, 1966.

CARDIOVASCULAR SYSTEM

"Biochemical and hematological reactions to noise in man," by V. Hrubes. ACTIV NERV SUP 9:245-248, August, 1967.

"The blood system reaction to the occupational effect of vibration and noise," by I. A. Gribova, et al. GIG SANIT 30:34-37, October, 1965.

"Changes in the bioelectric activity of the brain and in some autonomic and vascular reaction under the influence of noise," by E. A. Drogichina, et al. GIG SANIT 30:29-33, February, 1965.

"Contribution to the investigations on the influence of noise on hearing and the circulatory system in ship-yard employes," by S. Klajman, et al. MED PRACY 16:380-384, 1965.

"Contribution to the study of acoustic stimulation on the blood eosinophil count," by A. Amorelli, et al. ARCH ITAL LARING 73:515-520, November-December, 1965.

"Effect of air noise on the human cardiovascular system," by I. A. Sapov, et al. VOEN MED ZH 6:53-54, June, 1970.

"The effect of defined noise exposure on the peripheral circulatory system," by A. Fuchs-Schmuck, et al. Z GES HYG 15:651-653, September, 1969.

"The effect of industrial noise in winding and weaving factories on the arterial pressure of operators," by A. I. Andrukovich. GIG TR PROF ZABOL 9:39-42, December, 1965.

"Effect of isolated mental stimuli on arterial pressure and body weight in Sprague-Dawley rats," by S. Campus, et al. BOLL SOC ITAL BIOL SPER 41:1087-1089, September 30, 1965.

"Effect of intensive noise and neuro-psychic tension on arterial blood pressure levels and frequency of hypertensive disease," by N. N. Shatalov, et al. KLIN MED 48:70-73, March, 1970.

"Effects of two types of noise on cardiac rhythm of a man held immobile with constant visual attention," by T. Meyer-Schwertz, et al. ARCH SCI PHYSIOL 22:195-228, 1968.

"Electrodermal and cardiac responses of schizophrenic children to sensory stimuli," by M. E. Bernal, et al. PSYCHOPHYSIOLOGY 7:155-168, September, 1970.

"Ethanol inhibition of audiogenic stress induced cardiac hypertrophy," by W. F. Geber, et al. EXPERIENTIA 23:734-736, September 15, 1967.

"Experimental studies on the influence of impelling noise on various blood circulatory parameters," by A. Fuchs-Schmuck, et al. DTSCH GESUNDHEITSW 25:1951-1954, October 8, 1970.

"Experimental studies on noise susceptibility in impairment of circulation regulating function," by S. Sugiyama. J OTORHINOLARYNG SOC JAP 68:715-731, June, 1965.

"Influence of noise on the cardiovascular system," by C. Gradina, et al. FIZIOL NORM PAT 16:357-367, 1970.

"The reaction of the cardiovascular system of working adolescents to sound and vibration stimuli," by A. I. Tsysar'," GIG SANIT 31:33-38, February, 1966.

"Reactions of the human nervous and cardiovascular systems to the effects of aviation noises," by V. G. Terent'ev, et al. VOENNOMED ZH 6:55-58, June, 1969.

CHILDREN

"Apropos of 2 cases of mycoionic petit mal precipitated by noise in children," by A. Lerique-Koechlin, et al. REV NEUROL 113: 269, September, 1965.

"Auditory figure-background perception in normal children; speech reception threshold," by B. M. Siegenthaler, et al. CHILD DEVELOP 38:1163-1167, December, 1967.

"Effect of community noises on school children," by V. I. Pal'gov. PEDIAT AKUSH GIENK 5:29-32, September-October, 1966.

CHILDREN

"The effect of hearing aid use on the residual hearing of children with sensorineural deafness," by J. H. Macrae, et al. ANN OTOL 74:408-419, June, 1965.

"Effect of noise on the functional status of the nervous system in children of preschool age," by Kh. V. Storoshchuk. GIG SANIT 31:44-48, January, 1966.

"Electrodermal and cardiac responses of schizophrenic children to sensory stimuli," by M. E. Bernal, et al. PSYCHOPHYSIOLOGY 7:155-168, September, 1970.

"High intensity sounds in the recreational environment. Hazard to young ears," by D. M. Lipscomb. CLIN PEDIAT 8:63-68, February, 1969.

"Noise disturbance due to infants' home," by H. Wiethaup. MED KLIN 62:393, March 10, 1967.

"The noise level in a childrens hospital and the wake-up threshold in infants," by R. Gadeke, et al. ACTA PAEDIAT SCAND 58: 164-170, March, 1969.

"Noise levels in infant oxygen tents," by R. League, et al. LANCET 2:978, November 7, 1970.

"Noise characteristics in the baby compartment of incubators. Their analysis and relationship to environmental sound pressure levels," by F. L. Seleny, et al. AMER J DIS CHILD 117:445-450, April, 1969.

"Noise measurement in a children's hospital and the awaking noise threshold of infant," by R. Gadeke, et al. MSCHR KINDERHEILK 116:374-375, June, 1968.

"Performance of cerebral palsied children under conditions of reduced auditory input," by J. Fassler. EXCEPT CHILD 37:201-209, November, 1970.

"Studies of children under imposed noise and heat stress," by D. P. Wyon. ERGONOMICS 13:598-612, September, 1970.

CIRCUIT-BREAKERS

"Silencing airblast circuit-breakers," ENGINEER 225:274-275, February 16, 1968.

CLEFT PALATE

"Production of cleft palate by noise and hunger," by S. Peters, et al. DEUTSCH ZAHNAERZTL Z 23:843-847, August, 1968.

CLEREGIL

see: CR-121

CLINICAL ASPECTS

"Acoustic trauma. Clinical presentation," by D. L. Chadwick. PROC ROY SOC MED 59:957-966, October, 1966.

"Acoustic trauma in regular army personnel. Clinical audiologic study," by A. Saimivalli. ACTA OTOLARYNG 222:Suppl:1-85, 1967.

"The clinical and audiological pattern due to acoustic trauma," by A. Salmivalli. ACTA OTOLARYNG 224:Suppl:239+, June 27, 1966.

"Clinical and experimental aspects of the effects of noise on the central nervous system," by F. Angeleri, et al. MED LAVORO 60:759-766, December, 1969.

"Clinical observations of the hearing loss in perceptive deafness," by I. Kirikae, et al. OTOLARYNGOLOGY 40:599-606, August, 1968.

"Clinical study of the behavior of auditory adaptation in textile workers," by L. Bertocchi, et al. MINERVA OTORINOLARING 16:106-109, May-June, 1966.

"Clinical trial of EU 4200 in the treatment of sound injuries," by J. Cazaubon, et al. MED TROP 30:403-408, May-June, 1970.

"Occupational acoustic trauma in otosclerotic patients (considerations on the clinical and medicolegal evaluation related to 4 personal observations)," by A. Scevola. ARCH ITAL OTOL

78:474-486, July-August, 1967.

"Problems of adaptation to fatigue and occupational acoustic trauma. Experimental, anatomoclinical and audiometric study," by L. Teodorescu, et al. OTORINOLARINGOLOGIE 15:9-24, January-February, 1970.

"Tolerable limit of loudness: its clinical and physiological significance," by J. D. Hood, et al. J ACOUST SOC AMER 40:47-53, July, 1966.

COMBUSTION NOISE

"Combustion roar of turbulent diffusion flames," by R. D. Giammar, et al. J ENG POWER 92:157-165, April, 1970.

"Infernal combustion," by A. Brien. NEW STATESMAN 74:318, September 15, 1967.

COMMUNICATIONS

"The problem of the statistical inference. I. Communication in the presence of noise," by R. F. Wrighton. ACTA GENET 17:178-192, 1967.

"The problem of statistical inference. 3. Statistical inference as communication in the presence of noise," by R. F. Wrighton. ACTA GENET 18:84-96, 1968.

COMMUNITY NOISE

see: Urban Noise

COMPRESSED AIR NOISE

"Auditory disorders caused by high speed equipment: data on 345 cases," by J. C. Mooney. REV ASOC ODONT ARGENT 55:327-337, August, 1967.

"Compressed air: on the quiet. ENGINEERING 206:775, November 22, 1968.

"Cutting the noise of an air compressor," ENGINEER 228:34, June 26, 1969.

COMPRESSED AIR NOISE

"Development of engineering practices in jet and compressor noise," by J. B. Large, et al. J AIRCRAFT 6:189-195, May, 1969.

"Evaluation of the effectiveness of measures for lowering the noise of pneumatic rapier looms," by L. I. Maksimova, et al. GIG SANIT 33:87-89, December, 1968.

"Hygienic evaluation of manual pneumatic perforators," by N. M. Paran'ko. VRACH DELO 1:94-98, January, 1965.

"Integration of aircraft auxiliary power supplies," by J. Wotton. AIRCRAFT ENG 40:22-24+, October, 1968.

"Mechanisms of noise generation in a compressor model," by B. T. Hulse, et al. J ENG POWER 89:191-197; Discussion 197-198, April, 1967.

"Noisy compressors traced to misguided attempt at economy," by A. D. Sullivan. HEATING-PIPING 41:83-85, December, 1969.

"Perforated torus ring quiets 25,000 cfm of compressed air," by S. Butler. PRODUCT ENG 41:74, June 8, 1970.

"Recurrent impact noise from pneumatic hammers," by A. M. Martin, et al. ANN OCCUP HYG 13:59-67, January, 1970.

"Reduction of compressor noise radiation," by M. V. Lowson. ACOUSTICAL SOC AM J 43:37-50, January, 1968.

"Silencing of hand-held percussive rock drills for underground operations," by B. H. Weber. CAN MIN & MET BUL 63:163-166, February, 1970.

"Silent drills; where they stand now," by J. J. Daly. ROCK PROD 68:62-63, November, 1965.

"Socially acceptable compressors," by E. C. Hinck. COMP AIR MAG 74:18-21, January, 1969.

"Sound code," by G. M. Diehl. COM AIR MAG 74:22, January, 1969.

COMPRESSED AIR NOISE

"Sound impulses occurring during drilling processes and their harmful effect on the hearing organs," by K. Dietzel, et al. VESTN OTORINOLARINGOL 32:55-57, November-December, 1970.

"Step closer to quiet," by R. J. Hayes. MIN CONG J 55:68-72, April, 1969.

"Study of cochlear reactions in workers handling pneumatic hammers," by G. Jullien, et al. ARCH MAL PROF 28:297-300, January-February, 1967.

"On the effect on the organism of the vibration of pneumatic tools with high rate of rotation," by L. Ia. Tartakovskaia, et al. GIG TR PROF ZABOL 13:16-19, April, 1969.

"On the hearing impairment of bolt drivers in the building industry," by A. Schurno, et al. Z GES HYG 14:161-165, March, 1968.

COMPRESSOR NOISE
see: Compressed Air Noise

COMPUTER NOISE

"Computer room noises seen as hearing threat," IND RES 12:24-25, August, 1970.

"Computer units pose hearing threats," AUTOMATION 17:5, August, 1970.

"Cutting noise in data sampling," by T. Kobylarz. ELECTRONICS 41:70-73, May 13, 1968.

"Hazardous noise levels in computer labs," J ACOUST SOC AM 48:860-861, October, 1970.

"How to keep down noise levels in computer facilities," by L. L. Boyer, Jr. ARCH REC 145:165-166, May, 1969.

"Noise in computer rooms and in mechanical data processing central stations," by B. Kvasnicka. PRAC LEK 17:112-115, April, 1965,

CONCRETE BREAKERS

"Concrete breakers must be quieter," ENGINEERING 203:1050, June 30, 1967.

CONSTRUCTION INDUSTRY

see: Industrial Noise: Construction

CYTOCHROME C

"Protective effect of cytochrome C and adenosinotriphosphoric acid in acoustic trauma," by W. Jankowski, et al. OTOLARYNG POL 23:133-140, 1969.

DENTAL NOISE

"Acoustic microtrauma due to dental drills. Contribution to the technic of early diagnosis of acoustic trauma," by W. Niemeyer. ARCH KLIN EXP OHREN NASEN KEHIKOPFHEILKD 196:227-231, 1970.

"Acoustic trauma caused by dental turbine drills," by W. Lorenz, et al. DTSCH ZAHN MUND KIEFERHEILKD 54:343-351, June, 1970.

"Adverse conditions found in the use of ultra high-speed equipment," by P. L. Terranova. NEW YORK DENT J 33:143-148, March, 1967.

"Auditory discomfort associated with use of the air turbine dental drill," by A. F. Smith, et al. J ROY NAV MED SERV 52:82-83, Summer, 1966.

"An ear plug designed to obtund the sound of high speed dental engines," by R. B. Sloane. DENT DIG 72:218-220, May, 1966.

"Hearing damage following noise from dental turbines," by L. Dunker, et al. DDZ 24:33-35, January, 1970.

"The hearing threshold levels of dental practitioners exposed to air turbine drill noise," by W. Taylor, et al. BRIT DENT J 118:206-210, March 2, 1965.

"High speed equipment and dentists' health," by R. Von Krammer. J PROSTH DENT 19:46-50, January, 1968.

DENTAL NOISE

"Is noise caused by dental turbine drills injurious to hearing?" by J. S. Lumio, et al. MSCHR OHRENHEILK 99:192-199, 1965.

"New dental turbines and the problem of disposition to acoustic trauma. 2," by L. Dunker, et al. DDZ 23:257-260, June, 1969.

"Radiated noise from high speed dental handpieces," by H. N. Cooperman, et al. DENT DIG 71:404-407, September, 1965.

"Sound level measurements and frequency analyses in dental working area," by R. Mayer, et al. DEUTSCH ZAHNAERZTL Z 23: 800-805, August, 1968.

"Studies on the problem of hearing damage caused by the use of high speed dental turbines," by O. Arentsschild, et al. ZAHNAERZTL RUNDSCH 75:217-223, June 8, 1966.

DENTISTS

see also: Dental Noise

"Deafness in dentists," by C. Wark. J OTOLARYNG SOC AUST 2:89, March, 1967.

"Dentists' hearing: the effect of highspeed drill noise," by V. Bulteaul. MED J AUST 2:1111, December 9, 1967.

"Dentists' hearing: the effect of high speed drill," by B. A. Skurr, et al. AUST DENT J 15:259-260, August, 1970.

"Effects of high-speed drill noise and gunfire on dentists' hearing," by W. D. Ward, et al. J AMER DENT ASS 79:1383-1387, December, 1969.

DIABETES

"Dynamics of glycemic reactions after repeated exposure to noise," by K. Treptow. ACTIV NERV SUP 8:215-216, June, 1966.

DIET

"The importance of certain foodstuffs for maintaining operative work capacity under conditions of intense noise," by Iu. F. Udalov, et al. GIG TR PROF ZABOL 11:18-23, May, 1967.

DIGESTIVE SYSTEM

"Christmas cracker or duogastrone symptoms," by C. C. Evans, et al. BRIT MED J 1:120-121, January 11, 1969.

"Effect of air-craft noise on gastric function," by C. Y. Kim, et al. YONSEI MED J 9:149-154, 1968.

DIRECTIONAL DISCRIMINATION

"Array gain for the case of directional noise," by B. F. Cron, et al. ACOUSTICAL SOC AM J 41:864-867, April, 1967.

"Differentiation of the direction of hearing in noise," by J. Laciak. OTOLARYNG POL 22:787-791, 1968.

"The direction of change versus the absolute level of noise intensity as a cue in the CER situation," by K. Zielinski. ACTA BIOL EXP 25:337-357, 1965.

"Directional audiometry. I. Directional white-noise audiometry," by F. M. Tonning. ACTA OTOLARYNG 69:388-394, June, 1970.

"Effect of stimulus duration on localization of direction of noise stimuli," by W. R. Thurlow, et al. J SPEECH & HEARING RES 13:826-838, December, 1970.

"Effects of intermittent noise on human target detection," by H. D. Warner. HUM FACTORS 11:245-250, June, 1969.

DIVERS

see: Barotrauma

DOMESTIC ANIMALS

"Offensive noise caused by domestic animals," by H. Wiethaup. MED KLIN 60:1377-1378, August 20, 1965.

EEG

"The EEG and the impairment of sleep by traffic noise during the night: a problem of preventive medicine," by H. R. Richter. ELECTROENCEPH CLIN NEUROPHYSIOL 23:291, September, 1967.

EEG

"EEG measures of arousal during RFT performance in 'noise'," by R. W. Hayes, et al. PERCEPT MOT SKILLS 31:594, October, 1970.

"Effects of noise in EEG latency changes in an auditory vigilance task," by E. Gulian. ELECTROENCEPH CLIN NEUROPHYSIOL 27:637, December, 1969.

EU-4200

"Clinical trial of EU 4200 in the treatment of sound injuries," by J. Cazaubon, et al. MED TROP 30:403-408, May-June, 1970.

EAR INJURY

"Ear injury," by R. Willis. J OTOLARYNG SOC AUST 2:75-80, March, 1967.

"Explosive lesions of the ear," by A. Risavi. VOJNOSANIT PREGL 22:155-161, March, 1965.

"Hearing injury in building workers," by S. Lindqvist. LAKARTIDNINGEN 67:4283-4292, September 16, 1970.

"Hearing injuries among construction workers in Skaraborg County," NORD MED 83:445-446, April 2, 1970.

"Injury to hearing in foundrymen," by G. R. Atherley. OCCUP HEALTH REV 19:14-16, 1967.

"Lesions of the acoustic organ caused by noise among engine-room crews," by R. Nowak. BULL INST AMR GDANSK MED 17:339-342, 1966.

"Patterns of hair cell damage after intense auditory stimulation," by C. W. Stockwell, et al. ANN OTOL 78:1144-1168, December, 1969.

EARPHONES

"Severe equilibrium disorders following strong sound impression through bilaterally worn earphones," by H. Rudert. ARCH KLIN EXP OHR NAS KEHIKOPFHEILK 188:316-319, 1967.

EARPHONES

"Variables that influence sound pressures generated in the ear canal by an audiometric earphone," by N. P. Erber. ACOUSTICAL SOC AM J 44:555-562, August, 1968.

EDUCATION

see also: Attention
Learning

"Effect of noise on the teaching process," by V. Lejska. CESK PEDIAT 23:357-361, April, 1968.

"How to design a quiet school; additions and new structures," AM SCH BD J 156:21-25, October, 1968.

"Noise conditions in normal school classrooms," by D. A. Sanders. EXCEPTIONAL CHILD 31:344-353, March, 1965.

"The noise factor and its hygienic evaluation in the vocational training of students in the 9-11th grade," by E. A. Timokhina. GIG SANIT 30:46-50, February, 1965.

"Noise is for learning: four types of noise created by junior high school students," by J. Reedy. CLEAR HOUSE 43:154-157, November, 1968.

"Occupational deafness in a vocational school," by J. Knops, et al. ARCH BELG MED SOC 24:330-338, May, 1966.

"Professors are skeptical of positive claims made for environmental control (windowless classrooms, negative ions, noise are discussed)," by F. J. Versagi. AIR COND HEAT & REFRIG N 108:42-43, June 6, 1966.

"Sound and student behavior," by C. E. Schiller, et al. AV INSTR 14:92, March, 1969.

ELECTRIC SHOCK

"Aural trauma caused by electric shock by lightning," by T. Kobayashi. OTOLARYNGOLOGY 40:525-529, July, 1968.

"Damage of the ear by thunderbolt," by A. Spirov. VOJNOSANIT

ELECTRIC SHOCK

PREGL 25:648-651, December, 1968.

"Relative aversiveness of noise and shock," by B. A. Campbell, et al. J COMP PHYSIOL PSYCHOL 60:440-442, December, 1965.

ELECTRICAL SYSTEMS

"Reducing acoustic noise in electrical systems," by D. P. Costa. AUTOMATION 16:82-88, May, 1969.

ELEVATOR NOISE

"Elevator noise; architectural and mechanical considerations," by J. E. Sieffert. ARCH REC 143:199, April, 1968.

EMPLOYEES

"Contribution to the investigations on the influence of noise on hearing and the circulatory system in ship-yard employees," by S. Klajman, et al. MED PRACY 16:380-384, 1965.

"Office noise and employee morale," ADM MGT 26:48-49, March, 1965.

"Tune in on your employees' hearing problems," MOD MANUF 3:88, October, 1970.

ENGINES

"New engine silencer can be retracted," PRODUCT ENG 38:57, September 25, 1967.

"RB-211 engine given outdoor noise tests," AVIATION W 90:115, April 14, 1969.

ENGINES: AUTOMOBILE

see: Automobile Noise

ENGINES: DIESEL

"Acoustic trauma caused by diesel engine noise," by K. Ogata, et al. OTOLARYNGOLOGY 38:279-288, March, 1966.

"Diesel engines don't have to be so noisy," by A. E. W. Austen, et al. SAE J 73:71-75, May, 1965.

ENGINES: DIESEL

"Hearing disturbance due to diesel engine noises," by H. Ogata. OTOLARYNGOLOGY 38:279-288, March, 1966.

"Hearing-loss trend curves and the damage-risk criterion in diesel-engineroom personnel," by H. D. Harris. J ACOUST SOC AMER 37:444-452, March, 1965.

"Identification of mechanical sources of noise in a Diesel engine; sound emitted from the valve mechanism," by B. J. Fielding, et al. INST MECH ENG PROC 181,19:437-446; Discussion 447-450; Reply 451, 1966-1967.

"Light diesels give promise in mail service," by G. C. Nield. SAE J 74:76-79, July, 1966.

"Medium speed diesel engine noise," by R. Bertodo, et al. INST MECH ENG PROC 183,2:129-138 , 1968-1969.

"Quieter diesels at CAV," ENGINEERING 204:872-873, December 1, 1967.

"Reducing diesel engine noise," by C. Gray. ENGINEERING 209: 237, March 6, 1970.

"Reducing the noise from diesel engines," EINGNEERING 696, November 18.

"Reduction of the noise of railway traffic and of rheostat tests of Diesel locomotives," by E. V. Bobin. GIG SANIT 34:94-97, January, 1969.

ENGINES: DUCTS

"Acoustic-lining concepts and materials for engine ducts," by R. A. Mangiarotty. ACOUSTICAL SOC AM J 48:783-794, pt 3, September, 1970.

"Dynamic ratings for duct silencers," ENGINEER 223:120, January 20, 1967.

ENVIRONMENTAL HEALTH

"The Sixth AMA Congress on Environmental Health, Chicago,

April 28-29, 1969. Welcoming remarks," by G. D. Dorman. ARCH ENVIRON HEALTH 20:610-611, May, 1970.

ETHANOL

"Ethanol inhibition of audiogenic stress induced cardiac hypertrophy," by W. F. Geber, et al. EXPERIENTIA 23:734-736, September 15, 1967.

EXHAUST SYSTEMS

"Reducing noise in plant exhaust systems," by L. Klein, et al. PLANT ENG 19:154-157, October, 1965.

FAA

see: Aircraft Noise

FAN NOISE

"Accuracy consideration in fan sound measurement," by P. K. Baade. ASHRAE J 9:94-102, January, 1967.

"Application of sound-rated fans and ventilators," by H. R. Bohanon. ASHRAE J 9:73-77, August, 1967.

"Blower whine silenced by tuning inlet and outlet," PRODUCT ENG 37:59+, May 23, 1966.

"Centrifugal fan sound power level prediction," by G. C. Groff, et al. ASHRAE J 9:71-77, October, 1967.

"Control of centrifugal fan noise in industrial applications," by F. Oran, et al. AIR COND HEAT & VEN 62:85-96, May, 1965.

"Damping the noise of 6500-12-1 blowers (smoke exhaust fans)," by B. Z. Shushkovskii. GIG TR PROF ZABOL 11:47-48, September, 1967.

"Discrete frequency noise generation from an axial flow fan blade row," by R. Mani. J BASIC ENG 92:37-43, March, 1970.

"Doughnuts of foam absorb fan noise," PRODUCT ENG 38:52-53, July 3, 1967.

FAN NOISE

"Fan convector noise," ENGINEERING 207:720-721, May 9, 1969.

"High bypass ratio fan noise research test vehicle," by C. A. Warden. J AIRCRAFT 7:437-441, September, 1970.

"Keep that fan quiet," by C. J. Trickler. PLANT ENG 21:139-141, June, 1967.

"Measuring fan noise in the lab and in the field," by J. B. Graham. HEATING-PIPING 38:150-157, June, 1966.

"Method of estimating the sound power level of fans," by J. B. Graham. ASHRAE J 8:71-74, December, 1966.

"New concept; fan/silencer package," by C. J. Trickler. AIR COND HEAT & VEN 62:76-80, December, 1965.

"New fan law for sound," by R. Parker. ASHRAE J 9:83-85, October, 1967.

"Noise generation in roots type blowers," by R. N. Arnold, et al. INST MECH ENG PROC 178,3:202-208 , pt 3.

"Quieter turbofans could hit cost snag," by M. L. Yaffee. AVIATION W 91:31+, October 27, 1969.

"Silencing fan noise in underground garages," by K. A. Traub. AIR COND HEAT & VEN 66:47-50, May, 1969.

"Simplify your calculations for quiet fan systems," by C. J. Trickler. AIR COND HEAT & VEN 64:69-76, January, 1967.

"Turbofan-engine noise suppression," by R. E. Pendley, et al. J AIRCRAFT 5:215-220, May, 1968.

FERTILITY

"Fertility in couples working in noisy factories," by L. Carosi, et al. FOLIA MED 51:264-268, April, 1968.

FETUS

"Effects of intense noise during fetal life upon postnatal adaptability (statistical study of the reactions of babies to aircraft noise)," by Y. Ando, et al. J ACOUST SOC AMER 47:1128-1130, April, 1970.

FROGMEN
see: Barotrauma

FURNANCE NOISE
see also: Combustion Noise

"Engineering control of furnance noise," by S. H. Judd. AMER INDUSTR HYG ASS J 30:35-40, January-February, 1969.

"Fundamentals of fuel oil firing; how to eliminate burner noise," by A. A. Brahams. HEATING-PIPING 37:134-136, April, 1965.

"Quieting the gas flame," MECH ENG 89:49, July, 1967.

"Reduction of noise from warm air furnace blowers," by A. C. Potter. ASHRAE J 7:112-116, January, 1965.

"Silencing flames," AM GAS ASSN MO 49:15, May, 1967.

GAMMA GLUTAMYL TRANSPEPTIDAS
"Effects of physical work and work under condition of noise and vibration on the human body. II. Behavior of aspartic and alanine aminotransferase, gamma glutamyl transpeptidas, lactic dehydrogenase and L-idotyl dehydrogenase activities," by J. Gregorczyk, et al. ACTA PHYSIOL POL 16:709-714, September-October, 1965.

GENERATOR NOISE
"Silent generator," by A. Scott. ENGINEERING 207:904, June 27, 1969.

GLANDS: ADRENAL
"Changes in the adrenal glands under the effect of noise," by V. P. Osintseva, et al. GIG SANIT 34:119-122, October, 1969.

GLOSSARIES
"Commonly-used terms and definitions for understanding noise regulation," ENVIRONMENTAL CONTROL MGT 139:26-27, January, 1970.

GLOSSARIES

"Glossary of noise control terms," ARCH REC 142:204, October, 1967.

"Noise rating, terminology, and usage," by D. H. Ball. AIR COND HEAT & REFRIG N 121:32, October 5; 26, October 26; 17, November 2; 10-12, November 16; 18+, December 14; 15, December 21, 1970.

GLYCOGEN METABOLISM

"Effect of the acoustic stimulation on energy metabolism of the inner ear, with special reference to glycogen metabolism," by K. Ogawa. J OTOLARYNG JAP 73:335-352, March, 1970.

"Glycogen in the inner ear after acoustic stimulation. A light and electron microscopic study," by D. Ishii, et al. ACTA OTOLARYNG 67:573-582, June, 1969.

"Regeneration of glycogen in the hair cells of the organ of Corti," by S. Chodynicki. FOLIA HISTOCHEM CYTOCHEM 3:211-216, 1965.

GRINDERS

"Grinders cut noise level by 50 percent," PLASTICS WORLD 23: 86, December, 1965.

HEAD INJURIES

"Studies on late development of deafness following skull injuries," by W. Wagemann, et al. HNO 13:256-264, September, 1965.

HEARING AIDS

"Can hearing aids damage hearing?" by C. Roberts. ACTA OTOLARYNG 69:123-125, January-February, 1970.

"The effect of hearing aid use on the residual hearing of children with sensorineural deafness," by J. H. Macrae, et al. ANN OTOL 74:408-419, June, 1965.

HEARING CONSERVATION

see also: Protection

"Conservation of hearing programmes in North America," by O. M. Drew. OCCUP HLTH 20:179+, July-August, 1968.

HEARING CONSERVATION

"Control of noise and hearing conservation," by B. B. Bauer. AUDIO ENG SOC J 17:450+, August, 1969.

"Hearing conservation in industry; an overview," by A. J. Murphy. AMER ASS INDUSTR NURSES J 16:15-16, May, 1968.

"Hearing conservation in noise," by W. van der Sandt. S AFR MED J 44:558-561, May 9, 1970.

"Long-term industrial hearing conservation results," by M. T. Summer, et al. ARCH OTOLARYNG 82:618-621, December, 1965.

"The identity of the nurse in an industrial hearing conservation program," by A. J. Murphy. OCCUP HLTH NURS 17:32+, May, 1969.

"Military noise induced hearing loss: problems in conservation programs," by C. T. Yarington, Jr. LARYNGOSCOPE 78:685-692, April, 1968.

"Noise and the conservation of hearing," by S. S. Keys. TRANS ASS INDUSTR MED OFFICERS 15:12-17, January, 1965.

"Noise and the conservation of hearing," by M. Robinson. RHODE ISLAND MED J 53:146-149, March, 1970.

"A survey of hearing conservation programs in representative aerospace industries. I. Prevalence of programs and monitoring audiometry," by W. F. Rintelmann, et al. AMER INDUSTR HYG ASS J 28:372-380, July-August, 1967.

"10 years of hearing conservation in the Royal Air Force," by G. Jacobs, et al. NEDERL MILIT GENEESK T 18:212-218, July, 1965.

HEARING FIELD

"The phantom hearing (acouphene) after amputation of auditory field by sound trauma," by P. Pazat, et al. REV OTONEUROOPHTAL 42:81-90, March, 1970.

"Recent activity in the noise and hearing field," by W. L. Baughn. ARCH ENVIRON HEALTH 12:474-479, April, 1966.

HEARING MEASUREMENT

"Absolute thresholds of human hearing," by E. R. Hermann, et al. AMER INDUSTR HYG ASS J 28:13-20, January-February, 1967.

"Binaural summation of thermal noises of equal and unequal power in each ear," by R. J. Irwin. AMER J PSYCHOL 78:57-65, March, 1965.

"Comparison between the sound threshold audiogram and a speech audiogram in subjects with hearing impairment due to noise," by G. Fabian. Z GES HYG 14:165-169, March, 1968.

"Ear as a measuring instrument," by H. Fletcher. AUDIO ENG SOC J 17:532-534, October, 1969.

"Effect of noise on the perception of sounds in subject with normal and pathological hearing," by S. G. Kristostur'ian. VESTN OTORINOLARING 27:3-9, July-August, 1965.

"Effects of noise on arousal level in auditory vigilance," by E. Gulian. ACTA PSYCHOL 33:381-393, 1970.

"Evaluation of the results of various hearing tests for noise deafness," by T. Kawabata. J OTOLARYNGOL JAP 73:1858-1873, December, 1970.

"Examination of hearing in workers exposed to constant noise," by I. Prvanov. SRPSKI ARH CELOK LEK 93:1041-1046, November, 1965.

"Experimental study of changes in auditory acuity levels by exposing to noise," by S. Funasaka, et al. J OTOLARYNG JAP 72:338-339, February 20, 1969.

"Hearing tests by means of audiometry using whispered and normal speech for the testing of subjects working in noise," by K. Szymczyk. OTOLARYNG POL 21:277-280, 1967.

"Hearing tests in industry," by M. M. Hipskind. INDUSTR MED SURG 36:393-402, June, 1967.

HEARING MEASUREMENT

"Importance of the color of the iris in the evaluation of resistence of hearing to fatigue," by G. Tota, et al. RIV OTONEUROOFTAL 43:183-192, May-June, 1967.

"Improvement of hearing ability by directional information," by M. Ebata, et al. J ACOUST SOC AMER 43:289-297, February, 1968.

"Influence of noise on hearing function," by D. Filipo, et al. BOLL MAL ORECCH 83:133-145, March-April, 1965.

"Noise intermodulation audiometry as a simple method for determination of the intermodulation behavior in normal hearing persons," by M. Hoke, et al. ARCH KLIN EXP OHR NAS KEHLKOPFHEILK 194:482-488, December 22, 1969.

HEATHROW JETS
see: Aircraft Noise

HELICOPTER NOISE

"Army quiet helicopter effort aimed at reduction in losses," AVIATION W 91:33, September 15, 1969.

"Correlation of vortex noise data from helicopter main rotors," by S. E. Widnall. J AIRCRAFT 6:279-281, May, 1969.

"Helicopter noise," by I. M. Davidson, et al. ROY AERONAUTICAL SOC J 69:325-336, May, 1965.

"Influences of research on building design. 2. Helicopters: a noise problem on the doorstep," by W. Allen. BUILDER 1046-1047, June 3.

"Time perception for helicopter vibration and noise patterns," by G. R. Hawkes, et al. J PSYCHOL 76:71-77, September, 1970.

HISTORY

"Control of noise from the viewpoint of history. Prehistoric periods, age of the ancient cultures etc.," by H. Wiethaup. ZBL ARBEITSMED 16:120-124, May, 1966.

HOARSENESS

"Approach to the objective diagnosis of hoarseness," by N. Isshiki, et al. FOLIA PHONIAT 18:393-400, 1966.

HOSPITAL NOISE

"Caring for the total patient. Noise in hospitals: its effect on the patient," by P. Haslam. NURS CLIN NORTH AM 5:715-724, December, 1970.

"Detecting and correcting noises in the hospital," by S. Sorenson, et al. HOSPITALS 42:74-80, November 1, 1968.

"Hospital cooling tower sound problem solved by using stainless fill," AIR COND HEAT & REFRIG N 108:28, May 30, 1966.

"Hospital noises disturb U.S. patients also," MOD HOSP 105:85, December, 1965.

"Hospital quietude," by Maheux. REV INFIRM ASSIST SOC 20:657, July-September, 1970.

"Hospital: silence," SA NURS J 36:5-6, July, 1969.

"How quiet is a private room?" by I. D. Snook, Jr. J PRACT NURS 17:33 passim, June, 1967.

"Is noise important in hospitals?" by T. W. Hurst. INT J NURS STUD 3:125-135, September, 1966.

"Lots of rest?" by C. K. Harper. NURS MIRROR 126:22, February 9, 1968.

"Noise disturbances by a hydrotherapeutic section of a hospital," by H. Wiethaup. MED KLIN 62:612-613, April 14, 1967.

"Noise, a hazard in the operating room," by D. Kane. AORN J 7:78-80, January, 1968.

"Noise in the hospital," by W. Schweisheimer. SCHWEST REV 4:29-31, October, 1966.

HOSPITAL NOISE

"Noise in the hospital. One of the main complaints of hospitalized patients," by W. Schweisheimer. MED KLIN 60:816-817, May 14, 1965.

"Noise in hospitals: Its effect on the patient," by P. Haslam. NURS CLIN N AM 5:715+, December, 1970.

"Noise in hospitals--a problem the purchasing agent can help eliminate," by I. H. Hunt. CANAD HOSP 45:80-81, October, 1968.

"The noise level in a childrens hospital and the wake-up threshold in infants," by R. Gadeke, et al. ACTA PAEDIAT SCAND 58: 164-170, March, 1969.

"Noise measurement in a children's hospital and the awaking noise threshold of infant," by R. Gadeke, et al. MSCHR KINDERHEILK 116:374-375, June, 1968.

"The problem of low intensity impulse noises in hospital," by R. Wojtowicz. PRZEGL LEK 25:255-258, 1969.

"The problem of noise in hospitals. Considerations on some studies done in hospitals in Naples," by R. DeCapoa. G IG MED PREV 7:124-132, April-June, 1966.

"The problem of noise in hospitals. I. Phonometric data in various environments and in various conditions," by G. Spaziante. RIV ITAL IG 26:468-511, September-December, 1966.

"The problem of noise in hospitals. II. Results of acoustic spectrum analysis for some noise sources in hospital environments," by G. Spaziante. RIV ITAL IG 29:32-48, January-April, 1970.

"On permissible levels of noise in hospitals," By V. I. Pal'gov, et al. VRACH DELO 11:119-124, November, 1965.

"Preferred loudness of recorded music of hospitalized psychiatric patients and hospital employees," by H. L. Bonny. J MUS THERAPY 5:44-52, 1968.

"Softly, softly--anti-noise campaign at Edinburgh Royal Infirmary and associated hospitals," NURS TIMES 61:401, March 19, 1965.

"Sound-proof unit at West Middlesex Hospital," NURS MIRROR 126: 30-31, February 16, 1968.

"Sound-proofing the surgery," by A. Rathbone. PRACTITIONER 6:19-24, March, 1968.

"A study of noise and its relationship to patient discomfort in the recovery room," by B. B. Minckley. NURS RES 17:247-250, May-June, 1968.

"3 recommendation inquiries to the Board of Health," by R. J. Kruisinga. T ZIEKENVERPL 22:220, March 1, 1969.

HOUSEHOLD NOISE

"Acoustical properties of carpets and drapes," by J. W. Simons, et al. HOSPITALS 43:125-127, July 16, 1969.

"Field measurement study of the sound levels produced outdoors by residential air-conditioning equipment," by W. E. Blazier, Jr. ASHREE J 9:35-39, May, 1967.

"Give your house a tranquilizer," by M. A. Guitar. AM HOME 72: 48+, October, 1969.

"How to quiet down a noisy house," BET HOM & GARD 44:120, March, 1966.

"Medical consequences of environmental home noises," by L. E. Farr. JAMA 202:171-174, October 16, 1967.

"New ideas for noise control at home," by A. Lees. POP SCI 197: 94-96, September, 1970.

"Noise as a health hazard at work, in the community, and in the home," by H. H. Jones, et al. PUBLIC HEALTH REP 83:533-536, July, 1968.

"Quiet; San Antonio parade shows advantages sound conditioned homes can offer," AIR COND HEAT & REFRIG N 105:6-7,

August 23, 1965.

"Sound absorption of draperies," by J. H. Batchelder, et al. ACOUSTICAL SOC AM J 42:573-575, September, 1967.

"Sound absorptive properties of carpeting," by M. J. Kodaras. INTERIORS 128:130-131, June, 1969.

"Study of sound levels in houses," by J. J. Mize, et al. J HOME ECON 58:41-45, January, 1966.

"What ever became of the big noise about the quiet home?" HOUSE & HOME 30:88-91, December, 1966.

HUNTERS

"Acoustic trauma in the sports hunter," by G. D. Taylor, et al. LARYNGOSCOPE 76:863-879, May, 1966.

HYDRAULIC ENGINEERING

"Fluid power engineers take aim at pollution," PRODUCT ENG 41:76, October 12, 1970.

HYGIENIC STUDIES

"Flight noise; measurements on airplanes used in competition and hygienic aspects," by H. G. Dermus, et al. Z GES HYG 11:1-24, January, 1965.

"Focal points in research activities in regional universities, from the viewpoint of hygiene," by N. Saruta. JAP J HYG 23:31-34, April, 1968.

"Physiological-hygienic assessment of pulsed noise," by E. Ts. Andreeva-Galanina, et al. GIG TR PROF ZABOL 13:8-12, September, 1969.

"Public health hazards and vertigo," by T. Naito. J OTOLARYNG JAP 72:638-643, February 20, 1969.

"A review of hearing damage risk criteria," by W. I. Acton. ANN OCCUP HYG 10:143-153, April, 1967.

HYGIENIC STUDIES

"The significance of noise to health," by W. Klosterkotter. ZBL BAKT 212:336-353, 1970.

"Some clinico-physiological studies of workers subjected to the effect of constant noise," by G. Z. Dumkina. GIG TR PROF ZABOL 10:23-27, December, 1966.

"Temporary and permanent hearing loss. A ten-year follow-up," by J. Sataloff, et al. ARCH ENVIRON HEALTH 10:67-70, January, 1965.

"A test for susceptibility to noise-induced hearing loss," by P. E. Smith, Jr. AMER INDUSTR HYG ASS J 30:245-250, May-June, 1969.

"3 recommendation inquiries to the Board of Health," by R. J. Kruisinga. T ZIEKENVERPL 22:220, March 1, 1969.

HYPACUSIS

"Certain aspects of hypacusia caused by acoustic trauma studies with Bekesy's audiometer," by M. Maurizi, et al. ARCH ITAL LARING 76:131-146, May-June, 1968.

"Clinical and medico-legal considerations on hypoacusias of metallurgic workers," by M. Ciulla, et al. BOLL SOC MEDICOCHIR CREMONA 19:129-141, January-December, 1965.

"Experience with the dispensary treatment of occupational hypacusia," by K. Huber. CESK OTOLARYNG 15:173-175, June, 1966.

"The noise from the high speed turbine as a cause of hypacusis," by G. Girardi, et al. RASS INT STOMAT PRAT 17:405-415, November-December, 1966.

"On the possibility of early detection of perceptive hypoacusia in workers exposed to noise," by S. Kossowski, et al. MED PRACT 17:252-253, 1966.

HYPOXIA

"The effect of noise on the resistance to acute hypoxia," by J. Davidovic, et al. VOJNOSANIT PREGL 22:625-627, October, 1965.

IMPACT NOISE

"The effect of impact noise with various background noises," by E. P. Oriovskaia. GIG SANIT 32:21-25, September, 1967.

"Guide to airborne, impact and structure borne noise: control in multifamily dwellings," by R. D. Berendt, et al. HEATING-PIPING 41:147, April, 1969.

"Hazardous exposure to intermittent and steady-state noise," by K. D. Kryter, et al. J ACOUST SOC AMER 39:451-464, March, 1966.

"Hazards and hurdles in developing standards: a case history - rating the impact-noise resistance of floors," by R. E. Fischer. ARCH REC 147:147-150, May, 1970.

"Impact," by M. Ragon. WORLD HEALTH 19:26+, February-March, 1966.

"Impact noise analyser," ENGINEER 220:433, September 10, 1965.

"Impact-noise rating of various floors," by T .Mariner, et al. ACOUSTICAL SOC AM J 41:206-214, January, 1967.

IMPULSE NOISE

"The biological action of impulse noises (a review of the literature)," by V. N. Morozov, et al. VOENNOMED ZH 8:53-58, August, 1969.

"Data for hygienic assessment of impulse noise," by E. Ts. Andreeva-Galanina, et al. GIG SANIT 33:24-29, August, 1968.

"The effect of impulse noise on man," by A. A. Arkad'evskii, et al. GIG SANIT 31:29-33, May, 1966.

"Effect of loud impulse noise on the hearing organ of animals," by N. I. Ivanov. VOEN MED ZH 7:24-27, July, 1970.

"The effect of noise impulses on the organism as a function of the frequency of sequence and duration of impulse," by G. A. Suvorov. GIG TR PROF ZABOL 13:4-8, September, 1969.

IMPULSE NOISE

"Evaluation of exposures to impulse noise," by K. D. Kryter. ARCH ENVIRON HEALTH 20:624-635, May, 1970.

"Experiments on classification of peak and instantaneous value with impulse-rich work noise," by G. Vorwerk, et al. ARCH KLIN EXP OHR NAS KEHLKOPFHEILK 193:259-276, 1969.

"Generating specified shock pulses," by R. O. Brooks. J ENVIRONMENTAL SCI 10:28-33, April, 1967.

"Hazardous exposure to impulse noise," by R. R. Coles, et al. J ACOUST SOC AMER 43:336-343, February, 1968.

"Hazards from impulse noise," by R. R. Coles, et al. ANN OCCUP HYG 10:381-388, October, 1967.

"Hearing disorders due to impulsive noise," by J. Kuzniarz, et al. OTOLARYNG POL 22:781-785, 1968.

"Impulse duration and temporary threshold shift," by M. Loeb, et al. ACOSUTICAL SOC AM J 44:1524-1528, December, 1968.

"Impulse noise and neurosensory hearing loss. Relationship to small arms fire," by R. J. Keim. CALIF MED 113:16-19, September, 1970.

"Loudness sensitivity, measurement and trauma caused by impulse noise I.," by H. Niese. Z LARYNG RHINOL OTOL 44:209-217, April, 1965.

"Loudness sensitivity, measurement and trauma caused by impulse noises. 2.," by H. Weissing. Z LARYNG RHINOL OTOL 44: 217:223, April, 1965.

"Masking of speech by means of impulse noise," by J. Kuzniarz. OTOLARYNG POL 22:421-425, 1968.

"On gauge level frequency measurements in impulse-increased industrial noise," by H. G. Dieroff, et al. Z LARYNG RHINOL OTOL 44:639-648, October, 1965.

IMPULSE NOISE

"The problem of low intensity impulse noises in hospital," by R. Wojtowicz. PRZEGL LEK 25:255-258, 1969.

"Reliability of TTS from impulse-noise exposure," by D. C. Hodge, et al. J ACOUST SOC AMER 40:839-846, October, 1966.

"The status of hearing in the chronic action of impulse noise on the organism," by N. I. Ivanov. VESTN OTORINOLARING 29:73-78, March-April, 1967.

"Temporary threshold shift from impulse noise," by J. G. Walker. ANN OCCUP HYG 13:51-58, January, 1970.

"Towards a criterion for impulse noise in industry," by R. R. Coles, et al. ANN OCCUP HYG 13:43-50, January, 1970.

"Validation of the single-impulse correction factor of the CHABA impulse-noise damage-rick criterion," by D. C. Hodge, et al. J ACOUST SOC AM 48:Suppl 2:1429+, December, 1970.

INDUSTRIAL HYGIENE

"Noise as an industrial hazard," by H. H. Botsford. WATER & SEWAGE WORKS 116:194-196, November 28, 1969.

"Preventive measures for maintaining the health of persons working under noisy conditions," by M. Kvaas, et al. GIG TR PROF ZABOL 10:56-58, June, 1966.

"Problems of industrial hygiene in vulcanizing processes in rubber production," by Z. A. Volkova, et al. GIG SANIT 34:33-40, September, 1969.

INDUSTRIAL NOISE

"Acoustical enclosures muffle plant noise," by S. Wasserman, et al. PLANT ENG 19:112-115, January, 1965.

"Air pollution control: industrial noise control," by L. L. Beranek. CHEM ENG 77:227-230, April 27, 1970.

"Application of integrated circuits to industrial control systems with high-noise environments," by A. Wavre. IEEE TRANS IND

& GEN APPLICATIONS 5:278-281, May, 1969.

"Auditory and subjective effects of airborne noise from industrial ultrasonic sources," by W. I. Acton, et al. BRIT J INDUSTR MED 24:297-304, October, 1967.

"Case findings in hearing disorders in industry: the 'audiometric car;' 1st results," by J. C. Lafon, et al. J MED LYON 49: 1831-1834, November 20, 1968.

"Community noise--the industrial aspect," by K. M. Morse. AMER INDUSTR HYG ASS J 29:368-380, July-August, 1968.

"Concrete breakers must be quieter," ENGINEERING 203:1050, June 30, 1967.

"Considerations on the problem of noise in relation to its influence on some aspects of modern life. I. Behavior of the temporary shift of hearing sensitivity due to high energy level stimulations in persons employed in work with noise," by S. Collatina, et al. CLIN OTORINOLARING 17:357-370, July-August, 1965.

"Contribution to the evaluation of industiral noise," by J. Mayer. MSCHR OHRENHEILK 101:462-467, 1967.

"Contribution to the investigations on the influence of noise on hearing and the circulatory system in ship-yard employees," by S. Klajman, et al. MED PRACY 16:380-384, 1965.

"Control of centrifugal fan noise in industrial applications," by F. Oran, et al. AIR COND HEAT & VEN 62:85-96, May, 1965.

"Control of industrial noise, through personal protection," by R. R. Coles. AM SOC SAFETY ENG J 14:10-15, October, 1969.

"Controlling plant noise," by J. M. Handley. MACH 75:81-84, May, 1969.

"Controlling process-plant noise," by J. K. Floyd. MECH ENG 90: 23-26, October, 1968.

INDUSTRIAL NOISE

"Damage due to industrial noise," by H. Guttich. MUNCHEN MED WSCHR 107:1397-1406, July 16, 1965.

"Detection of hearing impairment in a large business in the Paris area. Initial results," by P. Housset, et al. ARCH MAL PROF 27:710-713, September, 1966.

"Drilling by ear," by J. Kuletz. AM MACH 110:63-65, July 4, 1966.

"Ear injuries and industry," by F. Montreuil. UN MED CANADA 95:1299-1306, November, 1966.

"Effect of loud industrial noise on hearing," by P. V. Kovalev. VOEN MED ZH 7:22-24, July, 1970.

"Effect of the noise level on the productivity of work," by S. D. Kovrigina, et al. GIG SANIT 30:28-32, April, 1965.

"Effect of noise on hearing observed in some small industrial plants," by I. P. Su. J FORMOSA MED ASS 64:171-175, March 28, 1965.

"The effect of noise on the organism and its control in industry," by R. I. Vorob'ev. FELDSH AKUSH 34:6-7, November, 1969.

"The effect of stable high frequency industrial noise on certain physiological functions of adolescents," by I. I. Ponomarenko. GIG SANIT 31:29-33, February, 1966.

"The effect of vibration and noise on protein metabolism of excavator-machinists using several biochemical indices," by A. M. Tambovtseva. GIG SANIT 33:58-61, January, 1968.

"Effects of noise measurements in the factories of Katowice Province," by J. Grzesik, et al. MED PRACY 16:489-496, 1965.

"Engineering can control industrial noise," by J. H. Botsford. AM SOC SAFETY ENG J 14:7-9, March, 1969.

"An epidemiologic approach to in-plant noise problems," by P. R.

Ebling, et al. INDUSTR MED SURG 34:508-512, June, 1965.

"Evaluation of the risks of hearing impairment due to industrial noise based on exposure parameters," by J. Grzesik. POL TYG LEK 25:1026-1028, July 6, 1970.

"Experience with creating an acoustic complex for studying the effect of industrial noise on the body," by E. Ts. Andreeva-Galanina, et al. GIG SANIT 30:44-47, January, 1965.

"Experimental-microscopic study on the problem of localization of industrial noise-conditioned hearing fatigue and of the later resulting permanent acoustic trauma," by H. G. Dieroff, et al. ARCH KLIN EXP OHR NAS KEHLKOPFHEILK 186:1-8, 1966.

"Experimental studies on the harmfulness of aircraft noise as opposed to industrial noise," by W. Lorenz. MSCHR OHREN-HEILK 103:492-498, 1969.

"Fertility in couples working in noisy factories," by L. Carosi, et al. FOLIA MED 51:264-268, April, 1968.

"Fitness for work of employees with hearing defects in noisy work places," by Z. Novotny. PRAC LEK 17:63-67, March, 1965.

"From experience in controlling noise and vibration in industrial plants in Kiev," by I. G. Guslits, et al. GIG TR PROF ZABOL 10:52-54, June, 1966.

"A further study on the temporary effect of industrial noise on the hearing of stapedectomized ears at 4,000 c.p.s.," by K. Ferris. J LARYNG 81:613-617, June, 1967.

"Government plans enforcement of industrial noise standards," AUTOMATION 16:18+, November, 1969.

"Health problems due to industrial noise. An environmental survey in three large plants," by M. el Batawi. J EGYPT PUBLIC HEALTH ASS 40:131-140, 1965.

"Hearing conservation in industry; an overview," by A. J. Murphy.

AMER ASS INDUSTR NURSES J 16:15-16, May, 1968.

"Hearing protection in industry," by R. E. Scott. SAFETY MAINT 129:40-41, January, 1965.

"Hirschorn says noise-control standards have industrial impact," AIR COND HEAT & REFRIG N 117:6, August 11, 1969.

"How to control industrial noise," by A. M. Teplitzky. AUTOMATION 17:70-74, March, 1970.

"How to estimate plant noises," by I. Heitner. HYDROCARBON PROCESS 47:67-74, December, 1968.

"How to plan a small plant hearing protection program," by R. J. Beaman. SAFETY MAINT 130:36-38, December, 1965.

"Hygeinic evlauation of manual pneumatic perforators," by N. M. Paran'ko. VRACH DELO 1:94-98, January, 1965.

"Hygienic evaluation of noise in a shop with curved designs," by N. M. Paran'ko, et al. GIG SANIT 31:81-82, May, 1966.

"The identity of the nurse in an industrial hearing conservation program," by A. J. Murphy. OCCUP HLTH NURS 17:32+, May, 1969.

"Incidence of noise as a cause of nuisance in an industrial city," by M. Braja, et al. IG MOD 60:26-36, January-February, 1967.

"Individual protection against damaging effect of industrial noise," by M. Prazic, et al. LIJECN VJESN 87:409-418, April, 1965.

"Industrial acoustics to amplify earnings with greater capacity," BARRONS 49:27, November 17, 1969.

"Industrial deafness," LAMP 25:13-14, November, 1968.

"Industrial deafness and the summed evoked potential," by T. G. Heron. S AFR MED J 42:1176-1177, November 9, 1968.

INDUSTRIAL NOISE

"Industrial hearing loss. Conservation and compensation aspects," by M. S. Fox. OCCUP HEALTH NURS 17:18-24, May, 1969.

"Industrial hygienic remarks on noise at the place of work," by W. Massmann, et al. ZBL ARBEITSMED 16:124-127, May, 1966.

"Industrial medicine and industrial physiology in the tropics," by J. Haas. MED KLIN 63:1001-1004, June 21, 1968.

"Industrial noise. Medicolegal considerations in prevention," by B. Testa, et al. FOLIA MED 52:311-317, May 5, 1969.

"Industrial noise and its control," by J. M. Handley. PLANT ENG 24:56-57, March 19, 1970.

"Industrial noise control is practical," by P. H. Hutton. AMER INDUSTR HYG ASS J 29:499-503, September-October, 1968.

"Industrial noise problems and solutions," by J. H. Botsford. MACHINE DESIGN 42:130+, August 20, 1970.

"Industrial noise; workers lose hearing," CHEM & ENG N 46:22, November 11, 1968.

"Industrial noises--hearing damage. Is prevention possible?" by B. Ingberg. SOCIALMED T 42:379-380, November, 1965.

"Industrial sudden deafness," by S. Kawata, et al. ANN OTOL 76: 895-902, October, 1967.

"Influence on auditory acuity of continuous industrial noise reaching 90 plus-or-minus 5 decibels within the range of an octave where it culminates," by P. Martinet. ARCH MAL PROF 30:323-335, June, 1969.

"The intermittent action of high frequency noise in industry," by V. V. Lipovoi. GIG TR PROF ZABOL 13:19-21, February, 1969.

"Investigations into noise and its effect on employees carried out in a manufacturing plant," by D. M. Cracknell. OCCUP HEALTH NURSE 20:184-193, July-August, 1968.

INDUSTRIAL NOISE

"Is industry facing up to its noise problems?" by R. K. Anderson. SAFETY MAINT 135:43-44, January, 1968.

"Knotty problem of industrial hearing loss," by L. W. Larson. SAFETY MAINT 135:39-42, May, 1968.

"Late development of acoustic trauma (studies on 500 workers in a noisy environment. II.," by W. Wagemann. MSCHR UNFALHEILK 69:23-37, January, 1966.

"Lighting, noise and floors in factories," ENGINEERING 201:530, March 18, 1966.

"Long-term industrial hearing conservation results," by M. T. Summer, et al. ARCH OTOLARYNG 82:618-621, December, 1965.

"The loudness determination by workers in noisy surroundings. On problems of getting accustomed to noise," by G. Linke, et al. Z LARYNG RHINOL OTOL 47:53-57, January, 1968.

"Medical aspects of industrial noise problem," by M. S. Fox. ENVIRONMENTAL CONTROL MGT 139:22-25+, January, 1970; also in INDUSTR MED SURG 39:241-244, June, 1970; NAT SAFETY CONGR TRANS 18:9-14, 1969.

"Minimizing danger in handling systems; noise control, the new necessity," MOD MATERIALS HANDLING 25:65, January, 1970.

"The modern industrial society, its illnesses and the maintaining of its health. 3. The effect of noise on the organ of hearing," by W. Wagemann. HIPPOKRATES 37:138-142, February 28, 1966,

"Noise and hearing in industry," by N. Williams. CANAD J PUBLIC HEALTH 58:514-517, November, 1967.

"Noise and its control in process plants," by B. G. Lacey. CHEM ENG 76:74-84, June 16, 1969.

"Noise as an environmental factor in industry," by W. Burns. TRANS ASS INDUSTR MED OFFICERS 15:2-11, January, 1965.

INDUSTRIAL NOISE

''Noise at the work place and its significance for human health,'' by G. Jansen. THERAPIEWOCHE 15:665-667, July, 1965.

''Noise control in chemical processing,'' by W. V. Richings. CHEM & PROCESS ENG 48:77-80, October; 66-68, November, 1967; 49:77-79+, January, 1968.

''Noise in industry,'' by S. J. Evans, et al. CHEM INDUSTR 9:275-281, March 2, 1968.

''Noise in industry,'' by M. M. Mackay. VIRGINIA MED MONTHLY 94:288-292, May, 1967.

''Noise levels in industry,'' by J. Morin. ARCH MAL PROF 28:517-522, June, 1967.

''Noise; modern industrial dilemma,'' by A. Teplitsky. SCI & TECH 36-39, October, 1969.

''Noise: new pressure on plants,'' CHEM W 105:17, December 3, 1969.

''Noise pollution in the molding room; what you should know about it,'' by A. R. Morse. PLASTICS TECH 15:51-54, July, 1969.

''Noise problems in industry,'' by D. C. Murphy. ANN OCCUP HYG 9:149-163, July, 1966.

''Noise program saves ears at Union Camp,'' PULP & PA 40:46, June 6, 1966.

''Noise protection in the industry,'' by H. Schmidt. MUNCH MED WOCHENSCHR 112:Suppl 52:3, December 25, 1970.

''Noise specification for industrial plant,'' by J. B. Erskine. ANN OCCUP HYG 10:407-414, October, 1967.

''Noise survey helps Japanese plant be a good neighbor,'' MOD MANUF 3:70-71, April, 1970.

''Noise transmission through shafts,'' by W. Sorge. Z GES HYG 13: 634-636, August, 1967.

INDUSTRIAL NOISE

"Observation of the effect of noise on the general health of workers in heavy industry; attempt at evaluation," by H. Jirkova, et al. PRAC LEK 17:147-148, May, 1965.

"Occupational deafness in theromelectric plants with Norberg methane motors," by N. Lo Martire, et al. ATTI ACCAD FISIOCR SIENA 16:202-212, 1967.

"Occupational deafness in young workers in a noisy environment with ear protection," by Y. Harada. ARCH ITAL OTOL 77:157-165, January-February, 1966.

"Occupational disease potentials in the heavy equipment operator," by F. Ottoboni, et al. ARCH ENVIRON HEALTH 15:317-321, September, 1967.

"Occupational hearing loss in relation to industrial noise exposure," by O. el-Attar. J EGYPT MED ASS 51:183-192, 1968.

"On changes in lipid metabolism in persons during prolonged action of industrial noise on the central nervous system," by P. S. Khomulo, et al. KARDIOLOGIIA 7:35-38, July, 1967.

"On the effect of industrial noise on the blood pressure level in workers in machine building plants," by N. N. Pokrovskii. GIG TR PROF ZABOL 10:44-46, December, 1966.

"On the effect of industrial noise on the functional status of the auditory analyzer in adolescents," by L. L. Kovaleva. GIG SANIT 32:52-56, January, 1967.

"On the effect of low frequency ultrasonic waves and high frequency sound waves on the organism of workers," by V. K. Dobroserdov. GIG SANIT 32:17-21, February, 1967.

"On the extent of hearing damage in workers exposed to noise," by F. Schwetz. MSCHR OHRENHEILK 102:663-668, 1968.

"On gauge level frequency measurements in impulse-increased industrial noise," by H. G. Dieroff, et al. Z LARYNG RHINOL OTOL 44:639-648, October, 1965.

INDUSTRIAL NOISE

"On the hygienic assessment of high-frequency intermittent occupational noise," by A. Z. Mariniako. GIG TR PROF ZABOL 10: 18-22, March, 1966.

"On noise evaluation at the working site in relation to hearing defects," by G. Wolff. ZBL ARBEITSMED 17:349-355, November 16, 1967.

"On the noise factor in some industrial plants (light and heavy industry) in Armenia," by Z. V. Babaian. GIG SANIT 30:98-99, June, 1965.

"On the pneumatization of the mastoid process in workers due to noise," by D. Kosa, et al. HNO 15:324-325, November, 1967.

"On the possibility of early detection of perceptive hypoacusia in workers exposed to noise," by S. Kossowski, et al. MED PRACY 17:252-253, 1966.

"On the problem of assessing the hearing and expert opinion on the work capacity of persons exposed to industrial noise," by N. I. Ponomareva, et al. GIG TR PROF ZABOL 11:50-53, October, 1967.

"On the problem of noise-induced lesions of the vestibular apparatus in the evaluation of workers in a noisy environment," by H. G. Dieroff, et al. Z LARYNG RHINOL OTOL 46:746-757, October, 1967.

"On the prolonged action of permissable parameters of noise on the hearing of workers," by L. I. Maksimova, et al. GIG SANIT 31: 11-16, February, 1966.

"On some hemodynamic changes due to the effects of industrial noise," by N. N. Shatalov. GIG TR PROF ZABOL 9:3-7, June, 1965.

"On standards of industrial noise," by A. P. Pronin. GIG SANIT 30:94-97, November, 1965.

"On the temporary effect of industrial noise on the hearing at

4,000 c/s of stapedectomized ears," by K. Ferris. J LARYNG 79:881-887, October, 1965.

"On the use of maximal value accumulator for the threshold frequency measurement in intermittent work noise," by H. G. Dieroff, et al. Z LARYNG RHINOL OTOL 47:58-63, January, 1968.

"Otology in industrial medicine," by R. L. Waston, Jr. INDUSTR MED SURG 36:731-734, November, 1967.

"Otosclerosis in workers in a noisy environment," by B. Gerth. HNO 14:205-208, July, 1966.

"Plant noise--the disturbing decibel dilemma," FACTORY 125:71-76, November, 1967.

"Preventing hearing loss in industry," by V. Hamilton. CANAD NURSE 66:37+, September, 1970.

"Prevention of deafness from industrial noise and acoustic trauma," by L. W. Benoay. J AMER OSTEOPATH ASS 68:161-167, October, 1968.

"The progression of hearing loss from industrial noise exposures," by E. J. Schneider, et al. AMER INDUSTR HYG ASS J 31:368-376, May-June, 1970.

"The psychological difficulties of control of traumatic hearing disorders due to industrial noise," by D. Hogger. INT ARCH GEWERBEPATH 22:306-314, August 17, 1966.

"Report on repeated audiometric examinations of industrial workers after having been exposed to harmful noise for a period of 5 years," by S. Podvinec, et al. J FRANC OTORHINOLARYNG 15:53-60, January-February, 1966.

"Research on changes in the auditory capacity in a noisy industrial environment," by J. Morin. ARCH MAL PROF 26:252-256, April-May, 1965.

INDUSTRIAL NOISE

"Research on industrial noise," by M. M. Woodhead. MANAGER 378-380, May.

"Results of some audiometric examinations performed in an iron and steel plant," by G. Sparacio, et al. G IG MED PREV 9:90-98, January-March, 1968.

"Review of noise in plants of the West Bohemian region," by J. Srutek. PRAC LEK 17:61-63, March, 1965.

"Role of sound-suppressing curtains in the control of industrial noise pollution," by L. Singer. WIRE & WIRE PROD 45:111-113, October, 1970.

"A serial study of noise exposure and hearing loss in a group of small and medium size factories," by D. E. Hickish, et al. ANN OCCUP HYG 9:113-133, July, 1966.

"Should your plant go underground," by C. E. Petak. SAFETY MAINT 132:46-49+, September, 1966.

"Standardization of high frequency industrial noise for adolescents," by I. I. Ponomarenko. GIG SANIT 33:34-38, August, 1968.

"State and evaluation of hearing in persons working under conditions of intense production of noise and vibration," by V. E. Ostapkovich, et al. KLIN MED 48:79-83, March, 1970.

"Statistical theory of extremal systems with dependent search under influence of multi-plicative and additive plant noise," by N. G. Gorelik. AUTOMATION & REMOTE CONTROL 1077-1083, July, 1967.

"The status of hearing and of arterial pressure in exposure to intense industrial noise," by N. N. Shatalov, et al. GIG TR PROF ZABOL 13:12-15, April, 1969.

"Studies on the action of the pulse after a given amount of industrial noise. Economic research program with the help of sequential analysis,I.," by A. Fuchs-Schmuck. INT Z ANGEW PHYSIOL 22:1-9, March 3, 1966.

INDUSTRIAL NOISE

"Studies on behavior of the pulse following definite industrial noise. Economical experimental design, using sequential analysis. II.," by A. Fuchs-Schmuck. INT Z ANGEW PHYSIOL 23: 345-353, March 7, 1967.

"Studies on the functional and perception efficiency in industrial workers working in noise," by B. Rogacka-Trawinska. POL TYG LEK 25:1023-1025, July 6, 1970.

"Subjective complaints in noise-deafness. (Studies on 500 workers in noisy environment. 8.," by W. Wagemann. MSCHR UNFAL-HEILK 68:546-553, December, 1965.

"A survey of hearing conservation programs in representative aerospace industries. I. Prevalence of programs and monitoring audiometry," by W. F. Rintelmann, et al. AMER INDUSTR HYG ASS J 28:372-380, July-August, 1967.

"Survey of industrial acoustic conditions," by J. Morin. ARCH MAL PROF 31:517-522, June, 1967.

"Systems approach key to reducing plant noise," PRODUCT ENG 40:102, December 1, 1969.

"Towards a criterion for impulse noise in industry," by R. R. Coles, et al. ANN OCCUP HYG 13:43-50, January, 1970.

"U.K. industry join in anti-noise research," by H. J. Coleman. AVIATION W 87:61+, August 14, 1967.

"U.S. Public Health Service field work on the industrial noise hearing loss problem," by A. Cohen. OCCUP HEALTH REV 17:3-10 passim, 1965.

"Variations of amylasuria and amylasemia under the effect of sound and vibratory 'stress' in industry," by S. Miclesco-Groholsky, et al. ARCH MAL PROF 31:277-282, June, 1970.

"Ways to reduce plant noises," by B. G. Golden. HYDROCARBON PROCESS 47:75-78, December, 1968.

INDUSTRIAL NOISE

"What can be done about plant noise," by H. S. Sharp. SUPERVISORY MGT 10:51-53, February, 1965.

"What to do about plant noise," by B. G. Golden. CHEM ENG PROG 65:47-52, April, 1969.

INDUSTRIAL NOISE: CHEMICAL PLANTS

"Hygienic assessment of the noise factor in the air dissociation workshop of a chemical plant," by M. I. Tsigel'nik, et al. GIG TR PROF ZABOL 13:43-45, October, 1969.

"Sound levels and development of audiometric technics in a chemical products firm," by Rety, et al. REV MED SUISSE ROM 88:59-63, February, 1968.

INDUSTRIAL NOISE: COMBUSTION

"Thermal efficiency, silencing and practicability of gas-fired industrial pulsating combustors," by D. Reay. COMBUSTION 41:16-24, January, 1970.

INDUSTRIAL NOISE: CONCRETE

"Noise and vibration in plants of prefabricated concrete structures," by Iu. K. Aleksandrovskii, et al. GIG SANIT 30:113-115, August, 1965.

"Noise disorders caused by a concrete plant," by H. Wiethaup. ZBL ARBEITSMED 16:128-130, May, 1966.

INDUSTRIAL NOISE: CONFECTIONERY

"The effect of some harmful factors of the confectionery industry on the organism of workers," by N. G. Prokof'eva, et al. GIG TR PROF ZABOL 11:47-49, December, 1967.

INDUSTRIAL NOISE: CONSTRUCTION

"The case of the unfortunate construction worker," by S. L. Shapiro. EYE EAR NOSE THROAT MON 49:383-386, August, 1970.

"The differential threshold of sound intensity perception in riveters," by N. Ia. Shalashov. ZH USHN NOS GORL BOLEZ 28:70-75, November-December, 1968.

INDUSTRIAL NOISE: CONSTRUCTION

"Hearing injuries among construction workers in Skaraborg County," NORD MED 83:445-446, April 2, 1970.

"Hearing injury in building workers," by S. Lindqvist. LAKARTIDNINGEN 67:4283-4292, September 16, 1970.

"A noise and hearing survey of earth-moving equipment operators," by P. LeBenz, et al. AMER INDUSTR HYG ASS J 28:117-128, March-April, 1967.

"On the hearing impairment of bolt drivers in the building industry," by A. Schurno, et al. Z GES HYG 14:161-165, March, 1968.

"On the status of meeting noise protection requirements in industrial housing construction," by W. Fasold. Z GES HYG 13:628-631, August, 1967.

"Steps to measure floor noises," by G. E. Marklew. ENGINEERING 674-675, May 13.

INDUSTRIAL NOISE: FLAX MILLS

"Hygienic assessment of working conditions at flax mills," by E. A. Krechkovskii. GIG TR PROF ZABOL 13:5-8, December, 1969.

INDUSTRIAL NOISE: FORGE SHOPS

"The effect of industrial noise on the organ of hearing in forging shop workers," by N. Ia. Shalashov. GIG TR PROF ZABOL 9: 41:43, July, 1965.

"Forge shop ventilation noise problem overcome," ENGINEER 225:956, June 21, 1968.

"Noise control in drop-forges," by W. Wiethaup. ZBL ARBEITSMED 16:227-228, August, 1966.

INDUSTRIAL NOISE: FOUNDRY

"Controlling foundry noise," by C. H. Borcherding, Jr. FOUNDRY 98:167-169, June, 1970.

"Environment control at Dayton foundry," by J. C. Miske. FOUNDRY 98:68-71, May, 1970.

INDUSTRIAL NOISE: FOUNDRY

"Experience with lowering noise of bar glazers in foundries," by J. Jerman, et al. PRAC LEK 17:245-248, August, 1965.

"Foundry noise and hearing in foundrymen," by G. R. Atherley, et al. ANN OCCUP HYG 10:255-261, July, 1967.

"Injury to hearing in foundrymen," by G. R. Atherley. OCCUP HEALTH REV 19:14-16, 1967.

"Need for inplant control," by P. S. Cowen. FOUNDRY 98:70-71, May, 1970.

"Plan before you biuld; planning a quiet foundry," by G. E. Warnaka. FOUNDRY 98:64-67, May, 1970.

"Quiet foundry," AM MACH 111;116, January 16, 1967.

"Some problems of improving working conditions in foundries," by V. N. Fomin. GIG TR PROF ZABOL 11:3-8, August, 1967.

INDUSTRIAL NOISE: GLASS BLOWERS

"Means of reducing industrial noise in the mechanical vat shops of glass container factories," by V. P. Goncharenko, et al. GIG TR PROF ZABOL 10:49-52, June, 1966.

"Noise-induced hearing loss in bench glass-blowers," by J. T. Sanderson, et al. ANN OCCUP HYG 10:135-141, April, 1967.

INDUSTRIAL NOISE: IRON WORKS

"Noise hazards in iron works," by V. Blaha, et al. PRAC LEK 17: 95-101, April, 1965.

"Noise in the iron and steel industry," by J. A. Adam, et al. IRON & STEEL INST J 205:701-713, July, 1967.

"Some notes on the provision of personal hearing protection for fettlers at an iron foundry," by D. B. Sugden. ANN OCCUP HYG 10:263-268, July, 1967.

INDUSTRIAL NOISE: KRASNODAR FACTORY

"Audiometric studies of hearing in workers in noisy shops of the

Krasnodar factory 'Traktorsel'khozzapchast'," by E. A. Melvnikova. GIG TR PROF ZABOL 10:34-38, July, 1966.

INDUSTRIAL NOISE: METAL CUTTING

"Hygienic importance of occupational factors in plasma-arc metal cutting," by A. V. Il'nitskaia. GIG TR PROF ZABOL 14:14-18, November, 1970.

INDUSTRIAL NOISE: METALLURGY

"Audiometric findings in a metallurgic industrial plant," by J. E. Fournier, et al. ARCH MAL PROF 28:523-529, June, 1967.

"Audiometric survey of a metallurgy enterprise," by J. E. Fournier, et al. ARCH MAL PROF 31:523-529, June, 1967.

"Clinical and medico-legal considerations on hypoacusias of metallurgic workers," by M. Ciulla, et al. BOLL SOC MEDICOCHIR CREMONA 19:129-141, January-December, 1965.

"New BISRA oxy-fuel burner test facility," METALLURGIA 74: 28, July, 1966.

"On the various degrees of stress of the hearing organ in the metal-processing industry," by H. G. Dieroff, et al. ARCH OHR NAS KEHLKOPFHEILK 185:485-488, 1965.

"Research on ear diseases caused by noise in workers in light metallurgical industry," by E. Manzo, et al. FOLIA MED 51: 720-731, September, 1968.

INDUSTRIAL NOISE: PAPER

"How Holmens Brukab reduced noise from a new paper machine," by P. Berg, et al. PAPER TR J 154:42-48, November 2, 1970.

"Noise exposures in pulp and paper production," by W. A. Ook, et al. AMER INDUSTR HYG ASS J 30:484-486, September-October, 1969.

"Solutions to the paper mill noise problem," by R. W. Gray. IND MED 35:257-258, April, 1966.

INDUSTRIAL NOISE: PLASTICS

"Certain problems of industrial hygiene in processing plastics by casting under pressure," by A. M. Dzhezhev. GIG SANIT 32: 19-22, March, 1967.

INDUSTRIAL NOISE: PLATE FIELD MILLS

"Elimination of Volta-potential noise from plate field mills," by K. Knott. R SCI INSTR 38:602-604, May, 1967.

INDUSTRIAL NOISE: PUNCH PRESSES

"Hygienic evaluation of the percussion noise of punch-presses," by E. P. Orlovskaia. VRACH DELO 5:89-92, May, 1966.

INDUSTRIAL NOISE: QUARRIES

"Dust- and soundproof stone-crushing plant protects workers and environment," PIT & QUARRY 62:134, February, 1970.

"Hygienic characteristisc of vibration and noise at work sites of crushing mills in ore enriching plants," by K. P. Antonova. GIG SANIT 34:116-118, November, 1969.

"An industrial hygiene study of flame cutting in a granite quarry," by W. A. Burgess, et al. AMER INDUSTR HYG ASS J 30:107-112, March-April, 1969.

"Maine's new dust-free crushed stone plant," by W. E. Trauffer. PIT & QUARRY 63:96-100, August, 1970.

INDUSTRIAL NOISE: REFINERY

"How noisy is a refinery?" by D. A. Tyler. HYDROCARBON PROCESS 48:173-174, July, 1969.

"Hygienic characteristics of working conditions in sugar refineries and ways in which they may be made healthier," by V. V. Paustovskaia, et al. GIG TR PROF ZABOL 14:44-45, May, 1970.

"Noise abatement in refineries," by R. C. Ewing. OIL & GAS J 67:83-88, October 13, 1969.

"Reduce those refinery noise levels," by D. A. Tyler. OIL & GAS J

67:140-141, July 28, 1969.

"Shushing refinery noises," CHEM ENG 76:84, March 24, 1969.

INDUSTRIAL NOISE: SHOE FACTORY

"Status of the organ of hearing in workers of the cutting shop of a shoe factory," by A. A. Tatarskaia. GIG TR PROF ZABOL 14: 45-46, February, 1970.

INDUSTRIAL NOISE: SYNTHETIC FIBER PLANT

"Hearing function in women workers under the effect of high-frequency noise in the twisting shop of a synthetic fiber plant," by Z. V. Babaian. ZH USHN NOS GORL BOLEZN 29:31-35, July-August, 1969.

INDUSTRIAL NOISE: TEXTILE

"Acoustic trauma in the textile industry," by R. Avellandea. ACTA OTORINOLARING IBER AMER 17:55-59, 1966.

"Assessment of the efficacy of periodic medical examinations of persons working in noisy shops of textile mills," by A. A. Tatarskaia, et al. GIG TR PROF ZABOL 9:45-46, July, 1965.

"Clinical study of the behavior of auditory adaptation in textile workers," by L. Bertocchi, et al. MINERVA OTORINOLARING 16:106-109, May-June, 1966.

"Exposure to noise in the textile industry of the U.A.R.," by M. H. Noweir, et al. AMER INDUSTR HYG ASS J 29:541-546, November-December, 1968.

"Noise abatement in textile mills," by P. H. R. Waldron. AM DYESTUFF REP 58:17-19, July 28, 1969; also in MOD TEXTILES MAG 50:49-50, July, 1969.

"Noise reduction in a textile weaving mill," by R. O. Mills. AMER INDUSTR HYG ASS J 30:71-76, January-February, 1969.

"On measures for controlling noise in textile industry plants," by F. S. Ravinskaia, et al. GIG TR PROF ZABOL 12:54, June, 1968.

INDUSTRIAL NOISE: TEXTILE

"On the use of individual means of protection from industrial noise during the vocational training of textile workers," by M. I. Krasil'shchikov. GIG SANIT 32:40-42, August, 1967.

INDUSTRIAL NOISE: WEAVING

"Damage caused by noise in weavers and preventive measures to reduce it," by G. Bologni, et al. ANN LARING 66:338-347, 1967.

"The effect of industrial noise in winding and weaving factories on the arterial pressure of operators," by A. I. Andrukovich. GIG TR PROF ZABOL 9:39-42, December, 1965.

"Evaluation of the effectiveness of measures for lowering the noise of pneumatic rapier looms," by L. I. Maksimova, et al. GIG SANIT 33:87-89, December, 1968.

"Hygienic evaluation of weaving loom machines without shuttles R-105," by E. V. Teterina, et al. GIG TR PROF ZABOL 10: 50-51, February, 1966.

"On the effect of noise on the functional status of the ear in weavers," by V. Ia. Kornev. GIG TR PROF ZABOL 12:26-30, October, 1968.

"A pilot study of hearing loss and social handicap in female jute weavers," by S. Taylor, et al. PROC ROY SOC MED 60:1117-1121, November 1, 1967.

"Study of noise and hearing in jute weaving," by W. Taylor, et al J ACOUST SOC AMER 38:113-120, July, 1965.

INDUSTRIAL NOISE: WELDING

"Noise measuring at workshops for autogenous welding and cutting and at a plasma-cutting device," by J. Gabelmann. ZBL ARBEITSMED 19:114-119, April, 1969.

INDUSTRIAL NOISE: YARN

"Noise problems connected with the manufacture of nylon and terylene yarn," by J. R. Kerr. PROC ROY SOC MED 60:1121-1126, November 1, 1967.

INFRASONIC NOISE

"Effects of low frequency and infrasonic noise on man," by G. C. Mohr, et al. AEROSPACE MED 36:817-824, September, 1965.

"Nearfield infrasonic noise generated by three turbojet aircraft during ground runup operations. AMRL-TR-65-132," by R. T. England, et al. US AIR FORCE AEROSPACE MED RES LAB 1-16, August, 1965.

INNER EAR

"Acoustic damage caused by noise as a cochlear anatomopathological manifestation and the prophylactic problem," by L. Ambrosio, et al. FOLIA MED 51:765-769, October, 1968.

"Adaptation behavior of Corti's organ following ephedrine and acoustic stimulation," by G. Stange, et al. ARCH OHR NAS KEHLKOPFHEILK 184:483-495, 1965.

"Biochemical observations of the alteration of DPN-diaphorase activity in the organ of Corti after noise exposure," by M. Tateda. J OTORHINOLARYNG SOC JAP 70:1312-1319, July, 1967.

"Certain aspects of hypacusia caused by acoustic trauma studies with Bekesy's audiometer," by M. Maurizi, et al. ARCH ITAL LARING 76:131-146, May-June, 1968.

"Cochleo-vestibular and general disorders induced by noise. The role of spas," by P. Molinery, et al. PRESSE THERM CLIMAT 103:5-9, 1966.

"Comparative Bekesy typing with broad and modulated narrow-band noise," by C. T. Grimes, et al. J SPEECH HEARING RES 12: 840-846, December, 1969.

"Comparison of Bekesy threshold for small band and sinus sounds in normal hearing and acoustic trauma," by R. Fischer, et al. ARCH KLIN EXP OHREN NASEN KEHLKOPFHEILKD 196: 223-227, 1970.

"Considerations on evaluation of acoustic and vestibular damage in pilots and specialists of military aeronautics,'' by C. Koch, et al. MINERVA MED 56:3832-3835, November 10, 1965.

''Effect of the acoustic stimulation on energy metabolism of the inner ear, with special reference to glycogen metabolism,'' by K. Ogawa. J OTOLARYNG JAP 73:335-352, March, 1970.

''The effect of trauma in causing cochlear losses after stapedectomy,'' by M. M. Paparella. ACTA OTOLARYNG 62:33-43, July, 1966.

''Effect on the labyrinth of intense acoustic stimulation,'' by P. Pialoux, et al. ANN OTOLARYNG 82:610-613, July-August, 1965.

''Electrocortical responses to a tonal stimulation and development of the cochlea,'' by R. Pujol, et al. J PHYSIOL 59:478, 1967.

''Experimental research on the cochlear function after exposure to intense noise. I. Variations of the microphonic potentials of the intact resected muscles of the middle ear,'' by W. Mozzo. BOLL SOC ITAL BIOL SPER 44:400-403, March 15, 1968.

''Experimental studies of the cochlear potential behaviour after stellatum blockade. II. Cochlear microphonic potential after stellatum blockade to the ear predamaged by noise,'' by D. Kleinfeldt, et al. ARCH KLIN EXP OHR NAS KEHLKOPFHEILK 190:398-406, 1968.

''Fluctuation of nucleic acid activity in the organ of corti resulting from noise exposure,'' by R. Nakamura. J OTORHINOLARYNG SOC JAP 70:1818-1827, November, 1967.

''Fluctuation of vitamin-B1 in blood and in the organ of Corti following exposure to noise,'' by S. Abiko. J OTORHINOLARYNG SOC JAP 69:1117-1133, June, 1966.

''Glycogen in the inner ear after acoustic stimulation. A light and electron microscopic study,'' by D. Ishii, et al. ACTA OTOLARYNG 67:573-582, June, 1969.

''Influence of an intensive acoustic stimulation on the cochlear response,'' by M. Aubry, et al. ACTA OTOLARYNG 60:191-196,

September, 1965.

"On the drug therapy of non-inflammatory diseases of the inner ear and balancing mechanism," by E. Ziegler. Z ALLGEMEINMED 45:927-929, July, 1969.

"Oxygen consumption in the organ of Corti based on observation of the dehydrogenase system following exposure to noise," by T. Takahaski. J OTORHINOLARYNG SOC JAP 70:1702-1715, October, 1967.

"Reaction of the hearing organ to sound stimulation in cochlear neuritis," by A. G. Rakhmilevich. VESTN OTORINORLARING 30:28-31, May-June, 1968.

"Recovery process and the extent of injuries to the organ of Corti (study by the surface specimen technic)," by A. Sugiura. J OTOLARYNGCL JAP 73:1770-1779, November, 1970.

"Regeneration of glycogen in the hair cells of the organ of Corti," by S. Chodynicki. FOLIA HISTOCHEM CYTOCHEM 3:211-216, 1965.

"Rehological understanding of origin of inner-ear acoustic fatigue and trauma," by H. Uchiyama, et al. BIORHEOLOGY 6:253, January, 1970.

"Scanning electron microscopy of the organ of Corti," by G. Bredberg, et al. SCIENCE 170:861-863, November 20, 1970.

"Sonic boom effects on the organ of Corti," by D. A. Majeau-Chargois, et al. LARYNGOSCOPE 80:620-630, April, 1970.

"Some problems concerning prevention of inner ear damage caused by noise," by L. Surjan. ORV HETIL 109:1365-1369, June 23, 1968.

INTERNAL EAR

"The audiologic picture and histologic substrate of noise-induced internal ear injury," by W. Lorenz. DEUTSCH GESUNDH 23:2423-2426, December 19, 1968.

INTERNAL EAR

"Effect of oxidoreductive agents and high-energy compounds on the efficiency of the internal ear during acoustic load," by W. Jankowski, et al. OTOLARYNG POL 23:141-144, 1969.

JET NOISE
see: Aircraft Noise

L-IDOTYL DEHYDROGENASE
"Effects of physical work and work under condition of noise and vibration on the human body. II. Behavior of aspartic and alanine aminotransferase, gamma glutamyl transpeptidas, lactic dehydrogenase and L-idotyl dehydrogenase activities," by J. Gregorczyk, et al. ACTA PHYSIOL POL 16:709-714, September-October, 1965.

LABORATORY NOISE
"Noise in the laboratory," BR MED J 3:662, September 19, 1970.

"Noise levels in a clinical chemistry laboratory," by P. D. Griffiths, et al. J CLIN PATHOL 23:445-459, July, 1970.

LACTIC DEHYDROGENASE
"Effect of phyiscal work and work under conditions of noise and vibration on the human body. I. Behavior of serum alkaline phosphatase, aldolase and lactic dehydrogenase activities," by J. Gregorczyk, et al. ACTA PHYSIOL POL 16:701-708, September-October, 1965.

"Effects of physical work and work under condition of noise and vibration on the human body. II. Behavior of aspartic and alanine aminotransferase, gamma glutamyl transpeptidas, lactic dehydrogenase and L-idotyl dehydrogenase activities," by J. Gregorczyk, et al. ACTA PHSYIOL POL 16:709-714, September-October, 1965.

LARYNGECTOMY
"The effects of environmental noise on pseudo voice after laryngectomy," by S. Drummond. J LARYNG 79:193-202, March, 1965.

LARYNX
"Occupational damages of the larynx," by E. Nessel. ARCH OHR NAS KEHLKOPFHEILK 185:474-477, 1965.

LAWS AND LEGISLATION
see also: Liability

"Aircraft noise law: a technical perspective," by J. J. Alekshun, Jr. AM BAR ASSN J 55:740-745, August, 1960.

"Are you prepared for noise laws?" by F. Haluska. STEEL 164:41+, February 17, 1969.

"Boom! here comes the noise rules explosion," by K. A. Kaufman. IRON AGE 205:19, April 16, 1970.

"Clinical and medico-legal considerations on hypoacusias of metallurgic workers," by M. Ciulla, et al. BOLL SOC MEDICOCHIR CREMONA 19:129-141, January-December, 1965.

"Coast noise verdict may set precedent," AVIATION W 88:35, January 1, 1968.

"Commonly-used terms and definitions for understanding noise regulation," ENVIRONMENTAL CONTROL MGT 139:26-27, January, 1970.

"Community noise ordinances," by A. E. Meling. ASHRAE J 9:40-43+, May, 1967.

"Companies warned: quieter, please!" BSNS W 28-29, July 26, 1969.

"Consumers and legislators shout down noisy products," MACHINE DESIGN 42:41, April 30, 1970.

"Contractors suggest revisions to proposed Los Angeles noise ordinance," AIR COND HEAT & REFRIG N 121:2, November 16, 1970.

"Control of neighborhood noise as a hygienic-legal problem," by A. Schubert. Z AERZTL FORTBILD 63:1056-1058, October 1, 1969.

"Cooperation, understanding make Coral Gables noise law workable," by P. B. Redeker. AIR COND HEAT & REFRIG N 112:1+, December 4, 1967.

LAWS AND LEGISLATION

"Coral Gables noise code gets grudging approval," by G. Duffy. AIR COND HEAT & REFRIG N 116:1+, April 21, 1969.

"Detroit studying ordinance to control noisy air conditioners," by G. M. Hanning. AIR COND HEAT & REFRIG N 106:1+, September 20, 1965.

"Disturbance by noise from the medical and legal viewpoint," by H. Wiethaup. MED KLIN 60:1218-1220, July 23, 1965.

"Excessive noise from the viewpoint of the penal law," by H. Wiethaup. MED KLIN 60:1254-1256, July 30, 1965.

"Government calls for quiet," MOD PLASTICS 46:80-81, December, 1969.

"How Ottawa's 'realistic' new traffic noise by law operates," FIN POST 64:27, February 14, 1970.

"The jet-set and the law: a summary of recent developments in noise law as it reltaes to airport and aircraft operations in California," by W. S. H. Hood, Jr. PACIFIC LAW JOURNAL 1:581-609, July, 1970.

"Law of noise," TIME 86:37-38, September 10, 1965.

"A legal action for noise deafness," by R. R. Coles. ANN OCCUP HYG 12:223-236, October, 1969.

"Legal aspects of noise control," by F. P. Houston. AUDIO ENG SOC J 17:321+, June, 1969.

"Legal aspects of noise in New York City," by J. J. Allen NEW YORK LAW JOURNAL 163:1+, June 12, 1970.

"Legal aspects of noise pollution," by H. A. Young. PLANT ENG 23:66-67, May 29, 1969.

"Medical opinion and lawful estimation of defective hearing caused by noise," by G. Kollmorgen. Z AERZTL FORTBILD 64:954-956, September 15, 1970.

LAWS AND LEGISLATION

"Medico-legal aspects of office hearing evaluations," by C. O. Istre, Jr., et al. ARCH OTOLARYNG 86:645-649, December, 1967.

"New fan law for sound," by R. Parker. ASHRAE J 9:83-85, October, 1967.

"New noise regulations under the Walsh-Healey Act," by F. A. Van Atta. J OCCUP MED 12:27-29, February, 1970.

"Noise bill draws opposing views," AM AVIATION 32:21, June 24, 1968.

"Noise bill limitation sought by airlines," AVIATION W 88:41, June 24, 1968.

"Noise can be costly now," by L. F. Yerges. HEATING-PIPING 42:66-69, February, 1970.

"Noise control: traditional remedies and a proposal for federal action," by J. M. Kramon. HARV J LEGIS 7:533, May, 1970.

"Noise pollution: an introduction to the problem and an outline for future legal research," by J. L. Hildebrand. COLUM L REV 70: 652, April, 1970.

"Nuisance: The law and economics," by P. Burrows. LLOYDS BANK REVIEW 95:36-46, January, 1970.

"Occupational health and safety: the government viewpoint," by F. A. Van Atta. FOUNDRY 98:109-113, February, 1970.

"On the law for protection from construction noise of September 9, 1965. (1)," by H. Wiethaup. ZBL ARBEITSMED 16:103-105, April, 1966.

"On the legality of anti-noise campaigns," by H. Wiethaup. MED KLIN 60:1095-1099, July 2, 1965.

"Proximity damages," by J. L. Sackman. APPRAISAL J 37:177-199, April, 1969.

LAWS AND LEGISLATION

"Regulating road traffic noise," ENGINEERING 205:823, May 31, 1968.

"Ssh! ordinances on the books in New York and other cities," NEWSWEEK 74:52, September 8, 1969.

"These guidelines suggested to limit noise pollution (digest of *Rational approach to legislative control of noise*)," by G. J. Thiessen. FIN POST 64:44, February 14, 1970.

"Triumph over traffic; New York's Court of appeals declares noise a compensable injury," TIME 92:74, July 12, 1968.

"2 LA suburbs plan noise ordinances," AIR COND HEAT & REFRIG N 121:29, October 5, 1970.

"Vehicle noise meter meets new law," ENGINEER 225:774, May 17, 1968.

"Walsh-Healey act changes spur design-out of noise to protect workers' hearing," by L. L. Beranek. POWER 114:52-53, May, 1970.

"Walsh-Healey, the first nine months; an evaluation of the noise control considerations," ENVIRONMENTAL CONTROL MGT 139:33-34, January, 1970.

"Walsh-Healey noise regulations cannot be ignored! interview with A. A. Fletcher.," SAFETY MAINT 138:34-39, August, 1969.

"Walsh-Healey revisited: Northrop reviews hearing loss prevention policies and procedures," by J. L. Morrow. ENVIRONMENTAL CONTROL & SAFETY MGT 139:32-34, June, 1970.

"War on noise; some small victories," ECONOMIST 214:1308, March 20, 1965.

"Who is going to do what about your noise problem?" by F. X. Worden. SAFETY MAINT 132:42-44, December, 1966.

"Work prohibition for expectant and breast feeding mothers in harmful noise,'' MUNCHEN MED WSCHR 31:Suppl:6, August 2, 1968.

LEARNING

see also: Attention

''Aversive stimulation as applied to discrimination learing in mentally retarded children,'' by P. S. Massey, et al. AM J MEN DEFICIENCY 74:269-272, September, 1969.

''Effects of achievement motivation and noise conditions on paired-associate learning,'' by K. Shrable. CALIF J ED RES 19:5-15, January, 1968.

''Effects of auditory stimulation on the performance of brain-injured and familial retardates,'' by L. S. Schoenfeld. PERCEPT MOTOR SKILLS 31:139-144, August, 1970.

''Effects of noise and difficulty level of imput information in auditory, visual, and audiovisual information processing,'' by H. J. Hsia. PERCEPT MOTOR SKILLS 26:99-105, February, 1968.

''Effects of noise on pupil performance,'' by B. R. Slater. J ED PSYCHOL 59:239-243, August, 1968.

''Effects of traffic noise on health and achievement of high school students of a large city,'' by G. Karsdorf, et al. Z GES HYG 14:52-54, January, 1968.

''Effects of white noise and presentation rate on serial learning in mentally retarded males,'' by J. R. Haynes. AMER J MENT DEFIC 74:574-575, January, 1970.

''Sound drive and paired-associates learning,'' by H. Levitt. J GEN PSYCHOL 72:343-357, April, 1965.

LIABILITY

see also: Laws and Legislation

''Clinical, social and insurance evaluation of professional deafness due to noise,'' by E. Vensi. ANN LARING 64:337-343, 1965.

LIABILITY

"Compensation claims for loss of hearing; impact of standards," by F. E. Frazier. AMER ASS INDUSTR NURSES J 15:17-19, passim, May, 1967.

"Compensation claims for loss of hearing; impact of standards," by F. E. Frazier. ARCH ENVIRON HEALTH 10:572-575, April, 1965.

"Damage risk criterion and contours based on permanent and temporary hearing loss data," by K. D. Kryter. AMER INDUSTR HYG ASS J 26:34-44, January-February, 1965.

"Damage-risk criterion for the impulsive noise of 'toys'," by K. Gjaevenes. J ACOUST SOC AMER 42:268, July, 1967.

"Hearing-loss trend curves and the damage-risk criterion in diesel-engineroom personnel," by J. D. Harris. J ACOUST SOC AMER 37:444-452, March, 1965.

"Indemnity in occupational deafness," by P. Mounier-Kuhn, et al. J MED LYON 48:1691-1696, November 20, 1967.

"Industrial hearing loss. Conservation and compensation aspects," by M. S. Fox. OCCUP HEALTH NURS 17:18-24, May, 1969.

"A new concept of damage risk criterion," by W. G. Noble. ANN OCCUP HYG 13:69-75, January, 1970.

"Ouch, that noise! Hearing loss gets compensation," FIN POST 64:4, September 19, 1970.

LIBRARIES

"Are academic libraries too noisy?" by J. A. McCrossan. AM LIB 1:396, April, 1970.

"I can't hear the flutes; excessive noise from nonbook learning devices in new Hume library at University of Florida," by L. Cassidy. AM LIB 1:888-889, October, 1970.

MACHINE DESIGN

"Design and decisions in machinery sound control," by G. M. Diehl.

COMP AIR MAG 75:7-9, August, 1970.

"Design for quiet," MACHINE DESIGN 39:174-224, September 14, 1967.

"Design for quiet: electrical systems," by J. Campbell. MACHINE DESIGN 39:192-199, September 14, 1967.

"Guidelines for designing quieter equipment," by C. H. Allen. MECH ENG 92:29-34, January, 1970.

"Internal noise attenuation of turbine auxiliary power unit must start during design stage," by J. J. Dias. SAE J 76:64-67, January, 1968.

"Noise controlled at design stage," by J. E. Seebold. OIL & GAS J 68:56-58, December 21, 1970.

"Noise must be designed out of machines; annual symposium, 6th, Cleveland," PRODUCT ENG 40:24+, October 20, 1969.

"Putting correlation to work on engineering test data," by G. T. Roberts. MACHINE DESIGN 42:108-113, January 8, 1970.

"When noise is a problem, design it out instead of in," by A. R. Gardner. PRODUCT ENG 37:86-91, April 11, 1966.

"Will you reduce noise on your designs for the 1970's?" by G. L. Charson. HYDRAULICS & PNEUMATICS 22:85-89, November, 1969.

MACHINERY NOISE

"Engineering outline; noise in machines," ENGINEERING 201: 1059-1062, June 3, 1966.

"Experimental investigation of discrete frequency noise generated by unsteady blade forces," by N. J. Lipstein, et al. J BASIC ENG 92:155-164, March, 1970.

"Fundamental nature of machine noise," by D. B. Welbourn. ENGINEERING 199:175-176, February 5, 1965.

MACHINERY NOISE

"Guidelines for noise control specifications for purchasing equipment," IRON & STEEL ENG 47:95-98, May, 1970.

"Iroquois plant; horsepower on a raft," MARINE ENG LOG 75:88-89 , October, 1970.

"Machinery hazards," by C. J. Moss. ANN OCCUP HYG 12:69-75, April, 1969.

"Machinery noise," AUTOMOBILE ENG 55:149, April, 1965.

"Machine tool computer resists noise," ELECTRONICS 38:190, October 4, 1965.

MACHINERY NOISE: CEMENT PLANTS

"Reducing machinery noise in cement plants," by G. L. Koonsman. IEEE TRANS IND & GEN APPLICATIONS 6:476-479, September, 1970.

MACHINERY NOISE: DATA PROCESSING

"Cutting noise in data sampling," by T. Kobylarz. ELECTRONICS 41:70-73, May 13, 1968.

"Noise in computer rooms and in mechanical data processing central stations," by B. Kvasnicka. PRAC LEK 17:112-115, April, 1965.

"Noise nuisance and noise protection in machine equipment for data processing facilities," by W. Winter. ZBL ARBEITSMED 19:201-204, July, 1969.

MACHINERY NOISE: ELECTRICAL

"Acoustic noise and vibration of rotating electric machines," by A. J. Ellison, et al. INST E E PROC 115:1633-1640, November, 1968; Discussion 117;127-129, January, 1970.

"Acoustic-noise measurements on nominally identical small electrical machines," by A. J. Ellison, et al. INST E E PROC 117: 555-560, March, 1970.

"Controlling noise from rotating electrical machines," ENGINEER 227:309, February 28, 1969.

MACHINERY NOISE: ELECTRICAL

"Design for quiet: electrical systems," by J. Campbell. MACHINE DESIGN 39:192-199, September 14, 1967.

"The effect of the noise of electric machines on certain physiological functions of the organism of machine test personnel," by N. T. Svistunov. GIG TR PROF ZABOL 13:15-19, February, 1969.

"Measurement and suppression of noise; with special reference to electrical machines, by A. J. King. A review," by K. A. Rose. ROY INST BRIT RRCH J 73:85, February, 1966.

--A review," by H. D. Parbrook. TOWN PLAN R 37:75-76, April, 1966.

"Methods of measurement of acoustic noise radiated by an electric machine," by A. J. Ellison, et al. INST E E PROC 116:1419-1431, August, 1969.

"Noise control for electric motors," by S. H. Judd, et al. AMER INDUSTR HYG ASS J 30:588-595, November-December, 1969.

"Noise in high-speed motors," by B. Brozek. MACHINE DESIGN 42:123-127, March 5, 1970.

MACHINERY NOISE: FARM EQUIPMENT

"Catch in the rye; it can make you deaf," by K. Mitchell. MACL MAG 80:3, August, 1967.

"Farm equipment noise exposure levels," by H. H. Jones, et al. AMER INDUSTR HYG ASS J 29:146-151, March-April, 1968.

"The hygienic characteristics of noise in agricultural machines," by K. Gruss. PRAC LEK 18:304-308, August, 1966.

MACHINERY NOISE: HYDRAULIC

"Can you hear the hydraulics system?" by J. S. Noss. MACHINE DESIGN 42:141-145, September 17, 1970.

"Design for quiet; hydraulic systems," by J. W. Sullivan. MACHINE DESIGN 39:210-215, September 14, 1967.

MACHINERY NOISE: HYDRAULIC

"Noise in oil-hydraulic systems," ENGINEERING 207:647-648, April 25, 1969.

MACHINERY NOISE: PAPER

"How Holmens Brukab reduced noise from a new paper machine," by P. Berg, et al. PAPER TR J 154:42-48, November 2, 1970.

"Paper machine and related noise abatement developments," by C. B. Dahl, et al. PAPER TR J 154:44-47, November 30, 1970.

MACHINERY NOISE: ROTATING

"Math quiets rotating machines," by J. H. Varterasian. SAE J 77: 53, October, 1969.

MACHINERY NOISE: TURBO

"Acoustic resonances and multiple pure tone noise in turbomachinery inlets," by F. F. Ehrich. J ENG POWER 91:253-262, October, 1969.

"Sound generation in subsonic turbomachinery," by C. L. Morfey. J BASIC ENG 92:450-458, September, 1970.

MACHINERY NOISE: TURBOCOMPRESSORS

"Frequency analysis of the noise of turbocompressors in an oxygen shop," by K. S. Rudakov. GIG SANIT 34:69-71, August, 1969.

MAN

see also: Attention, Behavior, Learning, Personality, Physiology, Psychology, Psychotics

"The effects of noise on man," by A. Glorig. JAMA 196:839-842, June 6, 1966.

"Man and his noises," CANAD MED ASS J 101:109-110, July 26, 1969.

"Standardization of vibration and noise and their combined action on man," by N. M. Paran'ko, et al. VRACH DELO 8:102-106, August, 1969.

MASKING NOISE

"Additivity of masking," by D. M. Green. J ACOUST SOC AMER 41:1517-1525, June, 1967.

"Alteration of the masking sound level during and after acoustic stimulation," by G. Muller. ARCH KLIN EXP OHR NAS KEHLKOPFHEILK 195:323-330, 1970.

"Audiologic examinations of the masking effect caused by flight noise," by W. Lorenz. MSCHR OHRENHEILK 103:438-444, 1969.

"Binaural masking of speech by periodically modulated noise," by R. Carhrat, et al. J ACOUST SOC AMER 39:1037-1050, June, 1966.

"Contralateral masking and the SISI-test in normal listeners," by B. Blegvad, et al. ACTA OTOLARYNG 63:557-563, June, 1967.

"Contralateral remote masking and the aural reflex," by K. Gjaevenes, et al. J ACOUST SOC AMER 46:918-923, October, 1969.

"Effect of background conversation and darkness on reaction time in anxious, hallucinating, and severely ill schizophrenics," by A. Raskin. PERCEPT MOTOR SKILLS 25:353-358, October, 1967.

"The effect of impact noise with various background noises," by E. P. Orlovskaia. GIG SANIT 32:21-25, September, 1967.

"Effect of masked noise on the bone conduction threshold," by D. A. Pigulevskii, et al. ZH USHN NOS GORL BOLEZ 27:38-43, March-April, 1967.

"Effect of masking noise upon syllable duration in oral and whispered reading," by M. F. Schwartz. J ACOUST SOC AMER 43:169-170, January, 1968.

"Masking noise: silence is golden, privacy is pink," by R. Farrell. PROGRES ARCH 48:152-155, November, 1967.

"Masking of speech by aircraft noise," by K. D. Kryter, et al. J ACOUST SOC AMER 39:138-150, January, 1966.

MASKING NOISE

"Masking of speech by continuous noise," by J. Kuzniarz. OTOLARYNG POL 21:401-407, 1967; also in POL MED J 7:1001-1008, 1968.

"Masking of speech by means of impulse noise," by J. Kuzniarz. OTOLARYNG POL 22:421-425, 1968.

"Masking of tones before, during, and after brief silent periods in noise," by L. L. Elliott. J ACOUST SOC AMER 45:1277-1279, May, 1969.

"Paced recognition of words masked in white noise," by C. M. Holloway. J ACOUST SOC AMER 47:1617-1618, June, 1970.

"Relations of the human vertex potential to acoustic input: loudness and masking," by H. Davis, et al. J ACOUST SOC AMER 43:431-438, March, 1968.

"Some effects of bone-conducted masking," by Z. G. Schoeny, et al. J SPEECH HEARING RES 8:253-261, September, 1965.

"Stimulation of efferent olivocochlear bundle causes release from low level masking," by P. Nieder, et al. NATURE (London) 227:184-185, July 11, 1970.

"Temporal effects in simultaneous masking and loudness," by E. Zwicker. J ACOUST SOC AMER 38:132-141, July, 1965.

MASS-TRANSIT SYSTEMS

"Noise in mass-transit systems," by V. Salmon. STANFORD RESEARCH INSTITUTE J 2-7, September, 1967.

MENTAL RETARDATION

"Aversive stimulation as applied to discrimination learning in mentally retarded children," by P. S. Massey, et al. AM J MED DEFICIENCY 74:269-272, September, 1969.

"Effects of auditory stimulation on the performance of brain-injured and familial retardates," by L. S. Schoenfeld. PERCEPT MOTOR SKILLS 31:139-144, August, 1970.

MENTAL RETARDATION

"Effects of white noise and presentation rate on serial learning in mentally retarded males," by J. R. Haynes. AMER J MENT DEFIC 74:574-577, January, 1970.

MIDDLE EAR

"Anticoagulation and gunshot concussion in eardrum bleeding," by N. Sonkin. RHODE ISLAND MED J 49:243-244, April, 1966.

"Atrophy of the long process of the incus in a patient with an occupational hearing disorder," by B. M. Gapanavichius. VESTN OTORINOLARINGOL 32:91-92, November-December, 1970.

"Dynamic response of middle-ear structures," by H. Fischler, et al. ACOUSTICAL SOC AM J 41:1220-1231, May, 1967.

"Effect of the acoustic reflex on the impedance at the eardrum," by A. S. Feldman, et al. J SPEECH HEARING RES 8:213-222, September, 1965.

"Experimental research on the cochlear function after exposure to intense noise. I. Variations of the microphonic potentials of the intact resected muscles of the middle ear," by W. Mozzo. BOLL SOC ITAL BIOL SPER 44:400-403, March 15, 1968.

"Relation of threshold shift to noise in the human ear," by H. Weissing. ACOUSTICAL SOC AM J 44:610-615, August, 1968.

MINING NOISE

"Acoustic trauma in depth miners," by A. Goubert, et al. J FRANC OTORHINOLARYNG 18:133-135, February, 1969.

"Control of mining noise exposure," by J. H. Botsford. MIN CONG J 53:22-24+, August, 1967.

"Effects of harmful noise of mechanized workings on workers at the bottom of the mine. Result of a study undertaken at the mines of Houilleres in the Lorraine Bassin," by A. Mas, et al. ARCH MAL PROF 27:815-818, October-November, 1966.

"A method of determining some parameters of the shock wave of explosions in mines and their hygienic assessment," by B. A. Shaparenko. GIG TR PROF ZABOL 9:9-13, July, 1965.

"Noise levels of underground mining equipment," by W. M. Ward. OCCUP HEALTH REV 19:16, 1967.

"Problems of protection from noise in the potassium mining industry," by H. Wolf. Z GES HYG 12:189-195, March, 1966.

"Studies on the determination of the true auditory threshold under the influence of experimental exposure to mine noise," by A. Dobrowolski. OTOLARYNG POL 21:43-48, 1967.

MONITORS

see: Noise Measurement Devices

MOTORS

"Controlling the tonal characteristics of the aerodynamic noise generated by fan rotors," by R. C. Mellin, et al. J BASIC ENG 92:143-154, March, 1970.

"Coping with noise from big motors," by R. L. Nailen. POWER ENG 72:54-57, October, 1968.

"Occupational deafness in thermoelectric plants with Norberg methane motors," by N. Lo Martire, et al. ATTI ACCAD FISIOCR SIENA 16:202-212, 1967.

"Silencing Impak metric motors," ENGINEER 227:233, February 14, 1969.

"Specifying motors for a quiet plant; absorption, acoustical barriers and distance can quiet airborne motor noise," by R. L. Nailen. CHEM ENG 77:157-161, May 18, 1970.

MOTORS: ELECTRICAL

"Electric motor noise: control of noise at the source," by B. L. Goss. AMER INDUSTR HYG ASS J 31:16-21, January-February, 1970.

MUSCLES

"Change in the electrical activity of muscles under the effect of noise and vibration," by A. F. Lebedeva. GIG SANIT 33:25-30, March, 1968.

"Importance of the individual factor of muscular exhaustibility of the sound-conducting apparatus in the genesis of the damage caused by continuous acoustic overstimulation," by T. Marullo, et al. CLIN OTORINOLARING 19:265-271, July-August, 1967.

MUSIC

"Acoustic trauma from rock and roll," HIGH FIDELITY 17:38+, November, 1967.

"Acoustic trauma from rock-and-roll music," by C. P. Lebo, et al. CALIF MED 107:378-380, November, 1967.

"Acoustical measurements in New York's Philharmonic Hall," by R. M. Schroeder. BELL LAB REC 43:38-45, February, 1965.

"Auditory fatigue and predicted permanent hearing defects from rock-and-roll music," by F. L. Dey. NEW ENG J MED 282:467-470, February 26, 1970.

"'Big-beat' music and acoustic traumas," by H. Kowalczuk. OTO-LARYNG POL 21:161-167, 1967.

"Dance bands can deafen," NEW ZEALAND NURS J 62:14, May, 1969.

"Deafening music," BRIT MED J 2:127, April 18, 1970.

"Does rock music damage hearing?" GOOD H 168:208, April, 1969.

"Ear damage from exposure to rock and roll music," by D .M. Lipscomb. ARCH OTOLARYNG 90:545-555, November, 1969.

"Effects of too-loud music on human ears. But, mother, rock'n roll has to be loud," by R. R. Rupp, et al. CLIN PEDIAT 8:60-62, February, 1969.

MUSIC

"Experimental study on sound perception--accuracy of musical scale with sound loading among members of a chorus group," by S. Sakurai, et al. J OTOLARYNGOL JAP 73:1527-1532, September, 1970.

"Going deaf from rock 'n' roll," TIME 92:47, August 9, 1968.

"Hearing loss in rock-and-roll musicians," by C. Speaks, et al. J OCCUP MED 12:216-219, June, 1970.

"Modern-day rock-and-roll music and damage-risk criteria," by J. M. Flugrath. J ACOUST SOC AMER 45:704-711, March, 1969.

"Music and noise," by J. C. Waterhouse. MUS & MUS 14:28-29+, January, 1966.

"Music as a source of acoustic trauma," by C. P. Lebo, et al. AUDIO ENG SOC J 17:535-538, October, 1969; also in LARYNGOSCOPE 78:1211-1218, July, 1968.

"Noise-induced hearing loss and pop music," by S. Hickling. NEW ZEAL MED J 71:94-96, February, 1970.

"Noise-induced hearing loss and rock and roll music," by W. F. Rintelmann, et al. ARCH OTOLARYNG 88:377-385, October, 1968.

"Noise reduction in electronic music," by R. M. Dolby. ELEC MUS R 6:33-37, April, 1968.

"Not exactly music to your ears; high sound levels of rock-and-roll music," CONSUMER REP 33:349-350, July, 1968.

"Pop music as noise trauma," by E. Fluur. LAKARTIDNINGEN 64: 794-796, February 22, 1967.

"Rock music and hearing," by R. L. Voorhees. POSTGRAD MED 48:108-112, July, 1970.

"Rock physically unsound. SCI DIGEST 63:67-68, June, 1968.

MUSIC

"Sounds of music," MED J AUST 1:865-866, April 26, 1969.

MUSICIANS

see also: Music

"On the question of occupationally related hearing difficulties in musicians," by M. Flach, et al. Z LARYNG RHINOL OTOL 45: 595-605, September, 1966.

"The risk of occupational deafness in musicians," by M. Flach, et al. GERMAN MED MONTHLY 12:49-54, February, 1967.

"Temporary threshold shift in rock-and-roll musicians," by J. Jerger, et al. J SPEECH HEARING RES 13:221-224, March, 1970.

MYOCIONUS

"Apropos of 2 cases of myocionic petit mal precipitated by noise in children," by A. Lerique-Koechlin, et al. REV NEUROL 113: 269, September, 1965.

NASA

see also: Aircraft Design
Aircraft Noise

"NASA acoustically treated nacelle program," by J. G. Lowry. ACOUSTICAL SOC AM J 48:780-782, pt 3, September, 1970.

"NASA begins major engine noise project," by M. L. Yaffee. AVIATION W 87:38-39+, August 21, 1967.

"NASA seeks quiet aircraft engine design," by W. S. Beller. TECH W 20:20-21, June 19, 1967.

"NASA to launch quiet engine program," by H. Taylor. AM AVIATION 32:23, August 19, 1968.

"NASA's quiet engine program focuses antinoise effort," AVIATION W 92:88-89, June 22, 1970.

"Review of research and methods for measuring the loudness and noisiness of complex sounds. NASA Contract Rep NASA CR-422," by K. D. Kryter. US NASA 1-57, April, 1966.

NATURAL GAS

"Quietening noisy natural gas," ENGINEERING 204:960-961, December 15, 1967.

NERVOUS SYSTEM

"Amplitude distribution of axon membrane noise voltage," by A. A. Verveen, et al. ACTA PHYSIOL PHARMACOL NEERL 15:353-379, August, 1969.

"Aspects of nervous fatigue in automated systems as a function of age of the operators," by R. Elias, et al. FIZIOL NORM PAT 13:447-454, September-October, 1967.

"Bioflavonoid therapy in sensorineural hearing loss: a double-blind study," by J. E. Creston, et al. ARCH OTOLARYNG 82: 159-165, August, 1965.

"Changes in the bioelectric activity of the brain and in some autonomic and vascular reaction under the influence of noise," by E. A. Drogichina, et al. GIG SANIT 30:29-33, February, 1965.

"Clinical and experimental aspects of the effects of noise on the central nervous system," by F. Angeleri, et al. MED LAVORO 60:759-766, December, 1969.

"Comparative study on steady and intermittent exposures to noise. ammonia content in the brain as an indicator," by K. Matsui, et al. JAP J HYG 23:225-228, June, 1968.

"Effect of extremely strong pulsed noise on certain sections of the central and peripheral nervous system," by O. S. Shemiakin. GIG TR PROF ZABOL 14:19-23, November, 1970.

"Effect of intensive noise and neuro-psychic tension on arterial blood pressure levels and frequency of hypertensive disease," by N. N. Shatalov, et al. KLIN MED 48:70-73, March, 1970.

"Effect of noise on the functional status of the nervous system in children of preschool age," by Kh. V. Storoshchuk. GIG SANIT

31:44-48, January, 1966.

"Effect of removing the neocortex on the response to repeated sensory stimulation of neurones in the mid-brain," by G. Horn, et al. NATURE (London) 211:754-755, August 13, 1966.

"On changes in lipid metabolism in persons during prolonged action of industrial noise on the central nervous system," by P. S. Khomulo, et al. KARDIOLOGIIA 7:35-38, July, 1967.

"Reactions of the human nervous and cardiovascular systems to the effects of aviation noises," by V. G. Terent'ev, et al. VOENNO-MED ZH 6:55-58, June, 1969.

NITROUS OXIDE

"Changes in the auditory pain threshold induced by nitrous oxide," by F. Fruttero, et al. MINERVA OTORINOLARING 16:135-137, July-August, 1966.

NOISE: CORAL GABLES

"Cooperation, understanding make Coral Gables noise law workable," by P. B. Redeker. AIR COND HEAT & REFRIG N 112: 1+, December 4, 1967.

"Coral Gables noise code gets grudging approval," by G. Duffy. AIR COND HEAT & REFRIG N 116:1+, April 21, 1969.

"Coral Gables would limit noise from air conditioners," by C. D. Mericle. AIR COND HEAT & REFRIG N 108:1+, May 9, 1966.

NOISE: ENGLAND

"The problem of noise in England," by F. Merluzzi. MED LAW 61:181-188, March, 1970.

NOISE: EUROPE

"Prevention of noise in Europe," by N. Simonetti. MINERVA MED 60:Suppl 26:22, March 11, 1969.

NOISE: FINLAND

"Noise and hearing ability: incidence of hearing defects induced by noise in Finland," by J. S. Lumio. IND MED 34:404-406, May, 1965.

NOISE: HONOLULU

"Antinoise campaign in Honolulu," ACOUSTICAL SOC AM J 45: 520-521, February, 1969.

NOISE: LOS ANGELES

"Contractors suggest revisions to proposed Los Angeles noise ordinance," AIR COND HEAT & REFRIG N 121:2, November 16, 1970.

"Noise governs transformer selection at UCLA health sciences complex in Los Angeles," by K. K. Leithold. ELEC CONSTR & MAINT 67:106-107, July, 1968.

NOISE: NEW YORK CITY

"Legal aspects of noise in New York City," by J. J. Allen. NEW YORK LAW JOURNAL 163:1+, June 12, 1970.

"Quieter city is goal of New York City study group," AUTOMOTIVE ENG 78:27-31, August, 1970.

"Sound of sounds that is New York," by H. C. Schonberg. N Y TIMES MAG 38+, May 23, 1965.

"Ssh! ordinances on the books in New York and other cities," NEWSWEEK 74:52, September 8, 1969.

"Urban noise control (focuses on efforts in New York city)," COLUMBIA J LAW & SOCIAL PROBLEMS 4:105-119, March, 1968.

NOISE: SAO PAULO

"Noise-induced health problems in Sao Paulo," by S. Marone. RESEN CLIN CIENT 38:223-234 conclusion, September-October, 1969.

NOISE ABATEMENT

"Abatement of control valve noise; Fisher Governor Co.," by E. E. Allen. GAS 45:53-56, November, 1969.

"Acoustical control (within the office)," by D. Anderson. ADM MGT 27:66+, March, 1966.

NOISE ABATEMENT

"Assessment and control of noise," by D. S. Gordon. INST E E PROC 113:775-776, May, 1966.

"Basics of noise control; sound power and sound pressure," by G. Diorio. HEATING-PIPING 39:194-195, April, 1967.

"Dust control and sound abatement," by D. Jackson, Jr. COAL AGE 71:66-70+, November, 1966.

"Effects of noise reduction in a work situation," by D. E. Broadbent, et al. OCCUPATIONAL PSYCHOLOGY 34,2:133-140.

"FAA fears political tug-of-war on noise abatement programs," AVIATION W 85:36-37, December 5, 1966.

"From experience in controlling noise and vibration in industrial plants in Kiev," by I. G. Guslits, et al. GIG TR PROF ZABOL 10:52-54, June, 1966.

"How noise control affects you," TEXTILE WORLD 119:43-46, June, 1969.

"How to keep the decibel level low," ADM MGT 28:90-91, May, 1967.

"Medical prevention of chronic acoustic trauma," by S. Kubik. PRAC LEK 18:176-179, May, 1966.

"Noise abatement by barriers," by M. Rettinger. PROGRES ARCH 46:168-169, August, 1965.

"Noise abatement in refineries," by R. C. Ewing. OIL & GAS J 67: 83-88, October 13, 1969.

"Noise abatement in textile mills," by P. H. R. Waldron. AM DYE-STUFF REP 58:17-19, July 28, 1969.

"Noise abatement in the Zetor-Super tractor prototype," by I. Seress. CESK HYG 10:81-85, March, 1965.

"Noise abaters," by H. Lawrence. AUDIO 51:82, May, 1967.

NOISE ABATEMENT

"Noise-breaks--a possible measure for the prevention of noise-induced deafness," by E. Lehnhardt, et al. INT ARCH GEWERBEPATH 25:65-74, 1968.

"Paper machine and related noise abatement developments," by C. B. Dahl, et al. PAPER TR J 154:44-47, November 30, 1970.

"Pressure could spur noise breakthrough," by H. J. Coleman. AVIATION W 85:49+, December 19, 1966.

"Preventing hearing loss in industry, " by V. Hamilton. CANAD NURSE 66:37+, September, 1970.

"The prevention of noise: a medico-social problem of today," by D. Terzuolo. MINERVA MED 61:3332-3336, August 3, 1970.

"Quiet: what you can do to preserve it; excerpts from The tyranny of noise," by R. A. Baron. HOUSE & GARD 138:128-129, October, 1970.

"Reducing machine sound levels," MACH 72:347-348, September, 1965.

"Rules lowering noise will increase plant costs," MOD MANUF 2: 98-100, September, 1969.

"Simple guidelines best for noise control," PRODUCT ENG 40: 75, June 2, 1969.

"Solution to the noise control problem," by H. Seelbach. AIR COND HEAT & VEN 64:87-90, April, 1967.

"Some particular problems of noise control," by R. E. Fischer. ARCH REC 144:185-192, September, 1968.

"Sound control in multi-purpose buildings," PARKS & REC 4:46-47, May, 1969.

"Sound control in office buildings," by R. C. Weber. SKYSCRAPER MGT 55:14-15, March; 20-24, April, 1970+.

NOISE ABATEMENT

"Sources of noise at camping areas and possibilities of their reduction," by E. Lange. Z GES HYG 14:143-145, February, 1968.

"The stability and control of conditioned noise aversion in the tilt cage," by M. Halpern, et al. J EXP ANAL BEHAV 9:357-367, July, 1966.

"Stopping all that racket," by K. A. Kaufman. IRON AGE 203:46, February 6, 1969.

"System keeps eye on noise," ELECTRONICS 41:197-199, June 10, 1968.

"Tape noise reduction center," by B. Whyte. AUDIO 54:10+, May, 1970.

"There's profit in noise control," IRON AGE 195:90-91, January 28, 1965.

"Tomorrow's markets: noise abatement; can marketers ignore the din?" SALES MGT 103:45-46+, November 10, 1969.

"U.S. mapping program to alleviate noise," by E. J. Bulban. AVIATION W 85:45+, October 24, 1966.

"Vibration and structure-borne noise control," by L. L. Eberhart. ASHRAE J 8:54-60, May, 1966.

"What you should know about noise control," MOD MATERIALS HANDLING 25:43-45, April, 1970.

NOISE MEASUREMENT

see also: Noise Measurement Devices

"Accuracy consideration in fan sound measurement," by P. K. Baade. ASHRAE J 9:94-102, January, 1967.

"Acoustic-noise measurements on nominally identical small electrical machines," by A. J. Ellison, et al. INST E E PROC 117: 555-560, March, 1970.

NOISE MEASUREMENT

"Acoustic power measuring device," by C. J. Moore, et al. J SCI INSTR series 2, 1:659-661, June, 1968.

"Acoustical measurements by time delay spectrometry," by R. C. Heyser. AUDIO ENG SOC J 15:370-382, October, 1967.

"Acosutical measurements in New York's Philharmonic Hall," by R. M. Schroeder. BELL LAB REC 43:38-45, February, 1965.

"Adaptive optimum detection: synchronous-recurrent transients," by L. W. Nolte. J ACOUST SOC AMER 44:224-239, July, 1968.

"Adaptive threshold detection of M-ary signals in statistically undefiend noise," by J. B. Millard, et al. IEEE TRANS COM TECH 14:601-610, October, 1966.

"Amplitude quantization; a new, more general approach," by M. Vinokur. IEEE PROC 57:246-247, February, 1969.

"Audio frequency analyzer measures auto noise pollution," ELECTRO-TECH 85:24-25, January, 1970.

"Audio measurements course, continued," by N. H. Crowhurst. AUDIO 50:28+, December, 1966; 51:36+, January; 36-8+, February; 44-46+, March; 44-45+, April; 34+, June; 26-28+, July, 1967.

"Audiometric configurations associated with blast truama," by D. I. Teter, et al. LARYNGOSCOPE 80:1122-1132, July, 1970.

"Audiometry and the medical department's function in management of noise problems," by A. J. Murphy. NAT SAFETY CONGR TRANS 18:14-17, 1969.

"Autonomic reactions to hearing impressions," by G. Lehmann. STUD GEN 18:700-703, 1965.

"Autonomic reactions to low frequency vibrations in man," by R. Coermann, et al. INT Z ANGEW PHYSIOL 21:150-168, September 13, 1965.

"Better noise measurement pins down trouble spots," by R. W. Carson.

PRODUCT ENG 37:28-30, October 24, 1966.

"Breakthrough in noise measurement," OCCUP HEALTH 22:183, June, 1970.

"Centrifugal fan sound power level prediction," by G. C. Groff, et al. ASHRAE J 9:71-77, October, 1967.

"Comparative study on steady and intermittent exposures to noise. Ammonia content in the brain as an indicator," by K. Matsui, et al. JAP J HYG 23:225-228, June, 1968.

"Complex laboratories for measuring vibration and noise," by O. E. Guzeev, et al. GIG SANIT 35:100-101, April, 1970.

"Concepts of perceived noisiness, their implementation and application," by K. D. Kryter. J ACOUST SOC AMER 43:344-361, February, 1968.

"Costly decibels," by R. O. Fehr. AUDIO ENG SOC J 18:110-111, February, 1970.

"Current methods of evaluation of noise," by J. Calvet, et al. ANN OTOLARYNG 86:113-126, March, 1969.

"Design for quiet; measuring noise," by J. Campbell. MACHINE DESIGN 39:216-224, September 14, 1967.

"Determination of sound pressure levels created by noise sources at standard points," by I. K. Razumov. GIG SANIT 34:29-32, November, 1969.

"Dichotic summation of loudenss," by B. Scharf. ACOUSTICAL SOC AM J 45:1193-1205, May, 1969.

"Direct test of the power function for loudness," by L. E. Marks, et al. SCIENCE 154:1036-1037, November 25, 1966.

"Down with decibles!" by O. Schenker-Sprungli. UNESCO COURIER 20:4-7, July, 1967.

NOISE MEASUREMENT

"The effects of ambient noise upon signal detection," by A. Mirabella, et al. HUM FACTORS 9:277-284, June, 1967.

"Effects of imaging of signal-to-noise ratio, with varying signal conditions," by S. J. Segal, et al. BRIT J PSYCHOL 60:459-464, November, 1969.

"Effects of low frequency and infrasonic noise on man," by G. C. Mohr, et al. AEROSPACE MED 36:817-824, September, 1965.

"Effects of noise and difficulty level of input information in auditory, visual, and audiovisual information processing," by H. J. Hsia. PERCEPT MOTOR SKILLS 26:99-105, February, 1968.

"Effects of noise measurements in the factories of Katowice Province," by J. Grzesik, et al. MED PRACY 16:489-496, 1965.

"Establishment of objective criteria reflecting subjective response to roller-bearing noise," by R. F. Lucht, et al. J ACOUST SOC AMER 44:1-4, July, 1968.

"Evaluation and rating of sound," by R. J. Wells. ASHRAE J 9:48-51, April, 1967.

"Evoked responses to clicks and tones of varying intensity in waking adults," by I. Rapin, et al. ELECTROENCEPH CLIN NEUROPHYSIOL 21:335-344, October, 1966.

"Frequency discrimination following exposure to noise," by J. F. Brandt. J ACOUST SOC AMER 41:448-457, February, 1967.

"Frequency discrimination in noise," by G. B. Henning. ACOUSTICAL SOC AM J 41:774-777, April, 1967.

"Fundamental audio; loudness and the decibel," by M. Leynard. AUDIO 51:12+, February, 1967.

"Gauging a noise's annoyance level," ELECTRONICS 40:202, February 6, 1967.

"Guidelines for noise control specifications for purchasing equipment," IRON & STEEL ENG 47:95-98, May, 1970.

NOISE MEASUREMENT

"Hearing and noise," by R. Chocholle. ARH HIG RADA 20:47-54, 1969.

"How to estimate plant noises," by I. Heitner. HYDROCARBON PROCESS 47:67-74, December, 1968.

"Hunam tolerance to low frequency sound," by B. R. Alford, et al. TRANS AMER ACAD OPHTHAL OTOLARYNG 70:40-47, January-February, 1966.

"Lets measure excessive noise," FIN POST 61:23, May 27, 1967.

"Loudness and the decibel," by M. Leynard. AUDIO 51:12+, February, 1967.

"Loudness, a product of volume times density," by S. S. Stevens, et al. J EXP PSYCHOL 69:503-510, May, 1965.

"Loudness recruitment and its measurement with special reference to the loudness discomfort level test and its value in diagnosis," by M. R. Dix. ANN OTOL 77:1131-1151, December, 1968.

"Measurement and suppression of noise; with special reference to electrical machines, by A. J. King. A review," by K. A. Rose. ROY INST BRIT ARCH J 73:85, February, 1966.

--A review," by H. D. Parbrook. TOWN PLAN R 37:75-76, April, 1966.

"Measuring noise exposure," by S. S. Meyers. INSTRUMENTATION TECH 16:66, December, 1969.

"Method of estimating the sound power level of fans," by J. B. Graham. ASHRAE J 8:71-74, December, 1966.

"Methods of measurement of acoustic noise radiated by an electric machine," by A. J. Ellison, et al. INST E E PROC 116:1419-1431, August, 1969.

"New data on acute acoustic trauma (explosion trauma) as based on new measurement technics," by J. Mayer. MSCHR OHRENHEILK

101:305-312, 1967.

"A new individual noise dosimeter," by S. Lagerholm, et al. ACTA OTOLARYNG 224:Suppl:234+, June 27, 1966.

"A new method for rating noise exposure," by J. H. Botsford. AMER INDUSTR HYG ASS J 28:431-446, September-October, 1967.

"Noise-damage criterion using A weighting levels," by D. M. Mercer. J ACOUST SOC AMER 43:636-637, March, 1968.

"The noise dosimeter for measuring personal noise exposure," by S. Lagerholm, et al. ACTA OTOLARYNG 263:Suppl:139-144, 1969.

"Noise figure measurement," by C. N. G. Matthews. WIRELESS WORLD 73:393-394, August, 1967; Discussion 73:451, 506-507, 554-555, September-November, 1967.

"Noise measurement," by C. H. G. Mills. AUTOMOBILE ENG 60: 111-113, March, 1970.

"Noise measurement in a children's hospital and the awaking noise threshold of infant," by R. Gadeke, et al. MSCHR KINDERHEILK 116:374-375, June, 1968.

"Noise rating terminology, and usage," by D. H. Ball. AIR COND HEAT & REFRIG N 121:32, October 5; 26, October 26; 17, November 2; 10-12, November 16; 18+, December 14; 15, December 21, 1970.

"On the evaluation of acute noise trauma," by P. Plath, et al. Z LARYNG RHINOL OTOL 44:754-762, November, 1965.

"On the practice of the measurement of noise and vibrations in factories," by G. Wolff. ZBL ARBEITSMED 15:181-184, August, 1965.

"On the use of maximal value accumulator for the threshold frequency measurement in intermittent work noise," by H. G. Dieroff, et al. Z LARYNG RHINOL OTOL 47:58-63, January, 1968.

NOISE MEASUREMENT

"Peripheral hearing organ and accidental factors," by T. Shida. J OTOLARYNG JAP 72:653-656, February 20, 1969.

"Primer on methods and scales on noise measurement," by W. Rudmose. AM SOC SAFETY ENG J 14:18-26, October, 1969.

"Primer on sound level meters," by B. Katz. AUDIO 53:22-24, July; 42+, August, 1969.

"Problems assoicated with measurement of acoustic transients," by G. J. Harbold, et al. AEROSPACE MED 36:767-773, August, 1965.

"Quantification of the noisiness of approaching and receding sound," by G. Rosinger, et al. ACOUSTICAL SOC AM J 48:843-853, pt 1, October, 1970.

"Rail test to evaluate equilibrium in low-level wideband noise. AMRL-TR-66-85," by C. W. Nixon, et al. US AIR FORCE AEROSPACE MED RES LAB 1-16, July, 1966.

"Recent noise measurement techniques," by P. L. Tanner. ANN OCCUP HYG 10:375-380, October, 1967.

"Significance of intermission in the strong sound loading test," by T. Shida, et al. J OTOLARYNG JAP 73:Suppl:976-977, July, 1970.

"Simple method for identifying acceptable noise exposures," by J. H. Botsford. J ACOUST SOC AMER 42:810-819, October, 1967.

"Sound level measurements and frequency analyses in dental working areas," by R. Mayer, et al. DEUTSCH ZAHNAERZTL Z 23:800-805, August, 1968.

"Sound meter," ENGINEER 223:686, May 5, 1967.

"Standard procedure for noise computation," SAFETY MAINT 136: 37, December, 1968.

"Steps to measure floor noises," by E. G. Marklew. ENGINEERING

674-675, May 13.

"Survey of noise measurement methods," by J. R. Ranz. MACHINE DESIGN 38:199-206, November 10, 1966.

"Toward more uniform noise ratings," by R. Warren. AIR COND HEAT & VEN 66:40-44, January, 1969; Discussion 66:4+, March; 7+, June, 1969.

"Use of probe microphones to measure sound pressures in the ear," by B. M. Johnstone, et al. ACOUSTICAL SOC AM J 46:1404-1405, December, 1969.

"Use of sensation level in measurements of loundess and of temporary threshold shift; discussion," ACOUSTICAL SOC AM J 41: 714-715, March, 1967.

"Variables that influence sound pressures generated in the ear canal by an audio-metric earphone," by N. P. Erber. ACOUSTICAL SOC AM J 44:555-562, August, 1968.

"What is a sound-level meter?" by F. Van Veen. SAFETY MAINT 130:45, August, 1965.

"Why, where, and when of noise measurement," by A. D. Woewucki, Jr. SAFETY MAINT 134:42-44, December, 1967.

NOISE MEASUREMENT DEVICES
see also: Noise Measurement

"Aircraft noise monitor," ENGINEER 219:195, January 22, 1965.

"Development of a personal monitoring instrument for noise," by F. W. Church. AMER INDUSTR HYG ASS J 26:59-63, January-February, 1965.

"Highway noise monitor," MECH ENG 92:52, June, 1970.

"Jet-noise monitor," ELECTRONICS 38:211, January 11, 1965.

"Noise monitor tells of impending failure," PRODUCT ENG 38:168+, May 8, 1967.

NOISE MEASUREMENT DEVICES

"Performance monitoring technique for arbitrary noise statistics," by G. D. Hingorani, et al. IEEE TRANS COM TECH 16:430-435, June, 1968.

NOISE RESEARCH

"Abolition of milieu-induced hyperlipemia in the rat by electrolytic lesion in the anterior hypothalamus," by M. Friedman, et al. PROC SOC EXP BIOL MED 131:288-293, May, 1969.

"Acoustic trauma in the guinea pig. I. Electrophysiology and histology," by H. A. Beagley. ACTA OTOLARYNG 60:437-451, November, 1965.

--II. Electron microscopy including the morphology of cell junctions in the organ of Corti," by H. A. Beagley. ACTA OTOLARYNG 60:479-495, December, 1965.

"Aero-engine noise research laboratory," ENGINEER 224:6-8, July 7, 1967.

"Age differences in the control of acquired fear by tone," by J. P. Frieman, et al. CANAD J PSYCHOL 23:237-244, August, 1969.

"Age factor in the response of the albino rat to emotional and muscular stresses," by W. F. Geber, et al. GROWTH 30:87-97, March, 1966.

"Analysis of the complex electric response of the guinea pig cochlea to click," by C. Vesely, et al. SBORN VED PRAC LEK FAK KARLOV UNIV 10:Suppl:453-460, 1967.

"Animal experiment studies of the readaptation behavior of acoustic organ," by J. Theissing, et al. Z LARYNG RHINOL OTOL 47: 64-70, January, 1968.

"Application of an amelioration technic of the signal-noise relationship to the measurement of the visual evoked potentials in the adult rabbit," by P. Magnien, et al. C R ACAD SCI (D) 266:929-932, February 26, 1968.

"Audiogenic seizure susceptibility induced in C57BL-6J mice by prior auditory exposure," by K. R. Henry. SCIENCE 158:938-940, November 17, 1967.

"Background light, temperature and visual noise in the turtle," by W. R. Muntz, et al. VISION RES 8:787-800, July, 1968.

"Behavior of glucose-6-phosphatase, glycogen and PAS-positive substances in the liver of guinea pigs following chronic exposure to noise," by J. Jonek, et al. Z MIKR ANAT FORSCH 72:256-263, 1965.

"The blood system reaction to the occupational effect of vibration and noise," by I. A. Gribova, et al. GIG SANIT 30:34-37, October, 1965.

"Cardiac hypertrophy due to chronic audiogenic stress in the rat, Rattus norvegicus albinus, and rabbit, Lepus cuniculus," by W. F. Geber, et al. COMP BIOCHEM PHYSIOL 21:273-277, May, 1967.

"A case of lethal outcome in severe acoustic trauma in the dog," by L. P. Rudenko. ZH VYSSH NERV DEIAT PAVLOV 15:105-108, January-February, 1965.

"Changes in acetylcholine concentration in cerebral tissue in rats repeatedly exposed to the action of mechanical vibration," by Z. Brzezinska. ACTA PHYSIOL POL 19:919-926, November-December, 1968.

"Changes in the serum level of unsaturated fatty acids in rats repeatedly exposed to noise," by M. Vondrakova. ACTIV NERV SUP 7:236-238, August, 1965.

"Click-evoked response patterns of single units in the medial geniculate body of the cat," by L. M. Aitkin, et al. J NEUROPHYSIOL 29:109-123, January, 1966.

"Cochlear hair-cell damage in guiena pigs after exposure to impulse noise," by L. B. Poche, Jr., et al. J ACOUST SOC AMER 46:947-951, October, 1969.

"Combined effects of chronic audio-visual stress and thiouracil administration on the cholesterol-fed rat," by T. A. Anderson, et al. J CELL COMP PHYSIOL 66:141-145, October, 1965.

"Combined effects of I-131 and noise on the cardiac activity in the dog," by T. Mukhamedov. BIULL EKSP BIOL MED 59:43-46, February, 1965.

"Comparative electrophysiological, histological and biochemical studies on guinea pigs following application of white noise," by E. A. Schnieder, et al. ARCH KLIN EXP OHR NAS KEHL-KOPFHEILK 194:579-583, December 22, 1969.

"Comparative examinations of the visual field in glaucoma and the influence of noise," by E. Ogielska, et al. KLIN OCZNA 36: 351-354, 1966.

"Continuity effects with alternately sounded tone and noise signals," by L. P. Elfner. et al. MED RES ENGIN 5:22-23, 1966.

"Continuity in alternately sounded tone and noise signals in a free field," by L. F. Elfner. J ACOUST SOC AMER 46:914-917, October, 1969.

"Contribution to the study of acoustic stimulation on the blood eosinophil count," by A. Amorelli, et al. ARCH ITAL LARING 73:515-520, November-December, 1965.

"Crossed centrifugal inhibitions and excitations in the nucleus cochlearis due to long click sequences in guinea pigs," by R. Pfalz. ARCH KLIN EXP OHR NAS KEHLKOPFHEILK 186: 9-19, 1966.

"Demonstration of nyctohemeral variations of susceptibility to audiogenic crisis in Swiss albino mice," by C. Poirel. C R SOC BIOL 161:4611-1465, 1967.

"The development of the startle reflex in the postnatal ontogenesis of the rat," by I. Friedrich, et al. ACTA BIOL MED GERMAN 19:605-607, 1967.

"The dilute locus, pyridoxine deficiency, and audiogenic seizures in mice," by K. R. Henry, et al. PROC SOC EXP BIOL MED 128:635-638, July, 1968.

"EEG measures of arousal during RFT performance in 'noise'," by R. W. Hayes, et al. PERCEPT MOT SKILLS 31:594, October, 1970.

"Effect of the caudate nucleus stimulation on audiogenic seizures in rats," by M. M. Oleshko. FIZIOL ZH 16:443-447, July-August, 1970.

"Effect of cochlear lesions upon audiograms and intensity discrimination in cats," by D. N. Elliott, et al. ANN OTOL 74:386-408, June, 1965.

"Effect of dietary syntehtic and natural chelating agents on the zinc-deficiency syndrome in the chick," by F. H. Nielsen, et al. J NUTR 89:35-42, May, 1966.

"Effect of isolated mental stimuli on arterial pressure and body weight in Sprague-Dawley rats," by S. Campus, et al. BOLL SOC ITAL BIOL SPER 41:1087-1089, September 30, 1965.

"Effect of loud impulse noise on the hearing organ of animals," by N. I. Ivanov. VOEN MED ZH 7:24-27, July, 1970.

"Effect of noise in the animal house on seizure susceptibility and growth of mice," by W. B. Iturrian, et al. LAB ANIM CARE 18: 557-560, October, 1968.

"Effect of noise on auditive localization of persons with normal hearing and with hearing disorders of the conductive and receptive type," by J. Laciak. MED PRACY 17:420-434, 1966.

"The effect of prolonged noise on the immunobiological reactivity of experimental animals," by Kh. V. Storoshchuk, et al. VRACH DELO 4:97-99, April, 1967.

"The effect of prolonged noise on oxidative processes of the rat brain," by N. F. Svadkovskaia, et al. GIG SANIT 32:19-23, July, 1967.

"Effect of short-lasting deafness on the microphonic potential of the guinea pig and an attempt at the pharmacological influencing of this effect," by L. Faltynek. SBORN VED PRAC LEK FAK KARLOV UNIV 8:269-289, 1965.

"The effect of stress stimuli on metabolism of laboratory animals. I. The effect of industrial noise on the behavior of DNA, RNA and soluble proteins in the liver, as well as the relative weight of this organ in the guinea pig," by J. Stanosek, et al. INT ARCH ARBEITSMED 26:216-223, 1970.

"Effects of high intensity impulse noise and rapid changes in pressure upon stapedetomiczed monkeys," by J. L. Fletcher, et al. ACTA OTOLARYNG 68:6-13, July-August, 1969.

"Effects of illumination and white noise on the rate of electrical self-stimulation of the brain in rats," by E. R. Venator, et al. PSYCHOL REP 21:181-184, August, 1967.

"Effects of negative air ions, noise, sex and age on maze learning in rats," by R. A. Terry, et al. INT J BIOMETEOR 13:39-49, June, 1969.

"Effects of noise on cortical evoked potentials in rats," by Z. Chaloupka. ACTIV NERV SUP 10:207-208, May, 1968.

"Effects of prolonged acoustic stimulation on adrenal glands of the rat," by M. Ameli, et al. CLIN OTORINOLARING 18:211-252, May-June, 1966.

"Effects of unilateral section of the facial nerve in the orientation of the initial motor syndrome of acoustic epilepsy in mice," by J. Requin, et al. C R SOC BIOL 160:1285-1290, 1966.

"Evaluation of the effect of acoustic stimuli on the rat central nervous system by means of the swimming test in the labyrinth," by J. Grzesik, et al. ACTA PHYSIOL POL 16:379-387, May-June, 1965.

"Experience with acoustic sensory stress. V. Beharior of sodium and potassium in the rat submandibular gland in audiogenic stress,'' by G. Croce, et al. BOLL MAL ORECCH GOLA NASO 87:300-306, 1969.

''Experimental data on assessing the effect of continuous whole-body vibration on warm blooded animals,'' by D. A. Mickhel'son. GIG SANIT 34:129-130, September, 1969.

''Experimental research on the cochlear function after exposure to intense noise. I. Variations of the microphonic potentials of the intact resected muscles of the middle ear,'' by W. Mozzo. BOLL SOC ITAL BIOL SPER 44:400-403, March 15, 1968.

''Experimental studies of the cochlear potential behaviour after stellatum blockade. II. Cochlear microphonic potential after stellatum blockade to the ear predamaged by noise,'' by D. Kleinfeldt, et al. ARCH KLIN EXP OHR NAS KEHIKOPFHEILK 190:398-406, 1968.

''Experimental studies on cochlear damage of rabbits following acoustic stimulation,'' by S. Mizuno. J OTOLARYNGOL JAP 73:1577-1594, October, 1970.

''Experimental study of the cochlear injury by sonic stimulation in rabbits,'' by H. Ouchi, et al. J OTOLARYNG JAP 72:372-373, February 20, 1969.

''Fluctuation of optic evoked potentials during conditioned avoidance behavior in cats: effects of attention and distraction on primary evoked potentials,'' by T. Ikeda. FOLIA PSYCHIAT NEUROL JAP 21:19-30, 1967.

''Genetics of autiogenic seizures. I. Relation to brain serotonin and norepinephrine in mice,'' by K. Schlesinger, et al. LIFE SCI 4:2345-2351, December, 1965.

--II. Effects of pharmacological manipulation of brain serotonin, norepinephrine and gamma-aminobutyric acid,'' by K. Schlesinger, et al. LIFE SCI 7:437-447, April 1, 1968.

''Harmonic configuration and audiological aspects of acoustic pictures evoked by auditory fatigue. Experimental study with low frequency generator tones,'' by F. Fruttero. MINERVA OTORLINOLARING 17:83-93, May-June, 1967.

''Heart rate response of anesthetized and unanesthetized dogs to noise and near-vacuum decompression,'' by J. P. Cooke, et al. AEROSPACE MED 37:704-709, July, 1966.

''Histochemical investigations of kidneys of guinea pigs following chronic noise influence,'' by J. Jonek, et al. Z MIKR ANAT FORSCH 73:1-13, 1965.

''Histochemical investigations on the behavior of some enzymes in the adrenal glands in guinea pigs following chronic noise influence,'' by J. Jonek, et al. Z MIKR ANAT FORSCH 73:174-186, 1965.

''Histochemical studies of the inner ear of guinea pigs subjected to industrial noise,'' by S. Chodynicki, et al. OTOLARYNG POL 22:831-838, 1968.

''Histologic study of guinea pig cochlea after acoustic trauma,'' by J. Calvet, et al. REV OTONEUROOPHTALMOL 42:150-152, April, 1970.

''Histological and histochemical aspects of the vaginal cycle of the mature rat during experiments with prolonged acoustic stimulation,'' by M. De Marini, et al. CLIN OTORINOLARING 18:380-402, July-August, 1966.

''Histological and histochemical changes in the dog brain under the effect of strong sound stimuli,'' by I. I. Tokarenko. FIZIOL ZH 15:272-279, March- April, 1969.

''Histopathologic changes of the hearing organ in cats caused by noise,'' By A. Andrevski, et al. GOD ZBORN MED FAK SKOPJE 15:23-35, 1969.

''Influence of acoustic stimuli on electroretinographic tracings of normal subjects,'' by G. Maffei, et al. RIV OTONEUROOFTAL 41:503-510, November-December, 1966.

"Influence of experimental vigil on the catecholamine content of rat suprarenal glands," by O. D. Kumanova. UKR BIOKHIM ZH 40:446-448, 1968.

"The influence of mechanical noise on the activity rhythms of finches," by M. Lohmann, et al. COMP BIOCHEM PHYSIOL 22: 289-296, July, 1967.

"Influence of rocket noise upon hearing in guinea pigs," by G. Gonzalez, et al. AEROSPACE MED 41:21-25, January, 1970.

"Influence of stress (electric and audiogenic) on the development of Walker 256 carcinosarcoma in rats," by S. M. Milcu, et al. STUD CERCET ENDOCR 19:131-137, 1968.

"Influence of 3 types of sound stimulation on the blood level of non-esterified fatty acids (FFA) in the sheep," by J. Bost, et al. C R SOC BIOL 160:2340-2343, 1966.

"The influence of various neuroleptic drugs on noise escape response in rats," by C. J. Niemegeers, et al. PSYCHOPHARMACOLOGIA 18:249-259, 1970.

"Instrumental escape conditioning to a low-intensity noise by rats," by A. K. Myers. J COMP PHYSIOL PSYCHOL 60:82-87, August, 1965.

"Interactions between synchronous neural responses to paired acoutsic signals," by D. C. Teas. J ACOUST SOC AMER 39: 1077-1085, June, 1966.

"Long-term observations on the auditory acuity of workers in a noisy environment," by T. Yokoyama, et al. J OTOLARYNG JAP 71:640-672, May, 1968.

"Masked tonal thresholds in the bottlenosed porpoise," by C. S. Johnson. J ACOUST SOC AMER 44:965-967, October, 1968.

"The mechanism of action of noise on the organism," by E. Ts. Andreeva-Galanina, et al. VESTN AKAD MED NAUK SSSR 24: 11-18, 1969.

"Metabolism of ammonia under some environmental conditions. 4. Mechanism of increase in the ammonia content of the brain and liver of animals exposed to noise," by C. Sakaguchi. JAP J HYG 21:33-37, April, 1966.

--5. Glutamine synthetase activity in the rat under noise," by C. Sakaguchi. JAP J HYG 21:296-298, October, 1966.

"Methodology for the use of primates in the exploration of hazardous environments," by D. N. Farrer. ANN NY ACAD SCI 162:635-645, July 3, 1969.

"Morphologic changes in the hearing organ of experimental animals under the effect of high frequency vibration and noise," by I. P. Enin. VESTN OTORINOLARING 27:25-29, January-February, 1965.

"Morphological and histochemcial characteristics of the rat endometrium after prolonged acoustic stimulation," by M. Ameli, et al. CLIN OTORINOLARING 18:354-379, July-August, 1966.

"Morphological changes in the hypothalamus in autonomic disorders caused by strong and auditory stimulus," by G. N. Krivitskaia, et al. ZH NEVROPAT PSIKHIAT KORSAKOV 66:1177-1183, 1966.

"Neuronal convergence of noxious, acoustic, and visual stimuli in the visual cortex of the cat," by K. Murata, et al. J NEUROPHYSIOL 28:1223-1239, November, 1965.

"Noise induced lesions with special reference to abortions in cattle," by E. Aehnelt. DTSCH TIERAERZTL WOCHENSCHR 77:543-547 continued, October 15, 1970.

"Observations upon the relationship of loudness discomfort level and auditory fatigue to sound-pressure level and sensation level," by J. D. Hood. J ACOUST SOC AMER 44:959-964, October, 1968.

"On the effect of repeated noise stress on rats," by V. Hrubes, et al. ACTA BIOL MED GERMAN 15:592, 1965.

"On the non-specific effect of noise on the human organism. Conclusion," by J. Kubik. CESK HYG 10:553-559, October, 1965.

"On the 'paradoxical' character of sound intensification in some forms of neurologic diseases associated with decreased hearing," by A. I. Lopotko. VESTN OTORINOLARING 27:63-68, May-June, 1965.

"On the problem of the effect of general vertical vibration and noise on several indices of protein, fat and carbohydrate metabolism in warm blooded animals," by G. I. Bondarev, et al. GIG TR PROF ZABOL 12:58-59, October, 1968.

"On the problem of studying the effect of noise on the organism," by E. Ts. Andreeva-Galanina, et al. GIG SANIT 34:70-75, May, 1969.

"On the separation of useful signal from noise in the neural impulse activity of the cochlear nucleus in cats," by E. A. Radionova. ZH VYSSH NERV DEIAT PAVLOV 15:481-490, May-June, 1965.

"On the time related differences in the hearing capacity with and without hearing protection in relation to sudden noise of various duration and intensity," by F. Pfander. ARCH OHR NAS KEHLKOPFHEILK 185:488-510, 1965.

"On variations of the responses to tests of liminal sensation decay: comparative evaluation," by G. Cervellera, et al. ARCH ITAL LARING 73:Suppl:123-136, 1965.

"An optic and electron microscopic study of the organ of Corti of a mice line having convulsive crises caused by some sound frequencies," by D. Usui, et al. ANN OTOLARYNG 87:167-182, March, 1970.

"Otitis and onset of audiogenic seizures induced with rapid stimulation in mice of a resistant strain," by M. M. Niaussat, et al. C R SOC BIOL 164:57-59, 1970.

"The pathological sensory cell in the cochlea," by H. Engstrom. ACTA OTOLARYNG 63:Suppl:20-26, 1967.

NOISE RESEARCH

"Pathology of noise in its medico-social aspects," by G. Salvadori. FOLIA MED 49:333-358, May, 1966.

"Pathomorphological changes in the organs of white rats under the prolonged effect of noise," by V. P. Osintseva, et al. GIG TR PROF ZABOL 11:23-27, May, 1967.

"Physiologic responses of the albino rat to chronic noise stress," by W. F. Geber, et al. ARCH ENVIRON HEALTH 12:751-754, June, 1966.

"Plasma lipid responses of rats and rabbits to an auditory stimulus," by M. Friedman, et al. AMER J PHYSIOL 212:1174-1178, May, 1967.

"Prelinimary results of the research of noise effect on some vegetative functions," by L. Blazekova. PRAC LEK 18:276-279, August, 1966.

"Progressive alterations in cochlear nucleus, inferior colliculus, and medial geniculate responses during acoustic habituation," by M. Kitzes, et al. EXP NEUROL 25:85-105, September, 1969.

"Prolonged effect of noise of moderate intensity on the functional state of the organism," by O. P. Kozerenko, et al. IZV AKAD NAUK SSSR 4:527-536, July-August, 1967.

"Protection against audiogenic crises in mice by nicotinamide," by A. Lehmann, et al. J PHYSIOL 59:Suppl:446, 1967.

"Protein fractions of serum in animals under the effects of prolonged noise," by N. N. Pushkina, et al. VRACH DELO 9:98-100, September, 1968.

"Quantitative studies on the spiral ganglion of guinea pigs after exposure to noise," by W. Wicke, et al. MONATSSCHR OHRENHEILKD LARYNGORHINOL 104:433-430, 1970.

"Reaction of the hearing organ to sound stimulation in cochlear neuritis," by A. G. Rakhmilevich. VESTN OTORINOLARING 30:28-31, May-June, 1968.

NOISE RESEARCH

"Reactions of neurons in cochlear nucleus to acoustic signals of varying duration," by E. A. Radionova. FED PROC 25:389-390, May-June, 1966.

"Readjustment reactions of cerebrovascular circulation to chronic noise," by E. Betz. ARCH PHYS THER 17:61-65, January-February, 1965.

"Regeneration of glycogen in the hair cells of the organ of Corti," by S. Chodynicki. FOLIA HISTOCHEM CYTOCEHM 3:211-216, 1965.

"The relation between motor activity and risk of death in audiogenic seizure of DBS mice," by I. Lieblich, et al. LIFE SCI 4:2295-2299, December, 1965.

"Relationship between hearing changes and wide range noise intensity and duration of its action," by S. V. Alekseev, et al. GIG TR PROF ZABOL 9:47-49, March, 1965.

"The relationship between the recovery pattern in auditory fatigue and the noise susceptibility," by K. Tsunoda. J OTORHINOLARYNG SOC JAP 69:2088-2096, December, 1966.

"Report on repeated audiometric examinations of industrial workers after having been exposed to harmful noise for a period of 5 years," by S. Podvinec, et al. J FRANC OTORHINOLARYNG 15:53-60, January-February, 1966.

"The residue phenomenon in animal experiments," by C. C. Leidbrandt. NEDERL T GENEESK 109:1781, September 18, 1965.

"Response of neurons of the dorsal and posteroventral cochlear nuclei of the cat to acoustic stimuli of long duration," by J. M. Goldberg, et al. J NEUROPHYSIOL 29:72-93, January, 1966.

"Responses of inferior colliculus neurons in the cat to binaural acoustic stimuli having wide-band spectra," by C. D. Geisler, et al. J NEUROPHYSIOL 32:960-974, November, 1969.

"Retention of conditioned noise aversion following medial geniculate lesions in the rat," by M. Lyon. EXP NEUROL 16:1-15, September, 1966.

"Review of research and methods for measuring the loudness and noisiness of complex sounds. NASA Contract Rep NASA CR-422," by K. D. Kryter. US NASA 1-57, April, 1966.

"Review of the research into the injurious effect of noise," by S. Kubik. PRAC LEK 18:224-225, June, 1966.

"Rocket blasts and guinea pigs," SCI DIGEST 64:63-64, October, 1968.

"Selection of strains of rabbits sensitive to an epileptogenic sound stimulus," by F. Horak. PHYSIOL BOHEMOSLOV 14:495-501, 1965.

"Sensory stimulation and rhesus monkey activity," by W. A. Draper. PERCEPT MOTOR SKILLS 21:319-322, August, 1965.

"Short-term changes in the threshold of hearing after stimulation with noise and autonomic equilibrium," by R. Tomanek. GIG TR PROF ZABOL 12:14-18, June, 1968.

"Sound effects: massive test vibrator," MECH ENG 87:66, November, 1965.

"Startle response of rats after the production of lesions at the junction of the mesencephalon and the diencephalon," by S. G. Carlsson. NATURE (Lodnon) 212:1504, December 24, 1966.

"Startling noise and resting refractive state," by N. Roth. BRIT J PHYSIOL OPT 23:223-231, 1966.

"Stomatologic research: high speed handpieces and loss of hearing," by H. Gelb. NEW YORK J DENT 35:353-354, December, 1965.

"Studies of audiogenic hypertension. I. Preventive effect of ethylcrotonylurea, maprobamate and pentobarbital on experimental audiogenic hypertension in rats,'' by F. Gesmundo. BOLL SOC ITAL BIOL SPER 43:647-651, June 15, 1967.

''Studies on the acoustic stress action with the simultaneous action of horizontal mechanical vibration of lower frequencies on biochemical changes in guinea pigs. II. Effect on the number of eosinophils in the blood on the ascorbic acid level in adrenals and on the behaviour of the relative organ weight,'' by A. Tokarz-Lewandowska, et al. MED PRACY 16:278-282, 1965.

''Studies on acoustic stress with the effect of simultaneous horizontal mechanical low-frequency vibrations on biochemical changes in guinea pigs,'' by J. Gregorczyk, et al. MED PRACY 16:124-129, 1965.

''Studies on the action of the pulse after a given amount of industrial noise. Economic research program with the help of sequential analysis. I.,'' by A. Fuchs-Schmuck. INT Z ANGEW PHYSIOL 22:1-9, March 3, 1966.

''Studies on behaviour of the pulse following definite industrial noise. Economical experimental design, using sequential analysis. II.,'' by A. Fuchs-Schmuck. INT Z ANGEW PHYSIOL 23: 345-353, March 7, 1967.

''Studies on the effect of acoustic and ultraacoustic fields on biochemical processes. IX. Effect on some blood components in workers under noise conditions,'' by S. Jozkiewicz, et al. ACTA PHYSIOL POL 16:727-737, September-October, 1965.

''Studies on the effects of acoustic stimuli on giunea pigs--experimental studies on the effects of long-term and repeated acoustic stimuli on the internal ear and pituitary--adrenal cortex system,'' by K. Sakashita. J OTORHINOLARYNG SOC JAP 70:1666-1701, October, 1967.

''Studies on the effect of acoustic stimuli on rats,'' by O. Ribari, et al. ACTA CHIR ACAD SCI HUNG 11:97-106, 1970.

''Studies on the effect of acoustic stress with simultaneous action of horizontal mechanical vibration of lower frequencies on the biochemical changes in guinea pigs. V. Effect on total lipids and phospholipids level and on total cholesterol and its esters content in guinea pigs blood serum," by J. Stanosek, et al. MED PRACY 16:434-437 continued, 1965.

"Studies on phosphorus metabolism in rabbit brain under some environmental factors. I. Changes in the concentrations of inorganic, organic and total phosphorus," by K. Nakao. JAP J HYG 21: 38-43, April, 1966.

"Studies on the polycystic ovaries of rats under continuous auditory stress," by K. B. Singh, et al. AM J OBSTET GYNECOL 108: 557-564, October 15, 1970.

"A study on the effect of repeated noise in rats," by V. Hrubes, et al. ACTIV NERV SUP 7:165-167, 1965.

"Temporary shifts in auditory thresholds of chinchilla after exposure to noise," by E. N. Peters. J ACOUST SOC AMER 37:831-833, May, 1965.

"Teratogenic effects of audiogenic stress in albino mice," by C. O. Ward, et al. J PHARM SCI 59:1661-1662, November, 1970.

"Threshold ultrasonic dosages for structural changes in the mammalian brain," by F. J. Fry, et al. J ACOUST SOC AM 48:Suppl 2:1413+, December, 1970.

"Timidity and metabolic elimination patterns in audiogenic-seizure susceptible and resistant female rats," by A. S. Weltman, et al. EXPERIENTIA 22:627-629, September 15, 1966.

"Tonal patterns of cochlear impairment following intense stimulation with pure tones," by F. Suga, et al. LARYNGOSCOPE 77:784-805, May, 1967.

"Total and free thiamine and thiamine diphosphate content of brains from rats with audiogenic epilepsy," by A. I. Goshev. VOP MED KHIM 15:581-583, November-December, 1969.

"A trial application of the labyrinth test in a water basin for studies on the combined action of phrenotropic drugs and strong acoustic stimuli in white rats," by J. Grzesik, et al. ACTA PHYSIOL POL 17:327-333, March-April, 1966.

"U.K., industry join in anti-noise research," by H. J. Coleman. AVIATION W 87:61+, August 14, 1967.

"Unilateral inhibition of sound-induced convulsions in mice," by R. L. Collins. SCIENCE 167:1010-1011, February 13, 1970.

"Water and alcohol consumption of mice sensitive or refractory to audiogenic crises," by A. Duveau, et al. C R SOC BIOL 160: 791-794, 1966.

NOISE STANDARDS

"American Conference of Governmental Industrial Hygienists' proposed threshold limit value for noise," by H. H. Jones. AMER INDUSTR HYG ASS J 29:537-540, November-December, 1968.

"Are you prepared for noise control?" by W. D. Huskonen. FOUNDRY 97:64-67, April, 1969.

"Compensation claims for loss of hearing: impact of standards," by F. E. Frazier. AMER ASS INDUSTR NURSES J 15:17-19 passim, May, 1967; also in ARCH ENVIRON HEALTH 10:572-575, April, 1965.

"Dofasco's program to save those eardrums," CAN BUS 43:29, June, 1970.

"Government plans enforcement of industrial noise standards," AUTOMATION 16:18+, November, 1969.

"Guidelines for noise exposure control," AMER INDUSTR HYG ASS J 28:418-424, September-October, 1967.

"Guidelines for noise exposure control," ARCH ENVIRON HEALTH 15:674-678, November, 1967.

"Guidelines for noise exposure control. Intersociety Committee Report-revised, 1970," J OCCUP MED 12:276-281, July, 1970.

NOISE STANDARDS

"Hirschorn says noise-control standards have industrial impact," AIR COND HEAT & REFRIG N 117:6, August 11, 1969.

"Hygienic assessment and establishment of standards for noise in the communications services," by A. P. Mikheev, et al. GIG TR PROF ZABOL 11:14-18, January, 1967.

"New medical standards for noise," by E. I. Denisov. GIG TR PROF ZABOL 14:47, May, 1970.

"Noise control also a medical question," by H. Wiethaup. THER GEGENW 107:1504-1506 passim, November, 1968.

"Noise control program is quiet success," by D. R. Carlson. MOD HOSP 105:82-85, December, 1965.

"Noise exposure control: guidelines," AMER ASS INDUSTR NURSES J 16:17-21, May, 1968.

"Noise study focuses on intakes, exhaust," AVIATION W 86:24, June 19, 1967.

"On hygienic standardization of medium- and high frequecny noise (experimental study)," by N. M. Paran'ko, et al. GIG TR PROF ZABOL 12:48-50, June, 1968.

"On noise and vibration exposure criteria," by H. E. von Gierke. ARCH ENVIRON HEALTH 11:327-339, September, 1965.

"On the problem of the basis for standard noise levels for adolescents," by E. A. Gel'tishcheva, et al. GIG SANIT 33:34-38, November, 1968.

"The possibility of modification of respiration by rhythmic acoustic stimuli," by H. J. Gerhardt, et al. Z LARYNG RHINOL OTOL 46:235-247, April, 1967.

"Professional risks in stomatological practice," by S. Bocskay, et al. STOMATOLOGIA 12:455-460, September-October, 1965.

"Progress towards standards for noise and audiometry," by C. N.

Davies. ANN OCCUP HYG 10:401-406, October, 1967.

"What belongs in acoustical specifications," by R. Farrell. ARCH REC 138:227-230+, September; 203-206, November, 1965.

"When standards are set, it will be thumbs down on noise," by K. Mansfield. PRODUCT ENG 40:56, May 5, 1969.

NOISE STUDIES

"Noise, air pollution study to be defined," PRODUCT ENG 39:29, April 22, 1968.

"The occurrence of noise disturbances in society. Two questionnaire studies," by E. Jonsson, et al. NORD HYG T 48:21-34, 1967.

"Relationship between auditory fatigue and noise exposure time," by Y. Katano. J OTORHINOLARYNG SOC JAP 69:1592-1602, September, 1966.

"Sieve-audiometric studies of workers exposed to noise," by M. Jonsson. DEUTSCH GESUNDH 22:2286-2289, November 30, 1967.

"Studies on the role of the state of dentition in the physiopathology of the auditory organ. IV. Studies on the effect of dentition on the etiology of acoustic trauma," by B. Semczuk. ANN UNIV CURIE SKLODOWSKA 22:173-178, 1967.

"Study of noise deafness," by T. Kawabata, et al. J OTOLARYNG JAP 72:396-397, February 20, 1969.

"A study of recruitment phenomenon of audition and its mechanism," by A. Watanabe. J OTORHINOLARYNG SOC JAP 68:1391-1403, November, 1965.

"Study of the residual noise of average evoked potentials," by D. Arnal, et al. ELECTROENCEPH CLIN NEUROPHYSIOL 27: 315-321, September, 1969.

NURSES

"The identity of the nurse in an industrial hearing conservation program," by A. J. Murphy. OCCUP HLTH NURS 17:32+, May, 1969.

"Nurse and noise," by P. N. Ghei, et al. NURS J INDIA 60:431 passim, December, 1969.

"Personal hearing protection: the occupational health nurse's challenge and opportunity," by R. B. Maas. OCCUP HEALTH NURS 17:25-27, May, 1969.

OCCUPATIONAL DEAFNESS

"Acute hearing loss: etiology, diagnosis and therapy," by H. Kricheldorff. Z AERZTL FORTBILD 59:562-564, May 15, 1965.

"The adaptation factor and aduitory rest in the appearance of occupational deafness," by L. Teodorescu, et al. OTORINOLARINGOLOGIE 10:109-121, April-June, 1965.

"The AMA and noise-induced hearing loss," by G. D. Taylor. ARCH OTOLARYNG 90:543, November, 1969.

"Atrophy of the long process of the incus in a patient with an occupational hearing disorder," by B. M. Gapanavichius. VESTN OTORINOLARINGOL 32:91-92, November-December, 1970.

"Behind the scenes; rise in hearing impairment attributed to increasing noise pollution," by B. Whyte. AUDIO 53:8+, March, 1969.

"Bilateral asymmetry in noise induced hearing loss," by J. E. Watson. ANN OTOL 76:1040-1042, December, 1967.

"A bio-physical law describing hearing loss," by E. R. Hermann. INDUSTR MED SURG 34:223-228, March, 1965.

"Chronic otitis media in protection against occupational deafness," by A. Monteiro, et al. HOSPITAL 70:1173-1178, November, 1966.

"Clinical, social and insurance evaluation of professional deafness due to noise," by E. Vensi. ANN LARING 64:337-343, 1965.

"Comparative provisions for occupational hearing loss," by M. S. Fox.

ARCH OTOLARYNG 81:257-260, March, 1965.

"The concept of susceptibility to hearing loss," by W. D. Ward. J OCCUP MED 7:595-607, December, 1965.

"Current prevention of occupational acoustic trauma," by W. Sulkowski. MED PRACY 18:51-59, 1967.

"Discovery and surveillance of occupational acoustic trauma in preventive medicine," by Y. Guerrier, et al. J FRANC OTORHINOLARYNG 14:237-247, May, 1965.

"Discussion on deafness," PRACTITIONER 194:691-693, May, 1965.

"Evaluation of the results of various hearing tests for noise deafness," by T. Kawabata. J OTOLARYNGOL JAP 73:1858-1873, December, 1970.

"Evaluation of the risks of hearing impairment due to industrial noise based on exposure parameters," by J. Grzesik. POL TYG LEK 25:1026-1028, July, 1970.

"Extra-auditory effects of noise as a health hazard," by J. R. Anticaglia, et al. AMER INDUSTR HYG ASS J 31:277-281, May-June, 1970.

"A few comments on hearing disorders caused by noise," by T. Takeuchi, et al. J OTOLARYNG JAP 73:Suppl:1002-1003, July, 1970.

"Further studies on the role of the dentition in the pathogenesis of occupational hearing defects caused by noise (acoustic trauma)," by B. Semczuk, et al. CZAS STOMAT 21:671-675, June, 1968.

"Hardness of hearing due to noise and noise deafness," by S. Mehmke. MED WELT 27:1595-1601, July 8, 1967.

"Hazards of the arc-air gouging process," by J. T. Sanderson. ANN OCCUP HYG 11:123-133, April, 1968.

OCCUPATIONAL DEAFNESS

"Hearing disorders caused by noise in loading personnel at a large civilian airport," by G. Pressel, et al. INT ARCH ARBEITSMED 26:231-249, 1970.

"Hearing loss from exposure to interrupted noise," by J. Sataloff, et al. ARCH ENVIRON HEALTH 18:972-981, June, 1969.

"Hearing loss from noise," by J. L. Konzen, et al. J OCCUP MED 8:388-389, July, 1966.

"Hearing loss in Canadian Army units," by R. W. Tooley. MED SERV J CANADA 21:173-176, March, 1965.

"Indemnity in occupational deafness," by P. Mounier-Kuhn, et al. J MED LYON 48:1691-1696, November 20, 1967.

"Industrial deafness," LAMP 25:13-14, November, 1968.

"Industrial deafness and the summed evoked potential," by T. G. Heron. S AFR MED J 42:1176-1177, November 9, 1968.

"Industrial noise; workers lose hearing," CHEM & ENG N 46:22, November 11, 1968.

"Industrial sudden deafness," by S. Kawata, et al. ANN OTOL 76: 895-902, October, 1967.

"Knotty problem of industrial hearing loss," by L. W. Larson. SAFETY MAINT 135:39-42, May, 1968.

"Military noise induced hearing loss: problems in conservation programs," by C. T. Yarington, Jr. LARYNGOSCOPE 78:685-692, April, 1968.

"Noise around us: report of Federal council for science and technology's task force on the problem of environmental noise," SCI N 94:541, November 30, 1968.

"Noise as a health hazard at work, in the community, and in the home," by H. H. Jones, et al. PUBLIC HEALTH REP 83:533-536, July, 1968.

OCCUPATIONAL DEAFNESS

"Noise induced hearing loss," by P. S. Rummerfield, et al. OCCUP HEALTH NURS 17:23-29 passim, November, 1969.

"Noise-induced hearing loss. Exposures to steady-state noise," by A. Cohen, et al. ARCH ENVIRON HEALTH 20:614-623, May, 1970.

"Noise--occupational hazard and public nuisance," INT NURS REV 13:42+, July-August, 1966.

"Noise: an occupational hazard and public nuisance," WHO CHRON 20:191-203, June, 1966.

"Noise an occupational hazard and public nuisance," by A. Bell. WHO PUBLIC HEALTH PAP 30:1-130, 1966.

"Observations on hearing loss in higher frequencies with chronic otitis media in excessively noisy environments," by T. Yokoyama, et al. J OTOLARYNG JAP 71:1428-1439, October, 1968.

"Occupational acoustic trauma in otosclerotic patients (considerations on the clinical and medicolegal evaluation related to 4 personal observations)," by A. Scevola. ARCH ITAL OTOL 78:474-486, July-August, 1967.

"Occupational damages of the ear," by E. Lehnhardt. ARCH OHR NAS KEHLKOPFHEILK 185:465-468, 1965.

"Occupational damages of the larynx," by E. Nessel. ARCH OHR NAS KEHLKOPFHEILK 185:474-477, 1965.

"Occupational deafness," by L. P. Sobrinho, et al. REV PAUL MED 70:259-269, June, 1967.

"Occupational deafness. Survey in a business concern in the Paris area," by M. J. Alibert. ANN OTOLARYNG 83:883-887, December, 1966.

"Occupational deafness and visual field," by H. Vynckier. ACTA OTORHINOLARYNG BELG 21:213-222, 1967.

OCCUPATIONAL DEAFNESS

"Occupational deafness in the German Democratic Republic," by H. Zenk. Z GES HYG 11:25-34, January, 1965.

"Occupational deafness in Luxembourg," by E. Faber. BULL SOC SCI MED LUXEMB 102:305-316, November, 1965.

"Occupational deafness in thermoelectric plants with Norberg methane motors," by N. Lo Martire, et al. ATTI ACCAD FISIOCR SIENA 16:202-212, 1967.

"Occupational deafness in a vocational school," by J. Knops, et al. ARCH BELG MED SOC 24:330-338, May, 1966.

"Occupational deafness in young workers in a noisy environment with ear protection," by Y. Harada. ARCH ITAL OTOL 77:157-165, January-February, 1966.

"Occupational disease potentials in the heavy equipment operator," by F. Ottoboni, et al. ARCH ENVIRON HEALTH 15:317-321, September, 1967.

"Occupational hearing loss and high frequency thresholds," JAMA 201:144, July 10, 1967.

"Occupational hearing loss and high frequency thresholds," by J. Sataloff, et al. ARCH ENVIRON HEALTH 14:832-836, June, 1967.

"Occupational hearing loss in relation to industrial noise exposure," O. el-Attar. J EGYPT MED ASS 51:183-192, 1968.

"Occupational hearing loss--recent trends and practices," by M. S. Fox. INDUSTR MED SURG 37:204-208, March, 1968.

"Occupational noise," by J. T. Sanderson. OCCUP HEALTH 18:61-71, March-April, 1966.

"Occupational noise-induced hearing loss," by R. Hinchcliffe. PROC ROY SOC MED 60:1111-1117, November 1, 1967.

"On the diagnosis and early detection of noise-induced deafness and on the prevention of acoustic trauma," by G. Fabian. Z GES HYG 14:508-510, July, 1968.

"On the question of occupationally related hearing difficulties in musicians," by M. Flach, et al. Z LARYNG RHINOL OTOL 45: 595-605, September, 1966.

"Ouch, that noise! Hearing loss gets compensation," FIN POST 64:4, September 19, 1970.

"Perceptive deafness in military aviation technicians caused by noise of F 104 G jet planes," by P. P. Castagliuolo. RIV MED AERO 29:Suppl:361-373, December, 1966.

"Prevalence of impaired hearing and sound levels at work," by J. H. Botsford. J ACOUST SOC AMER 45:79-82, January, 1969.

"Proactive inhibition, recency, and limited-channel capacity under acoustic stress," by D. Eldredge, et al. PERCEPT MOTOR SKILLS 25:85-91, August, 1967.

"Problems in the expert testimony of the occupational disease noise-related hearing disorders and deafness," by H. G. Boenninghaus. Z LARYNG RHINOL OTOL 44:578-582, September, 1965.

"Problems of deafness due to noise," by H. G. Dieroff. Z GES HYG 11:352-361, May, 1965.

"Professional hearing disorder during the influence of strong interrupted noise," by P. S. Kublanova, et al. ZH USHN NOS GORL BOLEZ 27:56-59, September-October, 1967.

"Remarks on the problem of diagnostic criteria of occupational hearing loss resulting from exposure to noise," by J. Borsuk, et al. OTOLARYNG POL 23:273-284, 1969.

"Reversible sudden deafness following a brief intensive sound," by M. Abrahamovic. CESK OTOLARYNG 17:200-205, August, 1968.

"The risk of occupational deafness in musicians," by M. Flach, et al. GERMAN MED MONTHLY 12:49-54, February, 1967.

OCCUPATIONAL DEAFNESS

"Statistical data from consultation in occupational deafness," by J. Bonnefoy. J MED LYON 48:1699-1700, November 20, 1967.

"Statistical data on consultation for occupational deafness," by J. Bonnefoy. J MED LYON 49:1835-1836, November 20, 1968.

"Traumatic deafness from nailing gun," by J. Delatour. J FRANC OTORHINOLARYNG 15:701-703, October, 1966.

OCCUPATIONAL HEALTH

"Effects of environmental noise," by D. A. Williams, et al. CANAD J PSYCHIAT NURS 10:9-12, October, 1969.

"Effects of noise on health," by G. Jansen. GERMAN MED MONTHLY 13:446-448, September, 1968.

"Environmental noise, hearing acuity, and acceptance criteria," by E. R. Hermann. ARCH ENVIRON HEALTH 18:784-791, May, 1969.

"Environmental noise pollution: a new threat to sanity," by D. F. Anthrop. BUL ATOM SCI 25:11-16, May, 1969.

"Implications of the changing environment to occupational health," by F. D. Yoder. OCCUP HEALTH NURS 18:23-25, July, 1970.

"Noise and the public health," by C. R. Bragdon. SCI & CIT 10: 183-184, September, 1968.

"Noise and your nerves," LIFE AND HLTH 84:6+, May, 1969.

"Noise from the viewpoint of public health, labor hygiene and economics," by A. Gotze, Jr. ORV HETIL 109:2153-2155, September 29, 1968.

"Noise, a major health problem: edited by A. Hamilton," by V. Knudsen. PARENTS MAG 45:66-68, February, 1970.

"Noise, the underrated health hazard," by S. Golub. RN 32:40-45, May, 1969.

OCCUPATIONAL HEALTH

"Report on the effects of noise on the worker. (Bristol Advisory Council on Occupational Health)," J ROY INST PUBLIC HEALTH 31:53-61, March-April, 1968.

OFFICE NOISE

"Acoustical control (within the office)," by D. Anderson. ADM MGT 27:66+, March, 1966.

"Bringing peace to the noisy office," by N. C. Crane. SUPERVISORY MGT 10:42-43, November, 1965.

"How to secure privacy of speech in offices," by E. H. Kone. OFFICE 62:14-15+, December, 1965.

"Noise conditioning shows results in office operation," FIN POST 62:1-23, May 25, 1968.

"Noise pollution in the office," by F. S. Burgen. OFFICE 71:120, January, 1970.

"Office noise and employee morale," ADM MGT 26:48-49, March, 1965.

"Sound control in office buildings," by R. C. Weber. SKYSCRAPER MGT 55:14-15, March; 20-24, April, 1970+.

OTOLOGY

"The environment in relation to otologic disease," by J. Sataloff, et al. ARCH ENVIRON HEALTH 10:403-415, March, 1965.

"Otology in industrial medicine," by R. L. Watson, Jr. INDUSTR MED SURG 36:731-734, November, 1967.

"Statistics in otology," by A. Morgon. ACTA OTOLARYNG 63:304-310, February-March, 1967.

OTOSCLEROSIS

"Otosclerosis in workers in a noisy environment," by B. Gerth. HNO 14:205-208, July, 1966.

OTOTOXICOSES

"Ototoxicoses under noise load," by M. Quante, et al. ARCH

OTOTOXICOSES

KLIN EXP OHREN NASEN KEHLKOPFHEILKD 196:233-237, 1970.

PC 63-14

"Acoustic trauma. Therapeutic trial of PC 63-14 (cogitum)," by P. J. Orsini, et al. ANN OTOLARYNG 86:209-212, March, 1969.

PERFORMANCE

see also: Psychology

"Reappraisal of the relationship between noise and human performance by means of a subsidiary task measure," by J. M. Finkelman, et al. J APPL PSYCHOL 54:211-213, June, 1970.

"Some effects of rhythmic distraction upon rhythmic sensori-motor performance," by J. J. Keenan. J EXP PSYCHOL 77:440-446, July, 1968.

PERSONALITY

"Noise-caused changes of fine motoricity and sensations of annoyance dependent on certain personality dimensions," by G. Jansen, et al. Z EXP ANGEW PSYCHOL 12:594-613, 1965.

"Personality and the slope of loudness function," by S. D. Stephens. Q J EXP PSYCHOL 22:9-13, February, 1970.

"Reactivity to noise and personality," by S. Dongier. REV MED PSYCHOSOM 9:283-285, October-December, 1967.

"Sound and the psyche," MECH ENG 90:40-41, August, 1968.

PERSPIRATION

"Studies of the effect of acoustic stimuli on the degree of acidification of sweat," by B. Semczuk, et al. OTOLARYNG POL 24:47-52, 1970.

PHYSIOLOGY

"Action of noise on the cerebral and peripheral rheogram and on the electroencephalogram," by M. Fusco, et al. FOLIA MED 48:88-98, February, 1965.

PHYSIOLOGY

"The c5 declivity; its interpretation on the basis of universally valid physiological concepts," by E. Lehnhardt. HNO 14:45-52, February, 1966.

"The effect of the addition of vitamin C to the diet on the human organism exposed to occupational hazards (in the presence of strong sound stimuli)," by A. M. Margolis, et al. VOP PITAN 25:34-38, March-April, 1966.

"The effect of noise and local vibration of permissible levels on the human organism," by N. M. Paran'ko, et al. GIG SANIT 32:25-30, September, 1967.

"The effect of noise on the human organism," by J. Kubik, et al. CESK STOMAT 69:339-343, May, 1969.

"The effect of octave noise bands on certain physiological functions of the organism," by S. V. Alekseev. GIG TR PROF ZABOL 12:27-31, June, 1968.

"Effect of oxidoreductive agents and high-energy compounds on the efficiency of the internal ear during acoustic load," by W. Jankowski, et al. OTOLARYNG POL 23:141-144, 1969.

"Effect of physical work and work under conditions of noise and vibration on the human body. I. Behavior of serum alkaline phosphatase, aldolase and lactic dehydrogenase activities," by J. Gregorczyk, et al. ACTA PHYSIOL POL 16:701-708, September-October, 1965.

"Effect of pulsed noise on the organism as a function of pulse periodicity," by G. A. Suvorov. GIG TR PROF ZABOL 13:23-27, December, 1969.

"Effect of supplementary ingestion of vitamins B1 and C on the function of the auditory organ in workers of 'noisy' trades," by Z. F. Nestrugina. ZH USHN NOS FORL BOLEZ 29:91-94, May-June, 1969.

"Fluctuation of nucleic acid activity in the organ of corti resulting from noise exposure," by R. Nakamura. J OTORHINOLAR-

YNG SOC JAP 70:1818-1827, November, 1967.

"Fluctuation of vitamin-B1 in blood and in the organ of Corti following exposure to noise," by S. Abiko. J OTORHINOLARYNG SOC JAP 69:1117-1133, June, 1966.

"Functional changes in cerebral blood supply and in hearing acuity occurring under the effect of noise," by L. N. Shkarinov, et al. GIG TR PROF ZABOL 14:23-26, November, 1970.

"Functional changes in the ear produced by high-intensity sound: 5.0-khz stimulation," by G. R. Price. ACOUSTICAL SOC AM J 44:1541-1545, December, 1968.

"Fundamental constraints to sensory discrimination imposed by two kinds of neural noise," by J. L. Stewart. BEHAV SCI 10: 271-276, July, 1965.

"Hemodynamic reactions during acoustic stimuli," by K. Klein, et al. WIEN KLIN WSCHR 81:705-709, October 8, 1969.

"The importance of the external ear for hearing in the wind," by H. Feldmann, et al. ARCH KLIN EXP OHR NAS KEHLKOPFHEILK 190:69-85, 1968.

"Inclusion of P32 in the internal organs during the effect of uninterrupted noise of low intensity," by N. F. Svadkovskaia, et al. VRACH DELO 2:105-107, February, 1969.

"Influence of fatigue on hearing in the presence of noise," by T. Bystrzanowska. OTOLARYNG POL 20:172-176, 1966.

"The influence of a loud acoustic stimulus on the ultra-low frequency acceleration ballistocardiogram in man," by P. J. Pretorius, et al. ACTA CARDIOL 22:238-246, 1967.

"Influence of noise on the cardiovascular system," by C. Gradina, et al. FIZIOL NORM PAT 16:357-367, 1970.

"Influence of noise on the heart rate and O2 consumption under moderate physical loading," by M. Quaas, et al. INT Z ANGEW

PHYSIOL 27:230-238, 1969.

"Influence of repeated 4-hour, intermittent, so-called 'muffled' noise on catecholamine secretion and pulse rate," by W. Hawel, et al. INT Z ANGEW PHYSIOL 24:351-362, 1967.

"Influence of systemic stresses on the development of an experimental inflammatory reaction. I. Effects of an auditory stress," by L. Thieblot, et al. J PHYSIOL 57:708-709, September-October, 1965.

"Influences of VARIABLE NOISE ON THE CEREBRAL ULTRASONIC ATTENUATION (CUSA) due to postural change and the function of concentration maintenance (TAF)," by E. Takakuwa, et al. JAP J HYG 23:527-529, February, 1969.

"Infrasound tests human tolerance," by H. M. David. MISS & ROC 17:31+, October 11, 1965.

"Inside every fat man," ECONOMIST 226:24+, March 2, 1968.

"Intermittent noise and the brain waves, especially alpha-wave," by E. Takakuwa, et al. JAP J HYG 23:370-373, October, 1968.

"Lateralization of sounds at the unstimulated ear opposite a noise-adapted ear," by E. C. Carterette, et al. SCIENCE 147:63-65, January 1, 1965.

"Magnitude of temporary threshold shift (TTS) in the audigram and 'response time' of noise-or detonation exposed ears as test of hearing organs endangered by detonation," by F. Pfander. ARCH KLIN EXP OHR NAS KEHLKOPFHEILK 191:586-590, 1968.

"A method of analyzing individual cortical responses to auditory stimuli," by C. W. Palmer, et al. ELECTROENCEPH CLIN NEUROPHYSIOL 20:204-206, February, 1966.

"Moderate acousti stimuli: the interrelation of subjective importance and certain physiological changes," by G. R. Atherley, et al. ERGONOMICS 13:536-545, September, 1970.

PHYSIOLOGY

"Morphologic changes in the central structures of the auditory analyzer under the prolonged effects of noise," by A. B. Strakhov, et al. BIULL EKSP BIOL MED 69:95-97, June, 1970.

"The morphology of the ganglion spirale cochleae," by B. Kellerhals. ACTA OTOLARYNG 226:Suppl:1-78, 1967.

"Noise and adrenal function," by H. Sakamoto, et al. JAP J CLIN MED 28:1621-1625, May, 1970.

"Noise and blood circulation," by R. Heinecker. DEUTSCH MED WSCHR 90:1107-1109, June 11, 1965.

"Noise and the ear," by L. Stein. PLANT ENG 23:54-55, May 1, 1969.

"Noise and emotional stress," by F. I. Catlin. J CHRONIC DIS 18:509-518, June, 1965.

"Noise and health," by G. Lehmann. UNESCO COURIER 20:26-27+, July, 1967.

"Noise as cause of disease," by G. Jansen. DEUTSCH MED WSCHR 92:2325-2328, December 15, 1967.

"Noise characteristics in the baby compartment of incubators. Their analysis and relationship to environmental sound pressure levels," by F. L. Seleny, et al. AMER J DIS CHILD 117:445-450, April, 1969.

"Noise in evoked cerebral potentials," by J. Woods, et al. ELECT TROENCEPH CLIN NEUROPHYSIOL 26:633, June, 1969.

"Noise pollution: how many decibels can we take?" by S. Sinclair. CAN BUS 43:22-24, 30+, June, 1970.

"Nucleic acid concentration in cerebral cortex nerve cells upon disruption of higher nervous activity," by I. I. Tokarenko. ZH VYSSH NERV DEIAT 19:692-697, July-August, 1969.

"Occlusion of the auditory canal and body-conducted sound," by

H. G. Dieroff. Z LARYNG RHINOL OTOL 44:417-426, June, 1965.

"On changes in lipid metabolism in persons during prolonged action of industrial noise on the central nervous system," by P. S. Khomulo, et al. KARDIOLOGIIA 7:35-38, July, 1967.

"On the character of changes in the enzyme activity in the brain tissue during reflex epilepsy," by A. A. Pokrovskii, et al. ZH VYSSH NERV DEIAT PAVLOV 15:120-127, January-February, 1965.

"On the effect of low frequency ultrasonic waves and high frequency sound waves on the organism of worker," by V. K. Dobroserdov. GIG SANIT 32:17-21, February, 1967.

"On the effect of noise on antibody formation," by N. N. Klemparskaia. GIG TR PROF ZABOL 10:54-56, June, 1966.

"On the occurrence of nervous disorders in the residential districts of Berlin. Preliminary report," by G. Feuerhahn, et al. PSYCHIAT NEUROL MED PSYCHOL 20:281-286, August, 1968.

"On the problem of the action of continuous spectrum noise on some physiological functions of the organism," by S. V. Alekseev, et al. GIG TR PROF ZABOL 9:8-11, June, 1965.

"On substantiation of the method of studying higher nervous activity during the effect of noise," by S. V. Alekseev, et al. GIG TR PROF ZABOL 11:35-39, May, 1967.

"On unification of physiological methods in studying the effect of noise on the human organism," by E. Ts. Andreeva-Galanina, et al. GIG TR PROF ZABOL 11:14-18, October, 1967.

"Oxygen consumption in the organ of Corti based on observation of the dehydrogenase system following exposure to noise," by T. Takahashi. J OTORHINOLARYNG SOC JAP 70:1702-1715, October, 1967.

"The physical and psychophysical characteristics of sound and noise,'' by A. C. Raes. ARCH BELG MED SOC 24:305-315, May, 1966.

''Physiological effects of audible sound; AAAS symposium, December 28-30, 1969,'' by B. L. Welch. SCIENCE 166:533-534, October 24, 1969.

''Psychological and physiological as well as organic damage due to the effect of noise and high sounds,'' by K. Jatho. OEFF GESUNDHEITSWESEN 29:293-298, July, 1967.

''Studies on the relation between cerebral ammonia content and intensity of noise exposure,'' by K. Matsui, et al. JAP J HYG 22:478-480, October, 1967.

''Studies on thiamine metabolism in the brain and liver under various environmental conditions,'' by K. Horio. JAP J HYG 22:487-495, October, 1967.

''Studies on the variations of monoamineoxidase and acetylcholinesterase activities in the brain under physical environmental conditions. 2. Influences of different physical environmental conditions on monoamineoxidase and acetylcholinesterase activities,'' by T. Kojima. JAP J HYG 21:20-26, April, 1966.

''Tolerable limit of loudness: its clinical and physiological significance,'' by J. D. Hood, et al. J ACOUST SOC AMER 40:47-53, July, 1966.

''Psychological and physiological reactions to noise of different subjective valence (TTS and EMG),'' by H. Hormann, et al. PSYCHOL FORSCH 33:289-309, 1970.

''The reaction of the cardiovascular system of working adolescents to sound and vibration stimuli,'' by A. I. Tsysar','' GIG SANIT 31:33-38, February, 1966.

''Reactions of the human nervous and cardiovascular systems to the effects of aviation noises,'' by V. G. Terent'ev, et al. VOENNO-MED ZH 6:55-58, June, 1969.

PHYSIOLOGY

"Readjustment reactions of cerebrovascular circulation to chronic noise," by E. Betz. ARCH PHYS THER 17:61-65, January-February, 1965.

"Regional vascular reactions during exposure to vibration and noise, individually and together," by N. M. Paran'ko. GIG SANIT 33: 19-24, November, 1968.

"Resolution of ballistographic records into cardiac and respiratory components," by P. M. Rautaharju, et al. CANAD J PHYSIOL PHARMACOL 44:691-700, September, 1966.

"Some physiological factors in noise-induced hearing loss," by M. Lawrence, et al. AMER INDUSTR HYG ASS J 28:425-430, September-October, 1967.

"Some physiological reactions found in persons experimentally exposed to sounds of 500 and 1000 Hz (80 plus or minus 3 db. and 90 plus or minus 2 db.)," by C. Gardina, et al. FIZIOL NORM PAT 14:453-460, September-October, 1968.

"Stresses in skin panels subjected to random acoustic loading," by B. L. Clarkson. AERONAUTICAL J 72:1000-1010, November, 1968.

"Studies of the effect of noise on the cardiorespiratory efficiency," by B. Semczuk, et al. POL TYG LEK 25:83-85, January 19, 1970.

"Studies on the dependency of the intraocular pressure from noise," by H. P. Vick. KLIN MBL AUGENHEILK 153:356-360, 1968.

"Studies on the influence of acoustic stimuli on respiratory movements," by B. Semczuk. POL MED J 7:1090-1096, 1968.

PILE DRIVERS

"Just a shimmy, no bang-bang: pile driver in Schenectady, N.Y.," AM CITY 80:24, September, 1965.

PILOTS

see also: Aircraft Noise

PILOTS

"Airplane cockpit noise levels and pilot hearing sensitivity," by K. J. Kronoveter, et al. ARCH ENVIRON HEALTH 20:495-499, April, 1970.

"Cockpit noise environment of airline aircraft," ACOUSTICAL SOC AM J 47:449, February, 1970.

"Cockpit noise environment of airline aircraft," by R. B. Stone. AEROSPACE MED 40:989-993, September, 1969.

"Cockpit noise intensity: fifteen single-engine light aircraft. AM 68-21," by J. V. Tobias. US FED AVIAT AGENCY OFFICE AVIAT MED 1-6, September, 1968.

"Cockpit noise intensity: fifteen single-engine light aircraft," by J. B. Tobias. AEROSPACE MED 40:963-966, September, 1969.

"Cockpit noise intensity: three aerial application (cropdusting) aircraft," by J. V. Tobias. J SPEECH HEARING RES 11:611-615, September, 1968.

"Considerations on evaluation of acoustic and vestibular damage in pilots and specialists of military aeronautics," by C. Koch, et al. MINERVA MED 56:3832-3835, November 10, 1965.

"Lateralization of hearing loss and vestibular nystagmus in test pilots," by A. Bruner, et al. AEROSPACE MED 41:684-687, June, 1970.

"Statistical evaluation of hearing losses in military pilots," by G. von Schulthess, et al. ACTA OTOLARYNG 65:137-145, January-February, 1968+.

"Studies on noise hazards for commerical aviators," by W. Lorenz, et al. Z GES HYG 13:10-17, January, 1967.

PIPELINES

"Noise abatement sprayed on pipelines," PIPELINE & GAS J 197: 80, September, 1970.

PLANTS

"Effect of random noise on plant growth," by C. B. Woodlief. ACOUSTICAL SOC AM J 46:481-482, pt 2, August, 1969.

"Rules lowering noise will increase plant costs," MOD MANUF 2:98-100, September, 1969.

PNEUMATIC MACHINERY AND TOOLS

see: Compressed Air Noise

POWER PLANTS

see also: Motors: Electrical

"Power plants abroad stress noise control," POWER 113:88, March, 1969.

"Some aspects of noise in an industrial environment: research performed on a group of hydroelectric power plants," by A. Lacquaniti, et al. MED LAVORO 58:41-59, January, 1967.

POWER STATION NOISE

see: Transformer Noise

PRODUCT ENGINEERING

"Acoustical theory being forced upon fluid power engineers," by G. W. Kamperman. PRODUCT ENG 40:54+, December 15, 1969.

"Quiet revolution in design: rooting out noise sources," by A. Hannavy. PRODUCT ENG 37:50-57, October 24, 1966.

PROTECTION

"Another tool for hearing conservation--an improved protector," by C. Zenz, et al. AMER INDUSTR HYG ASS J 26:187-188, March-April, 1965.

"Attenuation provided by fingers, palms, tragi, and V51R ear plugs," by H. H. Holland, Jr. ACOUSTICAL SOC AM J 41:1545, June, 1967.

"Basics of hearing protection," ENVIRONMENTAL CONTROL MGT 138:60-62, December, 1969.

PROTECTION

"Block that noise! try earmuffs," PULP & PA 41:58, September 11, 1967.

"The challenge of hearing protection," by R. Maas. INDURST MED SURG 39:124-128, March, 1970.

"Choosing ear protectors," SUPERVISORY MANAGEMENT 15:32-35, January, 1970.

"Clinico-physiologic assessment of the efficacy of certain types of modern antinoise ear caps," by N. Ia. Shalashov. GIG TR PROF ZABOL 14:46-47, May, 1970.

"Dampening properties of hearing protective devices with the view to the degree of inconvenience in their use," by J. Stikar, et al. CESK HYG 10:304-313, June, 1965.

"Design factors and use of ear protection," by G. G. Rice, et al. BRIT J INDUSTR 23:194-203, July, 1966.

"An ear defender with peak limited sound transmission--Erdefender," by P. J. Barker. OCCUP HLTH 20:67+, March-April, 1968.

"An ear plug designed to obtund the sound of high speed dental engines," by R. B. Sloane. DENT DIG 72:218-220, May, 1966.

"Ear plugs--or--damaged hearing," by R. R. Coles. J ROY NAV MED SERV 51:19-22, Spring, 1965.

"Ear protectors," by W. Melnick. OCCUP HEALTH NURS 17:28-31, May, 1969.

"Ear protectors; their usefulness and limitations," by P. L. Michael. ARCH ENVIRON HEALTH 10:612-618, April, 1965.

"Ear stoppies. A defense against snoring," by M. H. Boulware. J FLORIDA MED ASS 57:36, May, 1970.

"East Range Symposium. V. Personal protection in noise exposure," by M. T. Summar. J OCCUP MED 7:279-280, June, 1965.

PROTECTION

"Effect of ambient-noise level on threshold-shift measures of ear-protector attenuation," by R. Waugh. J ACOUST SOC AM 48: 597, August, 1970.

"Effect of hearing protectors on speech comprehensibility," by Z. Novotny. CESK OTOLARYNG 18:260-264, December, 1969.

"Effect of resonance hearing protectors in lowering acoustic trauma in the army," by J. Dominik, et al. VOJ ZDRAV LISTY 34:146-149, August, 1965.

"Effects of ear protection on communication," by W. I. Acton. ANN OCCUP HYG 10:423-429, October, 1967.

"The efficiency of several types of personal noise protectors and their selection as a function of the conditions of use," by L. N. Shkarinov, et al. GIG TR PROF ZABOL 10:38-43, June, 1966.

"Evaluation of devices for personal hearing protection against noise with special regard to their functional efficiency," by V. Psenickova, et al. PRAC LEK 17:313-317, September, 1965.

"Experience with glass fiber earplugs," by C. Zenz, et al. AMER INDUSTR HYG ASS J 28:499-500, September-October, 1967.

"Hearing protection in industry," by R. E. Scott. SAFETY MAINT 129:40-41, January, 1965.

"Hearing protection: keeping up with developments," SAFETY MAINT 137:35-40, February, 1969.

"Hearing protection pitfalls and how to avoid them," by F. P. Haluska. SAFETY MAINT 131:32-34+, January, 1965.

"How to plan a small plant hearing protection program," by R. J. Beaman. SAFETY MAINT 130:36-38, December, 1965.

"Keeping up with developments; hearing protection," ENVIRONMENTAL CONTROL MGT 139:47-50, February, 1970.

"Keeping up with developments; hearing protection," SAFETY MAINT 133:43-48, February, 1967.

PROTECTION

''Modern protection of hearing,'' by H. Gronemann. ZBL ARBEITS-MED 16:371-372, December, 1966.

''Noise protection in the industry,'' by H. Schmidt. MUNCH MED WOCHENSCHR 112:Suppl:52-53, December 25, 1970.

''On the use of sound-protection cotton (Billesholm),'' by F. Schwetz, et al. MSCHR OHRENHEILK 103:260-263, 1969.

''Personal ear protection,'' by W. I. Acton. OCCUP HEALTH 22: 315-320, October, 1970.

''Personal hearing protection: the occupational health nurse's challenge and opportunity,'' by R. B. Maas. OCCUP HEALTH NURS 17:25-27, May, 1969.

''Practical ear enclosure with selectively coupled volume,'' by A. L. DiMattia. AUDIO ENG SOC J 15:295-298, July, 1967.

''Practical experiences with individual noise protection devices,'' by E. Glock, et al. Z GES HYG 14:413-418, June, 1968.

''Problems of communication and ear protection in the Royal marines,'' by M. R. Forrest, et al. J ROY NAV MED SERV 56: 162-169, Spring, 1970.

''Problems of protection from noise in the potassium mining industry,'' by H. Wolf. Z GES HYG 12:189-195, March, 1966.

''Protecting hearing at Vulcan Mold & Iron Co.,'' by J. P. Donnelly. FOUNDRY 98:49-51, July, 1970.

''Protecting residual hearing in hearing aid user,'' by M. Ross, et al. ARCH OTOLARYNG 82:615-617, December, 1965.

''Protection against immission,'' by H. Wiethaup. ZBL ARBEITS-MED 15:237-243, October, 1965.

''Protection from bumps and noise,'' ENGINEERING 204:446, September 22, 1967.

''Protection of ground personnel against the noise of jet aircraft engines," by J. V. Quercy. ARCH MAL PROF 27:537-541, June, 1966.

"Recent developments in ear protection," by R. R. Coles. PROC R SOC MED 63:1016-1019, October, 1970.

"Safety helmet tunes out noise (blast cleaning)," IRON AGE 205: 114N, April 23. 1970.

"Sound attenuation provided by perforated earmuffs. SAM-TR-68-86," by H. C. Sutherland, Jr., et al. US AIR FORCE SCH AEROSPACE MED 1-4, September, 1968.

"Testing and evaluating hearing protectors," by F. P. Beguin. AM ASS INDUSTR NURSES J 13:11+, April, 1965.

"Trials and measurement results of a plastic ear mould as individual protection from noise," by C. J. Partsch, et al. HNO 14: 107-109, April, 1966.

"Wearing ear protectors in metal factories," by A. Lefort. ARCH MAL PROF 30:357-360, June, 1969.

"What goes into a hearing protection program?" SAFETY MAINT 132:37-40, July, 1966.

PROTEINS

"The effect of aviation noise on some indices of protein and vitamin metabolism," by Iu. F. Udalov, et al. VOENNOMED ZH 7:61-64, July, 1966.

"The effect of vibration and noise on protein metabolism of excavator-machinists using several biochemical indices," by A. M. Tambovtseva. GIG SANIT 33:58-61, January, 1968.

PROPELLERS

"Propeller research gains emphasis," by M. L. Yaffee. AVIATION W 91:56-57+, November 24, 1969.

PSYCHOLOGY

"Channel-capacity, intelligibility and immediate memory," by P. M.

Rabbitt. QUART J EXP PSYCHOL 20:241-248, August, 1968.

"Correlations between the noise level and conditioned reflex performance," by K. Hecht, et al. ACTA BIOL MED GERMAN 23: 133-143, 1969.

"Determination of simple reaction time to visual and acoustic stimuli under conditions of noise interference," by K. Witecki. POL TYG LEK 25:1012-1015, July 6, 1970.

"Differential effect of noise on tasks of varying complexity," by D. H. Boggs, et al. J APPL PSYCHOL 52:148-153, April, 1968.

"Distraction and Stroop Color-Word performance," by B. K. Houston, et al. J EXP PSYCHOL 74:54-56, May, 1967.

"Effect of the noise level on the productivity of work," by S. D. Kovrigina, et al. GIG SANIT 30:28-32, April, 1965.

"Effect of noise on psychological state," by A. Cohen. AM SOC SAFETY ENG J 14:11-15, April, 1969.

"Effect of noise on the reaction time to auditive stimuli," by S. Barbera, et al. BOLL MAL ORECCH 84:202-206, May-June, 1966.

"Effects of environmental factors on performance; abstract," by F. J. Vilardo. MACHINE DESIGN 41:150+, November 27, 1969.

"Effects of noise on a complex task," by S. J. Samtur. GRAD RES ED 4:63-81, Spring, 1969.

"How much does noise bother apartment dwellers? reprint," by D. R. Prestemon. ARCH REC 143:155-156, February, 1968.

"Human performance as a function of changes in acoustic noise levels," by R. W. Shoenberger, et al. J ENGIN PSYCHOL 4:108-119, 1965.

"Human response to intense low-frequency noise and virbation," by J. C. Guignard. INST MECH ENG PROC 182,3:55-59;

Discussion 78-83; Reply 84, 1967-1968.

"Human response to measured sound pressure levels from ultrasonic devices," by C. P. Skillern. AMER IDNUSTR HYG ASS J 26:132-136, March-April, 1965.

"Human responses to sonic boom," by C. W. Nixon. AEROSPACE MED 36:399-405, May, 1965.

"Human tolerance to low frequecny sound," by B. R. Alford, et al. TRANS AMER ACAD OPHTHAL OTOLARYNG 70:40-47, January-February, 1966.

"Individual reaction to noise," by A. Hedri. PRAXIS 57:1168-1169, August 27, 1968.

"The influence of noise on emotional states," by R. K. Mason. J PSYCHOSOM RES 13:275-282, September, 1969.

"Interaction of adversive stimuli: summation or inhibition?" by B. A. Campbell. J EXP PSYCHOL 78:181-190, October, 1968.

"The interaction of noise and personality with critical flicker fusion performance," by C. D. Frith. BRIT J PSYCHOL 58:127-131, May, 1967.

"An investigation of the effects of various noise levels as measured by psychological performance and energy expenditure," by D. W. Lehmann, et al. J SCH HEALTH 35:212-214, May, 1965.

"Louder, please: noiseless products distracting," TIME 95:92, May 4, 1970.

"Measurements of reaction time in intelligibility tests," by M. H. Hecker, et al. J ACOUST SOC AMER 39:1188-1189, June, 1966.

"Mental-hospital admissions and aircraft noise," by I. Abey-Wickrama, et al. LANCET 2:1275-1277, December 13, 1969+

"Mental-hospital admissions and aircraft noise," by R. H. Chowns. LANCET 1:467, February 28, 1970.

PSYCHOLOGY

"Mental hygiene aspects of urbanization and noise," by C. Bitter. T ZIEKENVERPL 20:535-537, August 1, 1967.

"Monitoring, activation, and disinhibition: effects of white noise masking on spoken thought," by P. S. Holzman, et al. J ABNORM PSYCHOL 75:227-241, June, 1970.

"Noise-caused changes of fine motoricity and sensations of annoyance dependent on certain personality dimensions," by G. Jansen, et al. Z EXP ANGEW PSHYCOL 12:594-613, 1965.

"Noise menace threatens man; hearing and sanity may be affected," by B. J. Culliton. SCI N 90:297-299, October 15, 1966.

"On the influence of attitudes to the source on annoyance reactions to noise. An experimental study," by E. Jonsson, et al. NORD HYG T 48:35-45, 1967.

"On the influence of attitudes to the source on annoyance reactions to noise. A field experiment," by R. Cederlof, et al. NORD HYG T 48:46-59, 1967.

"On the influence of continuous noise on the organization of the memory content," by H. Hormann, et al. Z EXP ANGEW PSYCHOL 13:31-38, 1966.

"On the influence of discontinuous noise on the organization of memory contents," by H. Hormann, et al. Z EXP ANGEW PSYCHOL 13:265-273, 1966.

"On noise annoyance and psychological disposition," by C. R. Johansson. NORD HYG T 47:19-25, 1966.

"Output, error, equivocation, and recalled information in auditory, visual, and audio-visual information processing with constraint and noise; with reply by R. E. Jester and rejoinder," by H. J. Hsia. J COMM 18:325-353, December, 1968.

"Psychic cost of adaptation to an environmental stressor," by D. C. Glass, et al. J PERSONALITY SOC PSYCHOL 12:200-210, July, 1969.

PSYCHOLOGY

"Scopolamine effects on go-no go avoidance discriminations; influence of stimulus factors and primacy of training," by N. Rosic, et al. PSYCHOPHARMACOLOGIA 17:203-215, 1970.

"Subjective matching of anxiety to intensities of white noise," by R. Sullivan. J ABNORM PSYCHOL 74:646-650, December, 1969.

PSYCHOTICS

"Behavioral and skin potential response correlations in chronic schizophrenic patients," by R. Wyatt, et al. COMPR PSYCHIAT 10:196-200, May, 1969.

"Effect of background conversation and darkness on reaction time in anxious, hallucinating, and severely ill schizophrenics," by A. Raskin. PERCEPT MOTOR SKILLS 25:353-358, October, 1967.

"Electrodermal and cardiac responses of schizophrenic children to sensory stimuli," by M. E. Bernal, et al. PSYCHOPHYSIOLOGY 7:155-168, September, 1970.

"Perceptual recognition in the presence of noise by psychiatric patients," by D. W. Stilson, et al. J NERV MENT DIS 142:235-247, March, 1966.

"Preferred loudness of recorded music of hospitalized psychiatric patients and hospital employees," by H. L. Bonny. J MUS THERAPY 5:44-52, 1968.

PULSE

see also: Physiology

"Effect of pulsed noise on the organism as a function of pulse periodicity," by G. A. Suvorov. GIG TR PROF ZABOL 13:23-27, December, 1969.

"Influence of repeated 4-hour, intermittent, so-called 'muffled' noise on catecholamine secretion and pulse rate," by W. Hawei, et al. INT Z ANGEW PHYSIOL 24:351-362, 1967.

"Studies on the action of the pulse after a given amount of industrial noise. Economic research program with the help of sequential analysis. I," by A. Fuchs-Schmuck. INT Z ANGEW PHYSIOL 22:1-9, March 3, 1966.

"Studies on behavior of the pulse following definite industrial noise. Economical experimental design, using sequential analysis. II," by A. Fuchs-Schmuck. INT Z ANGEW PHYSIOL 23:345-353, March 7, 1967.

RB.211

"RB.211 engine given outdoor noise tests," AVIATION W 90:115, April 14, 1969.

RFT

"EEG measures of arousal during RFT performance in 'noise'," by R. W. Hayes, et al. PERCEPT MOT SKILLS 31:594, October, 1970.

"RFT stability or failure to arouse?" by G. M. Vaught, et al. PERCEPT MOTOR SKILLS 28:378, April, 1969.

RADIO-TELEGRAPHERS

"Occupational damages in radio-telegraphers," by D. Petrovic, et al. VOJNOSANIT PREGL 25:123-126, March, 1968.

RADIOMEN

"The code copying capabilities of radiomen under simulated atmospheric noise conditions. Report No. 523," by A. M. Richards. US NAVAL SUBMAR MED CENT 523:1-4, May 15, 1968.

RAILWAY NOISE

"Attenuation of noise and ground vibrations from railways," by P. Grootenhuis. J ENVIRONMENTAL SCI 10:14-19, April, 1967.

"The development and results of the study of noise and vibration in railroad transportation," by A. M. Volkov. GIG TR PROF ZABOL 11:58-60, November, 1967.

"Isolation of railroad/subway noise and vibration," by L. N. Miller. PROGRES ARCH 46:203-208, April, 1965.

RAILWAY NOISE

"Problems of noise and its effect on hearing in some railroad establishments," by F. Singer, et al. CESK OTOLARYNG 17: 13-21, February, 1968.

"Rapid transit system environment," by W. V. Braktowski. J ENVIRONMENTAL SCI 10:20-27, April, 1967.

REFRIGERATOR NOISE

"Laboratory techniques useful in the design of refrigerator compressor valves," by R. Cohen, et al. ASHRAE J 7:106-111, January, 1965.

"Measurement and reduction of refrigerator noise," by R. J. Sabine. ASHRAE J 7:117-121, January, 1965.

SST

see: Sonic Boom

SAFETY HELMETS

"Safety helmet tunes out noise (blast cleaning)," IRON AGE 205: 114N, April 23, 1970.

SAILING

"What do you know about silencing aluminum spars," by D. B. Hoisington. YACHTING 125:83+, April, 1969.

SCHOOLS

"Professors are skeptical of positive claims made for environmental control (windowless classrooms, negative ions, noise are discussed)," by F. J. Versagi. AIR COND HEAT & REFRIG N 108:42-43, June 6, 1966.

"A review of research on noise, with particular reference to schools," by P. Wrightson. GREATER LONDON RESEARCH QUARTERLY BULLETIN 20-27, September, 1969.

"Sound polluted schools," by S. Hammon. SCH MGT 14:14-15, November, 1970.

"Study on acoustic trauma in a vocational school," by P. Van de Calseyde, et al. ACTA OTORHINOLARYNG BELG 22:664-666, 1968.

SCIENTIFIC AND TECHNICAL PERSONNEL

"Acoustic trauma in scientific and technical workers," by W. Sulkowski, et al. MED PRACY 17:515-518, 1966.

SERUM ALKALINE PHOSPHATASE

"Effect of physical work and work under conditions of noise and vibration on the human body. I. Behavior of serum alkaline phosphatase, aldolase and lactic dehydrogenase activities," by J. Gregorczyk, et al. ACTA PHYSIOL POL 16:701-708, September-October, 1965.

SERVICEMEN

"Acoustic trauma in regular army personnel. Clinical audiologic study," by A. Salmivalli. ACTA OTOLARYNG 222:Suppl:1-85, 1967.

"An audiometric survey of a Canadian armoured regiment. A follow-up to the 1963 report," by G. G. Jamieson. MED SERV J CANADA 23:1313-1320, December, 1967.

"Considerations on evaluation of acoustic and vestibular damage in pilots and specialists of military aeronautics," by C. Koch, et al. MINERVA MED 56:3832-3835, November 10, 1965.

"Detonation-traumatic load during military service," by C. J. Partsch, et al. ARCH KLIN EXP OHR NAS KEHLKOPFHEILK 191:581-586, 1968.

"Effect of resonance hearing protectors in lowering acoustic trauma in the army," by J. Dominik, et al. VOJ ZDRAV LISTY 34:146-149, August, 1965.

"Hearing loss in Canadian Army units," by R. W. Tooley. MED SERV J CANADA 21:173-176, March, 1965.

"High-intensity noise problems in the Royal Marines," by R. R. Coles, et al. J ROY NAV MED SERV 51:184-192, Summer, 1965.

"Military noise induced hearing loss: problems in conservation programs," by C. T. Yarington, Jr. LARYNGOSCOPE 78:685-692, April, 1968.

SERVICEMEN

"Noise--an increasing military problem," by R. H. Meyer. MILIT MED 133:550-556, July, 1968.

"Pathology characteristic of armored troups," by A. C. Benitte. CONCOURS MED 87:2539-2545, April 10, 1965.

"Problems of communication and ear protection in the Royal marines," by M. R. Forrest, et al. J ROY NAV MED SERV 56: 162-169, Spring, 1970.

"Results of the studies of hearing damages in military tank and infantry personnel," by A. Spirov. VOJNOSANIT PREGL 26: 78-80, February, 1969.

"Some cases of acoustic trauma observed in military service," by T. Stern. PRAXIS 58:898-900, July 15, 1969.

"Statistical evaluation of hearing losses in military pilots," by G. von Schulthess, et al. ACTA OTOLARYNG 65:137-145, January-February, 1968+.

"A study of the acoustic reflex in infantrymen," by M. H. Hecker, et al. ACTA OTOLARYNG 207:Suppl:1-16, 1965.

"10 years of hearing conservation in the Royal Air Force," by G. Jacobs, et al. NEDERL MILIT GENEESK T 18:212-218, July, 1965.

SHIP-YARD EMPLOYEES

see: Employees

SHIPS

"Hygienic assessment of noise and vibration on Diesel-powered hydrofoil boats of the 'Meteor' and 'Raketa' type," by V. I. Petrov, et al. GIG TR PROF ZABOL 12:23-27, June, 1968.

"Noise and vibration on vessels and their standardization," by L. Ia. Skratova. BULL INST MAR MED GDANSK 17:151-154, 1966.

"Noise in Polish ships," by A. Vent, et al. BULL INST MAR MED GDANSK 17:157-162, 1966.

SHIPS

"The problem of noise in the Royal Navy and Royal Marines," by R. R. Coles, et al. J LARYNG 79:131-147, February, 1965.

"The problem of noise on board the vessels of the Polish Merchant Marine," by C. Szczepanski. POL TYG LEK 25:760-762, June, 1970.

"The sanitary-hygienic characteristics of noise on dry-cargo Diesel ships of 1,200 to 1,800 tons capacity and the efficacy of anti-noise measures," by V. I. Petrov, et al. GIG TR PROF ZABOL 12:19-22, May, 1968.

"A survey of engine room noise in the Royal New Zealand Navy," by N. Roydhouse. NEW ZEAL MED J 67:133-140, January, 1968.

SINGERS

see also: Music
Musicians

"Acoustic trauma in singers," by A. Profazio, et al. OTORINOLARINGOL ITAL 37:337-346, August, 1969.

SLEEP

"Analysing effect of noise on sleep," ENGINEER 226:233, August 16, 1968.

"Characteristics of man's sleep under conditions of continuous protracted effect of broad-band noise of average intensity," by B. I. Miasnikov, et al. IZV AKAD NAUK SSSR 1:89-98, January-February, 1968.

"The EEG and the impairment of sleep by traffic noise during the night: a problem of preventive medicine," by H. R. Richter. ELECTROENCEPH CLIN NEUROPHYSIOL 23:291, September, 1967.

"Fractionated sleep. Nocturnal sleep disturbances provoked by noise: electroencephalographic aspects of a preventive medicine problem," by H. R. Richter. REV NEUROL 115:592-595, September, 1966.

SLEEP

"The incidence of snoring as a sleep problem in parents," by E. R. Seller. J ROY COLL GEN PRACT 19:247, April, 1970.

"Repetitive auditory stimuli and the development of sleep," by B. Tizard. ELECTROCNEEPH CLIN NEUROPHYSIOL 20:112-121, February, 1966.

SMALL BOREARMS

"Acoustic trauma caused by shooting at shooting ranges. Attempts at prevention," by A. Rigaud. REV CORPS SANTE ARMEES 9:39-60, February, 1968.

"Acoustic trauma due to firearms in the army," by P. Pazat, et al. REV CORPS SANTE ARMEES 9:213-230, April, 1968.

"Anticoagulation and gunshot concussion in eardrum bleeding," by N. Sonkin. RHODE ISLAND MED J 49:243-244, April, 1966.

"Auditory effects of acoustic impulses from firearms," by K. D. Kryter, et al. ACTA OTOLARYNG 211:Suppl:1-22, 1965.

"Auditory hazards of sport guns," by R. R. Coles, et al. LARYNGOSCOPE 76:1728-1731, October, 1966.

"Effects of high-speed drill noise and gunfire on dentists' hearing," by W. D. Ward, et al. J AMER DENT ASS 79:1383-1387, December, 1969.

"Hearing hazard from small-bore weapons," by W. I. Acton, et al. J ACOUST SOC AMER 44:817-818, September, 1968.

"Hearing defects in gunners," by B. Drettner, et al. FORSVARSMEDICIN 1:115-122, July, 1965.

"Impulse noise and neurosensory hearing loss. Relationship to small arms fire," by R. J. Keim. CALIF MED 113:16-19, September, 1970.

"Sensorineural hearing loss associated with firearms," by R. J. Keim. ARCH OTOLARYNG 90:581-584, November, 1969.

SNORING

"Ear stoppies. A defense against snoring," by M. H. Boulware. J FLORIDA MED ASS 57:36, May, 1970.

"The incidence of snoring as a sleep problem in parents," by E. R. Seller. J ROY COLL GEN PRACT 19:247, April, 1970.

"Snoring--the listeners' disease," LIFE AND HLTH 82:6+, April, 1967.

"Snoring: theory of compensative resortia," by M. H. Boulware. EYE EAR NOSE THROAT MONTHLY 47:664-668, December, 1968.

SOCIOLOGY

see: Behavior
Personality
Physiology
Psychology
Psychotics

SONIC BOOM

"Aerodynamics, noise, and the sonic boom," by W. R. Sears. AIAA J 7:577-586, April, 1969.

"The 'bang' of supersonic airplanes," ADM 25:341-351 continued, July-August, 1968.

"Boom nobody wants," NATIONS BSNS 56:76-78, September, 1968.

"Boom problem still clouds SST future," by C. M. Plattner. AVIATION W 86:28-29, January 9, 1967.

"The boom that's brewing a storm: sonic boom generated by Boeing's SST has stirred up critics who seek to clip the airplane's wings: before craft flies in mid-1970s, airlines and government face some tough decisions," BUS WEEK 64-65+, October 28, 1967.

"Booms," LANCET 2:295-296, August 5, 1967.

"Booms banned; exhausts exempted," NATURE (London) 226:889-890, June 6, 1970.

SONIC BOOM

"Breaking the sound barrier," SR SCHOL 97:15, December 7, 1970.

"Conveying through a sound barrier," AUTOMATION 16:57, July, 1969.

"Ears need protection from ultrasonics too," SAFETY MAINT 133: 41-42, January, 1967.

"Effect of shock waves from supersonic planes," by J. Calvet, et al. J FRANC OTORHINOLARYNG 18:79-85, February, 1969.

"The effect of ultrasonics and high-frequency noise on the blood sugar level," by Z. Z. Ashbel'. GIG TR PROF ZABOL 9:29-33, February, 1965.

"Effects of sonic booms and subsonic jet flyover noise on skeletal muscle tension and a paced tracing task. NASA CR-1522," by J. S. Lukas, et al. US NASA 1-35, February, 1970.

"Effects of the sonic boom on man," by R. E. Bouille. REV CORPS SANTE ARMEES 7:659-688, October, 1966.

"Effects of sonic boom on people: review and outlook," by H. E. Von Gierke. J ACOUST SOC AMER 39:Suppl:43-50, May, 1966.

"Effects of sonic boom on people: St. Louis, Missouri, 1961-1962," by C. W. Nixon, et al. J ACOUST SOC AMER 39:Suppl:51-58, May, 1966.

"Experience in the United Kingdom on the effects of sonic bangs," by C. H. Warren. J ACOUST SOC AMER 39:Suppl:59-64, May, 1966.

"Ground runup silencers for Concorde supersonic transport," ACOUSTICAL SOC AM J 41:1558, June, 1967.

"Human responses to sonic boom," by C. W. Nixon. AEROSPACE MED 36:399-405, May, 1965.

"Is supersonic transport worth the noise?" NATURE (London) 227: 873-874, August 29, 1970.

SONIC BOOM

"Laboratory tests of physiological-psychological reactions to sonic booms," by K. D. Kryter. J ACOUST SOC AMER 39:Suppl:65-72, May, 1966.

"Laboratory tests of subjective reactions to sonic booms. NASA CR-187," by K. S. Pearsons, et al. US NASA 1-34, March, 1965.

"Minimum sonic boom shock strengths and overpressures," by R. Seebass. NATURE (London) 221:651-653, February 15, 1969.

"Relative annoyance and loudenss judgments of various simulated sonic boom waveforms. NASA CR-1192," by L. J. Shepherd, et al. US NASA 1-52, September, 1968.

"The sonic boom," by R. Caporale. RIV MED AERO 28:199-212, April-June, 1965.

"Sonic boom claims (recommended techniques and procedures for investigating alleged sonic booms by military aircraft in such a manner as to give prompt satisfaction for legitimate damages while protecting the government against spurious claims)," by R. C. Smith. JAG J 23:23-26, July-August, 1968.

"Sonic boom effects on the organ of Corti," by D. A. Majeau-Chargois, et al. LARYNGOSCOPE 80:620-630, April, 1970.

"Sonic boom tests seek forecast data," by C. M. Plattner. AVIATION W 84:55-57, June 20, 1966.

"Sonic boom: what it is...what it will and will not do," TODAY'S HLTH 43:89, May, 1965.

"Sonic booms," by A. B. Lowenfels, et al. SCIENCE 158:313-315, October 20, 1967.

"SST noise critical to airport compatibility," AVIATION W 92:83-84, January 5, 1970.

"SST: why don't they ban the boom? (caused by supersonic flights: United States)," by A. R. Karr. WALL ST J 175:18, April 30, 1970.

SONIC BOOM

"SST's bag of mischief," by W. D. Lynn. SCIENCE 164:129, April 11, 1969.

"Supersonic blues," MED J AUST 1:1142-1143, May 31, 1969.

"Supersonic boom," LIFE 67:51-52, November 7, 1969.

"Supersonic 'boom'. Its nature and its effects," by J. Lavernhe, et al. PRESSE MED 74:1973-1975, September 17, 1966.

"Supersonic booms. A current problem," by J. Causse, et al. ANN OTOLARYNG 85:419-423, July-August, 1968.

"Testing sonic booms," by R. Clark. NATURE (London) 215:1122-1123, September 9, 1967.

SOUND STAGES

"Reminiscences; noise; sound stages," by V. O. Knudsen. AUDIO ENG SOC J 18:436-439, August, 1970.

SOUNDPROOFING

"Acoustic lining reduces jet noise," ENGINEERING 207:596, April 18, 1969.

"Acoustic shelters meet needs as noise pollution grows," by N. P. Chironis. PRODUCT ENG 41:160-161, April 27, 1970.

"Acoustical casing cuts engine room gear noise," SAFETY MAINT 136:47-48+, October, 1968.

"Acoustical enclosures muffle plant noises," by S. Wasserman, et al. PLANT ENG 19:112-115, January, 1965.

"Acoustical glass," GLASS IND 46:229-230, April, 1965.

"Acoustical locks set quiet mood for visitors to Bell telephone exhibit at N.Y. world's fair," AIR COND HEAT & REFRIG N 106:39, September 13, 1965.

"Acoustical panels control mechanical noise pollution," SAFETY MAINT 136:39-40, July, 1968.

SOUNDPROOFING

"Acoustical properties of carpets and drapes," by J. W. Simons, et al. HOSPITALS 43:125-127, July 16, 1969.

"Chrysler pavilion," by W. R. Farrell, et al. AUDIO 49:28+, April, 1965.

"Criterion for estimating spacecraft shroud acoustic field reductions," by S. M. Kaplan. J ENVIRONMENTAL SCI 12:27-29, February, 1969.

"Damping material cuts noise, vibration (called Deadbeat)," PURCHASING 61:76, December 1, 1966.

"Do-it-yourself acoustic structures can be designed, assembled by the purchaser," AIR COND HEAT & REFRIG N 105:12, July 19, 1965.

"Do-it-yourself style: sound attenuation (air diverter roof-mounted to residential air conditioner)," by F. J. Versagi. AIR COND HEAT & REFRIG N 107:24, February 7, 1976.

"Doughnuts of foam absorb fan noise," PRODUCT ENG 38:52-53, July 3, 1967.

"Hidamets; metals to reduce noise and vibration," by D. Birchon. ENGINEER 222:207-209, August 5, 1966.

"How Oregon school keeps out airport noise," by J. E. Guerusey. NATIONS SCH 80:65, November, 1967.

"How to keep down noise levels in computer facilities," by L. L. Boyer, Jr. ARCH REC 145:165-166, May, 1969.

"How to noiseproof a room," MECH ILLUS 65:100-102, March, 1969.

"How to quiet down a noisy house," BET HOM & GARD 44:120, March, 1966.

"How to save dollars when treating noise," MOD MANUF 2:88-89, October, 1969.

SOUNDPROOFING

"How to secure privacy of speech in offices," by E. H. Kone. OFFICE 62:14-15+, December, 1965.

"How to select acoustical materials," ARCH REC 138:187-188, July, 1965+.

"Integrated ceiling system handles air, light and sound," AIR COND HEAT & VEN 65:59-62, April, 1968.

"LAX studies house insulation as way to decrease jet noise," by F. S. Hunter. AM AVIATION 32:25, September 16, 1968.

"Lead for sound control doesn't impress experts," AIR COND HEAT & REFRIG N 121:30, December 7, 1970.

"Lead sheet for sound absorption: detail sheet," AIR COND HEAT & VEN 65:71-74, April, 1968.

"Lead sound barrier," MECH ENG 90:63, March, 1968.

"Look at lead, a versatile and flexible noise-cutting aid," POWER 113:62-63, August, 1969.

"The merits of double glazing," by J. Walton. PRACTITIONER 10: Suppl:13-17, March, 1970.

"National gypsum (co) offers record in push for acoustical tile," ADV AGE 36:94, December 6, 1965.

"New association of acoustical and insulating material manufacturing," by J. E. Nolan. OFFICE 70:35-36, September, 1969.

"New steel laminates can banish noise," PRODUCT ENG 39:100-101, January 15, 1968.

"Noise abatement by barriers," by M. Rettinger. PROGRES ARCH 46:168-169, August, 1965.

"Noise abaters," by H. Lawrence. AUDIO 51:82, May, 1967.

"Noise level control," TIMES ED SUP 2727:324, August 25, 1967.

SOUNDPROOFING

"Noise reduction by isolation," by J. Walton. INSTRUMENTS & CONTROL SYSTEMS 39:109-111, February, 1966.

"Parametric studies of the acoustic behavior of duct-lining materials," by J. Atvars, et al. ACOUSTICAL SOC AM J 48: 815-825, pt 3, September, 1970.

"Perforated stainless may cloak jet noise," STEEL 164:31, February 10, 1969.

"Perforated torus ring quiets 25,000 cfm of compressed air," by S. Butler. PRODUCT ENG 41:73, June 8, 1970.

"Plastic foam and the big noise," PLASTICS WORLD 23:74, July, 1965.

"Porous steel liner muffles jet whines," PRODUCT ENG 37:82, November 7, 1966.

"Practical designs for noise barriers based on lead," by B. Fader. AMER INDUSTR HYG ASS J 27:520-525, November-December, 1966.

"Products: soft core insulation," PROGRES ARCH 49:52, August, 1968.

"Quiet: San Antonio parade shows advantages sound conditioned homes can offer," AIR COND HEAT & REFRIG N 105:6-7, August 23, 1965.

"Reduction of aircraft noise measured in several school, motel and residential rooms," by D. E. Bishop. J ACOUST SOC AMER 39:907-913, May, 1966.

"Role of sound-suppressing curtains in the control of industrial noise pollution," by I. Singer. WIRE & WIRE PROD 45:111-113, October, 1970.

"Room noise is eliminated by using acoustical sealant," ADHESIVES AGE 11:32, April, 1968.

SOUNDPROOFING

"Sheet lead for soundproofing," AIR COND HEAT & VEN 65:46, December, 1968.

"Singing flame stills whine in jet engine," PRODUCT ENG 40:22, December 15, 1969.

"Sound absorption of draperies," by J. H. Batchelder,et al. ACOUSTICAL SOC AM J 42:573-575, September, 1967.

"Sound absorptive properties of carpeting," by M. J. Kodaras. INTERIORS 128:130-131, June, 1969.

"Sound control in office buildings," by R. C. Weber. SKYSCRAPER MGT 55:14-15, March; 20-24, April, 1970.

"Sound insulation and new forms of construction," BUILD RES STA DIGEST 96:1-8, August, 1968.

"Sound insulation of traditional dwellings," BUILD RES STA DIGEST 102:1-7, February; 103:1-8, March, 1969.

"Sound-insulation of windows," by E. Sonntag. Z GES HYG 13:632-633, August, 1967.

"Soundproof audiometry room," NT 61:728, May 28, 1965.

"Sound-proof shop office," IND FINISHING 42:106-107, May, 1966.

"Sound-proof unit at West Middlesex Hospital," NURS MIRROR 126:30-31, February 16, 1968.

"Soundproofing: architect defends house against the jet," HOUSE & HOME 33:8, February, 1968.

"Sound-proofing of buildings," by E. F. Stacy. ROY SOC HEALTH J 88:82-86, March-April, 1968.

"Sound-proofing the surgery," by A. Rathbone. PRACTITIONER 6:19-24, March, 1968.

"Soundwall soothes substation's neighbors," ELEC WORLD 173:29,

March 23, 1970.

"Specifying motors for a quiet plant; absorption, acoustical barriers and distance can quiet airborne motor noise," by R. L. Nailen. CHEM ENG 77:157-161, May 18, 1970.

"Sprayed mineral fiber manufacturers' association formed," AIR COND HEAT & REFRIG N 107:23, March 7, 1966.

"Stop that noise!" by J. R. Pritchard. CAN PERS 17:17-21, May, 1970.

"Thousands of tiny holes quiet noise from jumbo jet engines," PRODUCT ENG 41:96, January 1, 1970.

"Time-rated acoustical ceiling assemblies," ARCH REC 137:197-198, February, 1965.

"Turbofan-engine noise suppression," by R. E. Pendley, et al. J AIRCRAFT 5:215-220, May, 1968.

"Underexpanded jet noise reduction using radial flow impingement," by D. S. Dosanjh, et al. AIAA J 7:458-464, March, 1969.

"Waffled cone cuts throttle-valve noise," MACHINE DESIGN 42:150, February 19, 1970.

SPACECRAFT

"Appraisal of Apollo launch noise," by B. O. French. AERO-SPACE MED 38:719-722, July, 1967.

"Criterion for estimating spacecraft shroud acoustic field reductions," by S. M. Kaplan. J ENVIRONMENTAL SCI 12:27-29, February, 1969.

"Some problems of the development of an optimal acoustic environment in space ship cabins," by E. M. Iuganov, et al. IZV AKAD NAUK SSSR 1:14-20, January-February, 1966.

SPECTROMETER

"Spectrometer performance based upon signal-to-noise toleration,"

SPECTROMETER

by D. J. Lovell. AM J OPTOM 47:650-656, August, 1970.

SPEECH

"Effect of a competing message on synthetic sentence identification," by C. Speaks, et al. J SPEECH HEARING RES 10: 390-395, June, 1967.

"The effect of differing noise spectra on the consistency of identification of consonants," by A. C. Busch, et al. LANG SPEECH 10:194-202, July-September, 1967.

"Effect of experimental impairment of sound conduction in the ear on understanding of speech in noise," by J. Kuzniarz. OTOLARYNG POL 22:531-536, 1968.

"Effect of hearing protectors on speech comprehensibility," by Z. Novotny. CESK OTOLARYNG 18:260-264, December, 1969.

"Effect of masking noise upon syllable duration in oral and whispered reading," by M. F. Schwartz. J ACOUST SOC AMER 43:169-170, January, 1968.

"The effect of noise on synthetic sentence identification," by C. Speaks, et al. J SPEECH HEARING RES 10:859-864, December, 1967.

"Effect of pulsed masking on selected speech materials," by D. D. Dirks, et al. J ACOUST SOC AMER 46:898-906, October, 1969.

"The effect of spatially separated sound sources on speech intelligibility," by D. D. Dirks, et al. J SPEECH HEARING RES 12:5-38, March, 1969.

"The effect of speech-type background noise on esophageal speech production," by W. M. Clarke, et al. ANN OTOL 79:653-665, June, 1970.

"Effect of stimulus duration on localization of direction of noise stimuli," by W. R. Thurlow, et al. J SPEECH & HEARING RES 13:826-838, December, 1970.

SPEECH

"Effects of random and response contingent noise upon disfluencies of normal speakers," by R. H. Brookshire. J SPEECH HEARING RES 12:126-134, March, 1969.

"Enhancement of speech intelligibility at high noise levels by filtering and clipping," by I. B. Thomas, et al. AUDIO ENG SOC J 16:412-415, October, 1968.

"Importance of higher frequencies of the speech spectrum for its comprehension in noise," by J. Kuzniarz. OTOLARYNG POL 22:427-435, 1968.

"Influence of the frequency of interruption of signals on the intelligibility of the spoken language," by R. Lehmann. C R ACAD SCI (D) 261:5653-5656, December 20, 1965.

"Masking of speech by aircraft noise," by K. D. Kryter, et al. J ACOUST SOC AMER 39:138-150, January, 1966.

"Masking of speech by continuous noise," by J. Kuzniarz. OTOLARYNG POL 21:401-407, 1967.

"Masking of speech by means of impulse noise," by J. Kuzniarz. OTOLARYNG POL 22:421-425, 1968.

"Passing the strongly voiced components of noisy speech," by C. M. Holloway. NATURE (Lodnon) 226:178-179, April 11, 1970.

"Personal experiences with a stutter-aid," by W. D. Trotter, et al. J SPEECH & HEARING DIS 32:270-272, August, 1967.

"Problems of speech communication in noise and work performance in noise," by K. D. Kryter. AM ASS INDUSTR NURSES J 13:13, April, 1965.

"Results of speech audiometry in acoustic trauma," by E. Vojacek. MSCHR OHRENHEILK 102:152-157, 1968.

"The significance of binaural hearing for speech communication under the effects of noise," by H. Feldmann. ACTA OTOLARYNG 59:133-139, February-April, 1965.

SPEECH

"Some spectral features of 'normal' and simulated 'rough' vowels," by F. W. Emanuel, et al. FOLIA PHONIAT 21:401-415, 1969.

"Spectral noise levels and roughness severity ratings for normal and simulated rough vowels produced by adult females," by M. A. Lively, et al. J SPEECH HEAR RES 13:503-517, September, 1970.

"Speech communication in very noisy environments," by C. Cherry, et al. NATURE (London) 214:1164, June 10, 1967.

"Speech communications as limited by ambient noise," by J. C. Webster. J ACOUST SOC AMER 37:692-699, April, 1965.

"Speech intelligibility in a background noise and noise-induced hearing loss," by W. I. Acton. ERGONOMICS 13:546-554, September, 1970.

"Speech intelligibility in presence of ambient noise," by G. Grisanti. VALSALVA 42:348-372, December, 1966.

"Tone and speech hearing in acoustic trauma," by F. Schwetz. MSCHR OHRENHEILK 103:105-110, 1969.

STAPEDECTOMY

"The effect of trauma in causing cochlear losses after stapedectomy," by M. M. Paparella. ACTA OTOLARYNG 62:33-43, July, 1966.

"A further study on the temporary effect of industrial noise on the hearing of stapedectomized ears, at 4,000 c.p.s.," by K. Ferris. J LARYNG 81:613-617, June, 1967.

"Noise trauma deafness after stapedectomy with recovery," by T. R. Bull. J LARYNG 80:631-633, June, 1966.

"On the temporary effect of industrial noise on the hearing at 4,000 c/s of stapedectomized ears," by K. Ferris. J LARYNG 79: 881-887, October, 1965.

"Stapedectomy and air travel," by J. Bastien. ANN OTOLARYNG

83:69-74, January- February, 1966.

"The temporary effects of 125 c.p.s. octave-band noise on stapedectomized ears," by K. Ferris. J LARYNG 80:579-582, June, 1966.

"Treatment with vascular aim of recent acoustic traumas, especially after stapedectomy," by J. Causse, et al. REV LARYNG 87:360-368, May-June, 1966.

STUDENTS

see: Education

SUBWAYS

"Isolation of railroad/subway noise and vibration," by L. N. Miller. PROGRES ARCH 46:203-208, April, 1965.

"The noise level in installations of the Moscow subway," by P. N. Matveev. GIG TR PROF ZABOL 10:58-61, June, 1966.

SURGERY

"The problem of acoustic injury during auricular surgery," by J. Calvet, et al. J FRANC OTORHINOLARYNG 14:807-810, December, 1965.

SYMPOSIA

"Aircraft noise symposium; proceedings," ACOUSTICAL SO AM J 48:799-842, pt 3, September, 1970.

"American Conference of Govermental Industrial Hygienists' proposed threshold limit value for noise," by H. H. Jones. AMER INDUSTR HYG ASS J 29:537-540, November- December, 1968.

"Conference on acoustic noise and its control, London, January 23-27; with list of authors and titles of contributed papers," ACOUSTICAL SOC AM J 41:1383-1384, May, 1967.

"Conferecne on noise," NAT PARKS 42:21, August, 1968.

"Design studies urged at (London) noise conference," AVIATION W 85:67-68, December 12, 1966.

SYMPOSIA

"East Range Symposium. I. Hearing loss: diagnosis and anatomic condiderations," by C. J. Holmberg. J OCCUP MED 7:138-144, April, 1965.

--II. Hearing conservation programs," by M. T. Summar. J OCCUP MED 7:145-146, April, 1965.

--IV. Principles of noise control," by R. Donley. J OCCUP MED 7:222-226, May, 1965.

--V. Personal protection in noise exposure," by M. T. Summar. J OCCUP MED 7:279-280, June, 1965.

--VI. Criteria for assessing risk of hearing damage," by J. L. Fletcher. J OCCUP MED 7:281-283, June, 1965.

--8. Hearing conservation: preliminary report on a 7-year program," by M. T. Summar. J OCCUP MED 7:334-340, July, 1965.

"Noise must be designed out of machines: annual symposium, 6th, Cleveland," PRODUCT ENG 40:24+, October 20, 1969.

"Noise pollution; symposium," UNESCO COURIER 20:4-31, July, 1967.

"Noise 2000; VI. Internationaler Kongress fur larmbekampfing," by L. Trbuhovie. WEEK 57:429, July, 1970.

"Physiological effects of audible sound; AAAS symposium, December 28-30, 1969," by B. L. Welch. SCIENCE 166:533-534, October 24, 1969.

"Putting up with transport; Society of environmental engineers 10th anniversary symposium, London," ENGINEERING 207:600-602, April 18, 1969.

"Symposium: environmental protection. Introduction. N. A. Rockerfeller; Water quality control: a modern approach to state regulation. F. E. Maloney, R. C. Ausness; Federal enforcement under the refuse act of 1899. J. T. B. Tripp, R. M. Hall; Comments: environmental control in New York City,

Introduction. Environmental problems of an expanding population. Problems of solid waste disposal. Noise pollution. Abandoned cars--the facts. Aesthetic considerations in land use planning," ALBANY L REV 35:23, 1970.

"Symposium: residuals and environmental quality management. Environmental quality management. B. T. Bower, et al.; Governmental responsibility for waste management in urban regions. R. T. Anderson; The community noise problem: factors affecting its management. C. Bragdon; Pesticide residues and environmental economics. W. F. Edwards, et al; Environmental litigation-where the action is? E. P. Grad, et al.," NATURAL RESOURCES J 10:655, October, 1970.

TAF

"Studies on the function of concentration maintenance (TAF). 7. CPT-swing degree and TAF under exposure to noise," by E. Takakuwa, et al. JAP J HYG 20:359-363, February, 1966.

TELEPHONE NOISE

"Loudness rating of telephone subscribers' sets by subjective and objective methods," by W. D. Cragg. ELEC COM 43,1:39-43; 3:228-232, 1968.

TELETYPE OPERATORS

"Audiometric studies in teletype operators," by G. Bocci, et al. BOLL MAL ORECCH 84:190-199, March-April, 1966.

TELETYPEWRITER

"Teletypewriter used near phones and dictators," OFFICE 68:32, August, 1968.

TEMPORARY THRESHOLD SHIFT

"Considerations on the problem of noise in relation to its influence on some aspects of modern life. I. Behavior of the temporary shift of hearing sensitivity due to high energy level stimulations in persons employed in work with noise," by S. Collatina, et al. CLIN OTORINOLARING 17:357-370, July-August, 1965.

"Effect of ambient-noise level on threshold-shift measures of ear-protector attenuation," by R. Waugh. J ACOUST SOC AM 48:597,

August, 1970.

"Impulse duration and temporary threshold shift," by M. Loeb, et al. ACOUSTICAL SOC AM J 44:1524-1528, December, 1968.

"Magnitude of temporary threshold shift (TTS) in the audiogram and 'response time' of noise-or detonation exposed ears as test of hearing organs endangered by detonation," by F. Pfander. ARCH KLIN EXP OHR NAS KEHLKOPFHEILK 191: 586-590, 1968.

"Predicting hearing loss from noise-induced TTS," by J. C. Nixon, et al. ARCH OTOLARYNG 81:250-256, March, 1965.

"Proposed threshold limit value for noise," by H. H. Jones. SAFETY MAINT 137:46-49, January, 1969.

"Psychological and physiological reactions to noise of different subjective valence (TTS and EMG)," by H. Hormann, et al. PSYCHOL FORSCH 33:289-309, 1970.

"Relation of threshold shift to noise in the human ear," by H. Weissing. ACOUSTICAL SOC AM J 44:610-615, August, 1968.

"Relationships for temporary threshold shifts produced by three different sources. Rep No. 633," by J. L. Fletcher, et al. US ARMY MED RES LAB 41-45, June 30, 1965.

"Reliability of TTS from impulse-noise exposure," by D. C. Hodge, et al. J ACOUST SOC AMER 40:839-846, October, 1966.

"Results from the retest of noise-induced temporary threshold shift (NI-TTS)," by T. Yokoyama, et al. J OTORHINOLARYNG SOC JAP 70:1421-1429, August, 1967.

"Slight differences in noise stimulation and NI-TTS (noise-induced temporary threshold shift)," by T. Yokoyama, et al. J OTORHIN-OLARYNG SOC JAP 70:1343-1357, August, 1967.

"Temporary changes of the auditory system due to exposure to noise for one or two days," by J. H. Mills, et al. J ACOUST SOC AM

48:524-530, August, 1970.

"Temporary changes of the auditory threshold after the exposure to noise at different frequencies," by S. Kubik, et al. PRAC LEK 17:240-243, August, 1965.

"The temporary effects of 125 c.p.s. octabe-band noise on stapedectomized ears," by K. Ferris. J LARYNG 80:579-582, June, 1966.

"Temporary hearing loss due to vibration and noise," by K. Yamamura, et al. JAP J HYG 25:472-478, December, 1970.

"Temporary shifts in auditory thresholds of chinchilla after exposure to noise," by E. N. Peters. J ACOUST SOC AMER 37: 831-833, May, 1965.

"Temporary threshold shift and damage-risk criteria for intermittent noise exposures," by W. D. Ward. J ACOUST SOC AM 48:561-574, August, 1970.

"Temporary threshold shift from impulse noise," by J. G. Walker. ANN OCCUP HYG 13:51-58, January, 1970.

"Temporary threshold shift in rock-and-roll musicians," by J. Jerger, et al. J SPEECH HEARING RES 13:221-224, March, 1970.

"Temporary threshold shift produced by exposure to high-frequency noise," by P. E. Smith, Jr. AMER INDUSTR HYG ASS J 28: 447-451, September-October, 1967.

"Temporary threshold shifts in hearing from exposure to combined impact-steady-state noise conditions," by A. Cohen, et al. J ACOUST SOC AMER 40:1371-1380, December, 1966.

"Tone height shifts under noise conditions," by H. G. Dieroff, et al. FOLIA PHONIAT 18:247-255, 1966.

"Use of sensation level in measurements of loudness and of temporary threshold shift; discussion," ACOUSTICAL SOC AM J 41:

714-715, March, 1967.

TEST RESULTS

"Initial notes on content in auditory projective testing," by I. Breger. J PROJ TECH PERS ASSESS 34:125-130, April, 1970.

"Sound tests show hardboard effective in controlling noise," FOREST IND 96:81, June, 1969.

"Temporary and permanent hearing loss. A ten-year follow-up," by J. Sataloff, et al. ARCH ENVIRON HEALTH 10:67-70, January, 1965.

TINNITUS

"Homolateral and contralateral masking of subjective tinnitus by broad spectrum noise, narrow spectrum noise and pure tones," by H. Feldmann. ARCH KLIN EXP OHR NAS KEHLKOPFH-EILK 194:460-465, December 22, 1969.

"Study of tinnitus induced temporarily by noise," by G. R. Atherley, et al. J ACOUST SOC AMER 44:1503-1506, December, 1968.

TOYS

"Acoustical hazards of children's 'toys'," by D. C. Hodge, et al. J ACOUST SOC AMER 40:911, October, 1966.

"Damage-risk criterion for the impulsive noise of 'toys'," by K. Gjaevenes. J ACOUST SOC AMER 42:268, July, 1967.

"Measurements on the impulsive noise from crackers and toy firearms," by K. Gjaevenes. J ACOUST SOC AMER 39:403-404, February, 1966.

TRACTOR NOISE

see also: Machinery: Farm Equipment

"Can we quiet big tractors?" by B. Coffman. FARM J 93:26F+, August, 1969.

"Deafness in agricultural tractor drivers," by J. P. Vallee. GAZ

MED FRANCE 72:3193-3194, October 10, 1965.

"Experimental application of the word association test in a study of the effect of tractor noise and vibration on the human organism," by V. N. Kozlov, et al. GIG TR PROF ZABOL 13:46-48, December, 1969.

"Noise abatement in the Zetor-Super tractor prototype," by I. Seress. CESK HYG 10:81-85, March, 1965.

"Reducing automobile and tractor noise," by V. P. Goncharenko. GIG TR PROF ZABOL 14:46-47, January, 1970.

TRAFFIC NOISE

"Annoyance reactions to traffic noise in Italy and Sweden," by E. Jonsson, et al. ARCH ENVIRON HEALTH 19:692-699, November, 1969.

"Coming to terms with traffic noise," ENGINEERING 201:374-375, February 25, 1966.

"Contribution to the study of the noise caused by street traffic," by G. Sparacio, et al. G IG MED PREV 7:385-394, October-December, 1966.

"The EEG and the impairment of sleep by traffic noise during the night: a problem of preventive medicine," by H. R. Richter. ELECTROENCEPH CLIN NEUROPHYSIOL 23:291, September, 1967.

"Effects of traffic noise on health and achievement of high school students of a large city," by G. Karsdorf, et al. Z GES HYG 14:52-54, January, 1968.

"Effects of traffic noise on human life," by O. Guthof, et al. OEFF GESUNDHEITSWESEN 30:1-6, January, 1968.

"Findings on traffic noise in the city of Imperia," by I. Murruzzu. NUOVI ANN IG MICROBIOL 18:503-510, November-December, 1967.

TRAFFIC NOISE

"Highway noise monitor," MECH ENG 92:52, June, 1970.

"How Ottawa's 'realistic' new traffic noise by law operates," FIN POST 64:27, February 14, 1970.

"Noise nuisance from motorways; effects on residential areas," by F. J. Langdon. ARCHITECTS J 141:1453-1455, June 23, 1965.

"Noise pollution efforts now focus on turbines and highways," PRODUCT ENG 40:105, April 7, 1969.

"The nuisances of traffic in residential areas," by D. M. Winterbottom. TRAFFIC Q 19:384-395, July, 1965.

"Problems in the field of urban traffic noise control in town buildings," by A. S. Perotskaia, et al. GIG SANIT 35:3-9, August, 1970.

"Regulating road traffic noise," ENGINEERING 205:823, May 31, 1968.

"Traffic noise," by B. H. Sexton. TRAFFIC Q 23:427-439, July, 1969.

"Traffic noise draws R & D attention," by W. J. Kalb. IRON AGE 201:21, June 20, 1968.

"Traffic noise from roundabouts," by M. A. McCormick, et al. ARCHITECTS J 146:861-862, October 4, 1967.

"Traffic noise: technique for assessing motorway noise and protecting buildings against it," by J. Simmons. ARCHITECTS J 144:239-244, July 27, 1966.

"Triumph over traffic; New York's Court of appeals declares noise a compensable injury," TIME 92:74, July 12, 1968.

TRANQUILIZERS

"Objective studies of the effects of tranquilizing agents," by G. Harrer, et al. ARZNEIM FORSCH 20:921-923, July, 1970.

TRANSFORMER NOISE

"Appearance combined with soundproofing in a transformer enclosure," ENGINEER 226:232, August 16, 1968.

"Bibliography on transformer noise," INST ELEC & ELECTRONICS ENG TRANS POWER APPARATUS & SYSTEMS 87:372-387, February, 1968.

"Keeping power-station noise within acceptable limits," by R. X. French. IEEE TRANS POWER APPARATUS & SYSTEMS 233-237+.

"Magnetostriction and transformer noise; abstracts," ULTRASONICS 6:77, April, 1968.

"Measures for lowering noise intensity at the Tashkent hydroelectric power stations," by P. E. Popov. GIG SANIT 32:94-95, April, 1967.

"Noise control of emergency power generating equipment," by P. J. Torpey. AMER INDUSTR HYG ASS J 30:596-606, November-December, 1969.

"Noise governs transformer selection at UCLA health sciences complex in Los Agneles," by K. K. Leithold. ELEC CONSTR & MAINT 67:106-107, July, 1968.

"Quieter transformers," ENGINEERING 204:547, October 6, 1967.

"Select an appropriate decibel rating when specifying transformers," ELEC CONSTR & MAINT 68:123-126, May, 1969.

"Transformer noise," ENGINEER 224:399, September 29, 1967.

TRANSPORTATION NOISE

"Location-design control of transportation noise," by A. Cohen. AM SOC C E PROC 93(UP 4 no 5693):63-86, December, 1967; Discussion by T. E. Parkinson. 94(UP 1 no 6052):95-96, August, 1968; Reply 95(UP 1 no 6487):102-103, April, 1969.

"Noise on transportation trunk lines of the Zakarpatskaia District," by N. F. Grishchenko, et al. GIG SANIT 35:101-103, April, 1970.

TRANSPORTATION NOISE

"Putting up with transport; Society of environmental engineers 10th anniversary symposium, London," ENGINEERING 207:600-602, April 18, 1969.

"Transportation noise sources," by R. C. Potter. AUDIO ENG SOC J 18:119-127, April, 1970.

"Transportation vehicle noise control: application and acceptability," by P. A. Franken. ARCHIVES OF ENVIRONMENTAL HEALTH 20:636-643, May, 1970.

TRUCKS

"Quest for quiet," by B. Swart. FLEET OWNER 60:85-92, February, 1965.

"Long-term observations on the effect of noise on drivers of large tanktrucks," by H. Herrmann, et al. ZBL ARBEITSMED 17:73-78, March, 1967.

TUG BOATS

"Acoustic conditions on harbour tug boats," by C. Szczepanski. BULL INST MAR MED GDANSK 21:215-222, 1970.

TULLIO PHENOMENON

"The Tullio phenomenon and a possibility for its treatment," by G. Lange. ARCH KLIN EXP OHR NAS KEHLKOPFHEILK 187:643-649, 1966.

"Unusual Tullio phenomena," by S. K. Kacker, et al. J LARYNG 84:155-166, February, 1970.

TURBINES

"Fundamental noise research emphasized; aircraft turbine engines," AVIATION W 92:90, June 22, 1970.

"Internal noise attenuation of turbine auxiliary power unit must start during design stage," by J. J. Dias. SAE J 76:64-67, January, 1968.

"New silencers quiet turbine exhaust noise," by L. S. Wirt. SAE J 74:88-89, February, 1966.

TURBINES

"The noise from the high speed turbine as a cause of hypacusis," by G. Girardi, et al. RASS INT STOMAT PRAT 17:405-415, November-December, 1966.

"The noise of the turbines," AN ESP ODONTOESTOMAT 27:253-254, July-August, 1968.

"Noise pollution efforts now focus on turbines and highways," PRODUCT ENG 40:105, April 7, 1969.

"Studies on turbine noise in the hearing and ultrasonic range. I.," by L. Dunker, et al. DDZ 23:211-218, May, 1969.

TURBINES: GAS

"Add-on quieting for gas turbine," by W. J. Pietrucha. POWER 114: 54-55, May, 1970.

"Internally generated noise from gas trubine engines; measurement and prediction," by M. J. T. Smith, et al. J ENG POWER 89: 177-185; Discussion 186-187; Reply 187-190, April, 1967.

"Silencing a gas-turbine generator," ENGINEERING 202:938, November 25, 1966.

"Turbine noise problem licked," GAS 44:60, September, 1968.

"Vibration and noise characteristics of an aircraft-type gas turbine used in a marine propulsion system," by R. E. Harper. NAVAL ENG J 81:103-110, December, 1969.

TURBINES: HYDRAULIC

"Reduction of noise and vibrations in a hydraulic turbine," by T. Sagawa. J BASIC ENG 91:722-737, December, 1969.

URBAN NOISE

"Aggravation of deafness caused by acoustic trauma after retreating from the noisy environment," by P. Pialoux, PROBL ACTUELS OTORHINOLARYNGOL 1-6, 1968.

"Assault on the ear; city dweller," NEWSWEEK 67:70, April 4, 1966.

URBAN NOISE

"Cities lend an ear to noise control," BSNS W 108, January 17, 1970.

"City; a challenge to engineering and political sciences," by P. R. Achenbach. ASHRAE J 11:33-38, March, 1969.

"City noise; designers can restore quiet, at a price," by H. W. Bredin. PRODUCT ENG 39:28-35, November 18, 1968.

"City noise--a sociological and psychological study in a defined environment. Experimental pilot studies on the occurrence of noise nuisances and their correlation with various psychological variables," by O. Arvidsson, et al. NORD HYG T 46:153-188, 1965.

"Citizens vs. noise," SCI & CIT 10:31, March, 1968.

"Communal hygienic studies of the noise problem in Warsaw," by A. Brodniewicz. Z GES HYG 13:760-764, October, 1967.

"Community care today and tomorrow," by A. W. Macara. DISTRICT NURS 9:210+, December, 1966.

"Community noise--the industrial aspect," by K. M. Morse. AMER INDUSTR HYG ASS J 29:368-380, July-August, 1968.

"Community noise ordinances," by A. E. Meling. ASHRAE J 9:40-43+, May, 1967.

"Community noise problems-origin and control," by L. S. Goodfriend. AMER INDUSTR HYG ASS J 30:607-613, November-December, 1969.

"Concrete breakers must be quieter," ENGINEERING 203:1050, June 30, 1967.

"Control of neighborhood noise as a hygienic-legal problem," by A. Schubert. Z AERZTL FORTBILD 63:1056-1058, October 1, 1969.

"Dangers of urbanization," DISTRICT NURS 10:125, September, 1967.

URBAN NOISE

"Effect of community noises on school children," by V. I. Pal'gov. PEDIAT AKUSH GINEK 5:29-32, September-October, 1966.

"Environmental hazards. Community noise and hearing loss," by J. D. Dougherty, et al. NEW ENG J MED 275:759-765, October 6, 1966.

"London noises," LANCET 2:1289, December 13, 1969.

"Man and his cities. Part 1.," QUEENSLAND NURSES J 8:38+, June, 1966.

--Part 2.," QUEENSLAND NURSES J 8:5+, July, 1966.

--Part 4.," QUEENSLAND NURSES J 8:5+, September, 1966.

"Man and his cities. The agression against man," INFIRMIERE 45:35-37 conclusion, June, 1967.

"Man and his cities--impact," AUST NURSES J 64:81+, April, 1966; also in JAMAICAN NURSE 6:17+, August, 1966; NEW ZEALAND NURS J 59:8+, April, 1966; NURS J INDIA 57:100+, April, 1966.

"Mental hygiene aspects of urbanization and noise," by C. Bitter. T ZIEKENVERPL 20:535-537, August 1, 1967.

"Muffling the clamor of urban construction," BSNS W 168-169, December 14, 1968.

"Noise and urban man," by B. A. Baron. AMER J PUBLIC HEALTH 58:2060-2066, November, 1968.

"Noise as an environmental disturbance," by O. Kitamura. NAIKA 21:903-906, May, 1968.

"Noise as a health hazard at work, in the community, and in the home," by H. H. Jones, et al. PUBLIC HEALTH REP 83:533-536, July, 1968.

"Noise: the audible pollutant; cities could do much to lessen noise levels," by R. A. Baron. NATIONS CITIES 7:28+, September, 1969.

"Noise: city dwellers, teenagers face deafness from noise," CONG Q W REPT 28:1035-1038, April 17, 1970.

"Noise nuisance from motorways; effects on residential areas," by F. J. Langdon. ARCHITECTS J 141:1453-1455, June 23, 1965.

"Noise: a syndrome of modern society," by C. R. Bragdon. SCI & CIT 10:29-37, March, 1968.

"The nuisances of traffic in residential areas," by D. M. Winterbottom. TRAFFIC Q 19:384-395, July, 1965.

"Old folks at home? life in a Manhattan apartment," by E. G. Smith. ATLAN 220:118-119, December, 1967.

"On the occurrence of nervous disorders in the residential districts of Berlin. Preliminary report," by G. Feuerhahn, et al. PSYCHIAT NEUROL MED PSYCHOL 20:281-286, August, 1968.

"Realistic assessment of the vertiport/community noise problem," by N. Shapiro, et al. J AIRCRAFT 5:407-411, July, 1968.

"Sound of sounds that is New York," by H. C. Schonberg. N Y TIMES MAG 38+, May 23, 1965.

"Sound pollution; another urban problem," by P. A. Breysse. SCI TEACH 37:29-34, April, 1970.

"Street noises in Naples," by R. De Capoa, et al. G IG MED PREV 7:20-47, January-March, 1966.

"Studies of air pollution and noise in urban Korea," by M. H. Kim, et al. YONSEI MED J 8:40-52, 1967.

"Symposium: residuals and environmental quality management. Environmental quality management. B. T. Bower, W. Spofford; Governmental responsibility for waste management in urban regions. R. T. Anderson; The community noise problem: factors affecting its management. C. Bragdon; Pesticide residues and environmental economics. W. F. Edwards, M. Langham, J. C.

Headley; Environmental litigation-where the action is? F. P. Grad, L. Rockett," NATURAL RESOURCES J 10:655, October, 1970.

"System's analyst's view of noise and urban planning," by M. Wachs, et al. AM SOC C E PROC 96(UP 2 no 7639):147-159, October, 1970.

"The uproar of our cities: electroencephalographic aspects," by H. R. Richter. CONFIN NEUROL 25:215-223, 1965.

"Urban noise control (focuses on efforts in New York city)," COLUMBIA J LAW & SOCIAL PROBLEMS 4:105-119, March, 1968.

URBAN PLANNING

see also: Urban Noise

"City; a challenge to engineering and political sciences," by P. R. Achenbach. ASHRAE J 11:33-38, March, 1969.

"City noise: designers can restore quiet, at a price," by H. W. Bredin. PRODUCT ENG 39:28-35, November 18, 1968.

"Definition of human requirements in respect to noise, to be used by town planners and builders," J SCI MED LILLE 87:328-329, April, 1969.

"Noise: the audible pollutant: cities could do much to lessen noise levels," by R. A. Baron. NATIONS CITIES 7:28+, September, 1969.

"Noise control on the local level," by H. M. Fredrikson. ARCH ENVIRON HEALTH 20:651-653, May, 1970.

"System's analyst's view of noise and urban planning," by M. Wachs, et al. AM SOC C E PROC 96(UP 2 no 7639):147-159, October, 1970.

VALVE NOISE

"Abatement of control valve noise; Fisher Governor Co.," by E. E.

VALVE NOISE

Allen. GAS 45:53-56, November, 1969.

"Combating valve noise in process control systems," by C. B. Schuder, et al. AUTOMATION 16:50-53, September, 1969.

VIBRATION

"Acoustic noise and vibration of rotating electric machines," by A. J. Ellison, et al. INST E E PROC 115:1633-1640, November, 1968; Discussion 117:127-129, January, 1970.

"Application of the Mossbauer method to ear vibrations," by P. Gilad, et al. ACOUSTICAL SOC AM J 41:1232-1236, May, 1967.

"Art and technique of noise and vibration control," by H. C. Carter. DOM ENG 205:68-74, May, 1965.

"Autonomic reactions to low frequecny vibrations in man," by R. Coermann, et al. INT Z ANGEW PHYSIOL 21:150-168, September 13, 1965.

"The blood system reaction to the occupational effect of vibration and noise," by I. A. Gribova, et al. GIG SANIT 30:34-37, October, 1965.

"Change in the electrical activity of muscles under the effect of noise and vibration," by A. F. Lebedeva. GIG SANIT 33:25-30, March, 1968.

"Complex laboratories for measuring vibration and noise," by O. E. Guzeev, et al. GIG SANIT 35:100-101, April, 1970.

"The control of vibration and noise," by T. P. Yin. SCI AMER 220: 98-106, January, 1969.

"The effect of noise and local vibration of permissible levels on the human organism," by N. M. Paran'ko, et al. GIG SANIT 32: 25-30, September, 1967.

"Effect of noise and vibration on man," by J. C. Guignard. NATURE 188:533-534.

VIBRATION

"The effect of pulsating noise and vibration on the body in straightening and hammering out work," by L. N. Shkarinov, et al. GIG SANIT 33:104-106, July, 1968.

"The effect of vibration and noise on protein metabolism of excavator-machinists using several biochemical indices," by A. M. Tambovtseva. GIG SANIT 33:58-61, January, 1968.

"The effect of vitamin B 1 and niacin on the course of biochemical processes in organisms exposed to the action of vibration," by D. A. Mikhel'son. VOP PITAN 28:54-58, November-December, 1969.

"Effects of physical work and work under condition of noise and vibration on the human body. II. Behavior of aspartic and alanine aminotransferase, gamma glutamyl transpeptidas, lactic dehydrogenase and L-idotyl dehydrogenase activities," by J. Gregorczyk, et al. ACTA PHYSIOL POL 16:709-714, September-October, 1965.

"Experimental application of the word association test in a study of the effect of tractor noise and vibration on the human organism," by V. N. Kozlov, et al. GIG TR PROF ZABOL 13:46-48, December, 1969.

"From experience in controlling noise and vibration in industrial plants in Kiev," by I. G. Guslits, et al. GIG TR PROF ZABOL 10:52-54, June, 1966.

"Hygienic characteristics of vibration and noise at work sites of crushing mills in ore enriching plants," by K. P. Antonova. GIG SANIT 34:116-118, November, 1969.

"Isolating equipment for vibration control," by W. E. Whale. HEATING-PIPING 38:122-125, January; 97-100, February; 118-120, March, 1966; Discussion 38:96, June; 94+, July, 1966.

"Noise and vibration are analyzed for designers," PRODUCT ENG 41:43, February 16, 1970.

"Noise and vibration; conference, Glasgow, Scotland, April 4-6," ULTRASONICS 6:191, July, 1968.

VIBRATION

"Noise and vibration exposure criteria," by H. E. von Gierke. ARCHIVES ENVIRONMENTAL HEALTH 11:327-330, September, 1965.

"Noise and vibration in moulding shops of iron concrete plants," by I. N. Ivatsevich, et al. GIG SANIT 33:105-107, January, 1968.

"Noise and vibration in plants of prefabricated concrete structures," by Iu. K. Aleksandrovskii, et al. GIG SANIT 30:113-115, August, 1965.

"Noise and vibration on vessels and their standardization," by L. Ia. Skratova. BULL INST MAR MED GDANSK 17:151-154, 1966.

"On the functional interrelationship of certain analyzers in the action of noise-vibration stimuli," by A. F. Lebedeva. GIG TR PROF ZABOL 10:22-28, June, 1966.

"On noise and vibration exposure criteria," by H. E. von Gierke. ARCH ENVIRON HEALTH 11:327-339, September, 1965.

"On the practice of the measurement of noise and vibrations in factories," by G. Wolff. ZBL ARBEITSMED 15:181-184, August, 1965.

"On the problem of assessing the functional status of the vestibular analyzer in persons subjected to the effects of vibration," by N. I. Ponomareva, et al. GIG TR PROF ZABOL 12:31-35, June, 1968.

"On the problem of the effect of vibration and noise on general morbidity," by S. S. Kangelari, et al. GIG TR PROF ZABOL 10:47-49, June, 1966.

"Regional vascular reactions during exposure to vibration and noise, individually and together," by N. M. Paran'ko. GIG SANIT 33: 19-24, November, 1968.

"Standarizationd of vibration and noise and their combined action on man," by N. M. Paran'ko, et al. VRACH DELO 8:102-106, August, 1969.

VIBRATION

"Variations of amylasuria under the effect of sound and vibration injuries (preliminary note)," by S. Miclesco-Groholsky. ANN OTOLARYNG 86:251-258, April-May, 1969.

"Vibration and structure-borne noise control," by L. L. Eberhart. ASHRAE J 8:54-60, May, 1966.

VISION

"Auditory thresholds during visual stimulation as a function of signal bandwidth," by E. J. Moore, II., et al. J ACOUST SOC AMER 47:659-660, February, 1970.

"Background light, temperature and visual noise in the turtle," by W. R. Muntz, et al. VISION RES 8:787-800, July, 1968.

"Comparative examinations of the visual field in glaucoma and the influence of noise," by E. Ogielska, et al. KLIN OCZNA 36: 351-354, 1966.

"Determination of simple reaction time to visual and acoustic stimuli under conditions of noise interference," by K. Witecki. POL TYG LEK 25:1012-1025, July 6, 1970.

"Effect of noise on the modulation transfer function of the visual channel," by H. Pollehn, et al. J OPT SOC AM 60:842-848, June, 1970.

"Effect of a specific noise on visual and auditory memory span," by S. Dornic. SCAND J PSYCHOL 8:155-160, 1967.

"Effects of noise and difficulty level of imput information in auditory, visual, and audiovisual information processing," by H. J. Hsia. PERCEPT MOTOR SKILLS 26:99-105, February, 1968.

"Effects of two types of noise on cardiac rhythm of a man held immobile with constant visual attention," by T. Meyer-Schwertz. ARCH SCI PHYSIOL 22:195-228, 1968.

"Experimental studies on the influence of electric and light stimulation upon the susceptibility to acoustic trauma," by T.

Nakamura. J OTORHINOLARYNG SOC JAP 69:1439-1454, August, 1966.

"Identification of form in patterns of visual noise," by H. Munsinger, et al. J EXP PSYCHOL 75:81-87, September, 1967.

"Importance of the color of the iris in the evaluation of resistence of hearing to fatigue," by G. Tota, et al. RIV OTONEUROOFTAL 43:183-192, May-June, 1967.

"The influence of noise on the visual field," by E. Ogielska, et al. ANN OCULIST 198:115-122, February, 1965.

"Is noise harmful to vision?" by R. Grandpierre, et al. PRESSE THERM CLIMAT 102:164-165, 1965.

"Occupational deafness and visual field," by H. Vynckier. ACTA OTORHINOLARYNG BELG 21:213-222, 1967.

"On alteration of color perception by the exposition to noise," by D. Broschmann. GRAEFE ARCH OPHTHAL 168:250-255, May 13, 1965.

"On the problem of visual field reduction due to acoustic trauma," by U. Grohmann. KLIN MBL AUGENHEILK 152:600-602, 1968.

"Relative effects of figrual noise and rotation on the visual perception of form," by E. A. Alluisi, et al. PERCEPT MOT SKILLS 31:547-554, October, 1970.

"Sensory interaction: perception of loudness during visual stimulation," by R. S. Karlovich. ACOUSTICAL SOC AM J 44:570-575, August, 1968.

"Static visual noise and the Ansbacher Effect," by G. Stanley. Q J EXP PSYCHOL 22:43-48, February, 1970.

VITAMINS

"The effect of the addition of vitamin C to the diet on the human organism exposed to occupational hazards (in the presence of strong sound stimuli)," by A. M. Margolis, et al. VOP PITAN

25:34-38, March-April, 1966.

"The effect of aviation noise on some indices of protein and vitamin metabolism," by Iu. F. Udalov, et al. VOENNOMED ZH 7:61-64, July, 1966.

"Effect of supplementary ingestion of vitamins B1 and C on the function of the auditory organ in workers of 'noisy' trades," by Z. F. Nestrugina. ZH USHN NOS GORL BOLEZ 29:91-94, May-June, 1969.

"The effect of vitamin B 1 and niacin on the course or biochemical processes in organisms exposed to the action of vibration," by D. A. Mikhel'son. VOP PITAN 28:54-58, November-December, 1969.

WALSH-HEALEY ACT
see: Laws and Legislation

WATER HAMMER NOISE
"Water hammer stops: join the anti-noise arsenal," by R. Zell. DOM ENG 206:72-73, September, 1965.

WELDING
"Acoustic trauma in welders exposed to strong noise in mechanical-technological workshops," by A. Brusin, et al. MED PREGL 22: 545-549, 1969.

WHITE NOISE
"Anesthesia and 'white noise'," by V. K. Sipko. STOMATOLOGIIA 45:91-92, January-February, 1966.

"Considerations on the phenomena of adaptation to white noise," by I. De Vincentiis, et al. VALSALVA 42:247-254, October, 1966.

WIND TUNNELS
"Radiated aerodynamics noise effects on boundary-layer transition in supersonic and hypersonic wind tunnels," by S. R. Pate, et al. AIAA J 7:450-457, March, 1969.

WOMEN

"Changes in hearing acuity of noise-exposed women," by S. Pell, et al. ARCH OTOLARYNG 83:207-212, March, 1966.

YOUTH

"Effect of the combined action of noise and vibration on the sensitivity to vibration in adolescents," by A. I. Tsysar'. GIG SANIT 30:30-36, June, 1965.

"The effect of noise of various levels and spectral composition on several functions of the organism in adolescents," by I. I. Ponomarenko. GIG TR PROF ZABOL 10:32-38, June, 1966.

"The effect of stable high frequency industrial nosie on certain physiological functions of adolescents," by I. I. Ponomarenko. GIG SANIT 31:29-33, February, 1966.

"Effects of noise on pupil performance," by B. R. Slater. J ED PSYCHOL 59:239-243, August, 1968.

"Effects of traffic noise on health and achievement of high school students of a large city," by G. Karsdorf, et al. Z GES HYG 14:52-54, January, 1968.

"Noise: city dwellers, teenagers face deafness from noise," CONG Q W REPT 28:1035-1038, April 17, 1970.

"The noise factor and its hygienic evaluation in the vocational training of students in the 9-11th grade," by E. A. Timokhina. GIG SANIT 30:46-50, February, 1965.

"Noise in students residences; an investigation," by D. C. R. Porter. ARCHITECTS J 141:971-979, September 21, 1965.

"Noise is for learning; four types of noise created by junior high school students," by J. Reedy. CLEAR HOUSE 43:154-157, November, 1968.

"On the effect of industrial noise on the functional status of the auditory analyzer in adolescents," by L. L. Kovaleva. GIG SANIT 32:52-56, January, 1967.

YOUTH

"On the problem of the basis for standard noise levels for adolescents," by E. A. Gel'tishcheva, et al. GIG SANIT 33:34-38, November, 1968.

"Pupil behavior in relation to various sensory stimuli," by N. Orzalesi, et al. BOLL OCULIST 36:284-291, April, 1967.

"Pupillographic changes induced by diverse acoustic stimulations in normal subjects," by B. Carenini, et al. BOLL OCULIST 45:75-82, February, 1966.

"The reaction of the cardiovascular system of working adolescents to sound and vibration stimuli," by A. I. Tsysar'. GIG SANIT 31:33-38, February, 1966.

"Sound and student behavior," by C. E. Schiller, et al. AV INSTR 14:92, March, 1969.

"Standardization of high frequency industrial noise for adolescents," by I. I. Ponomarenko. GIG SANIT 33:34-38, August, 1968.

"The work of adolescents in environments exposed to noise," by M. Fisarova. CESK NEUROL 29:396-401, November, 1966.

AUTHOR INDEX